Autorenkollektiv

Structural Chemistry of Boron and Silicon

Pyrazole Derivatives of Boron

K. Niedenzu and S. Trofimenko

A Survey of Structural Types of Borates and Polyborates

G. Heller

Molecular and Electronic Structure of Penta- and Hexacoordinate Silicon Compounds

S. N. Tandura, M. G. Voronkov and N. V. Alekseev

AKADEMIE-VERLAG BERLIN

Autorenkollektiv

Structural Chemistry of Boron and Silicon

Autorenkollektiv

Structural Chemistry of Boron and Silicon

With 3 Figures and 28 Tables

AKADEMIE-VERLAG BERLIN
1986

Die Originalausgabe erscheint im Springer-Verlag Berlin · Heidelberg · New York · Tokyo
in der Schriftenreihe „Topics in Current Chemistry“, Volume 131

Vertrieb ausschließlich für die DDR und die sozialistischen Länder

ISBN 3-05-500 196-6

Erschienen im Akademie-Verlag Berlin, DDR-1086 Berlin, Leipziger Straße 3—4
Lizenznummer: 202 · 100/510/86
Printed in the German Democratic Republic
Gesamtherstellung: VEB Druckerei „Thomas Müntzer“, 5820 Bad Langensalza
Umschlaggestaltung: Karl Salzbrunn
LSV 1275
Bestellnummer: 763 595 1 (3072/131)
11800

Table of Contents

Pyrazole Derivatives of Boron

Kurt Niedenzu[1] **and Swiatoslaw Trofimenko**[2]

[1] Department of Chemistry, University of Kentucky, Lexington, Kentucky 40506, USA
[2] Polymer Products Department, E. I. DuPont de Nemours & Co., Parkersburg, West Virginia 26102, USA

Table of Contents

I Introduction

Despite the multitude of well characterized boron compounds, the knowledge of azole derivatives of boron is still rather limited. For example, only three C-borylated pyrazoles are known. Although N-borylated pyrazole derivatives are considerably more abundant, until most recently these were restricted to compounds containing only four-coordinate boron. Of these, the poly(1-pyrazolyl)borate anions have been a bonanza for the coordination chemist, since the steric and electronic features of these ions render them as extremely useful (polydentate and chelating) ligands. However, recent studies of the chemistry of boron derivatives of pyrazoles have provided for some noteworthy developments and a survey of such compounds and their chemistry appears to be a timely subject.

In the following, the known boron derivatives of pyrazoles are discussed but without consideration of the historical development. Specific properties are highlighted which illustrate not only the usefulness of many of these compounds but also their potential for studying fundamental aspects of contemporary chemistry.

Historically, C-borylated pyrazoles were first described in 1962. A few years later, N-borylated species containing four-coordinate boron were prepared, i.e., the poly(1-pyrazolyl)borate anions and the (neutral) dimeric pyrazol-1-ylboranes = pyrazaboles. The first monomeric pyrazol-1-ylborane containing trigonal boron was reported in 1980. In addition, several cationic species in which two or more boron atoms or a boron and a carbon atom are bridged by pyrazolyl groups are known, but their chemistry has not yet been explored at all.

II C-Borylated Pyrazoles

Pyrazole derivatives in which a boron is bonded directly to a pyrazole-carbon atom are exceedingly scarce. Only three such compounds have been described in the literature [1], all of which were obtained by cycloaddition reactions. For example, $(n\text{-}C_4H_9O)_2BC{\equiv}CH$ was found to react with ethyl diazoacetate to yield the ester *1* as the initial product, but which was isolated as the corresponding dihydroxyborane. The latter has been hydrolyzed to yield the acid *2*.

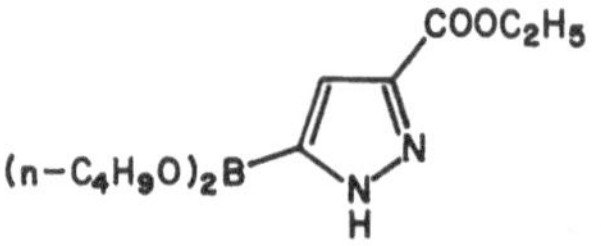

1

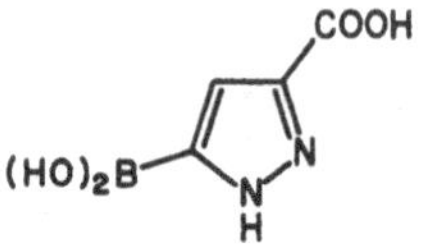

2

An analogous reaction of $(n\text{-}C_4H_9O)_2BC{\equiv}CH$ with diphenyldiazomethane yielded, with rearrangement and after hydrolysis, the dihydroxyborane *3*.

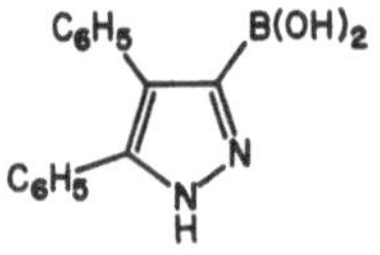

3

No additional C-borylated pyrazoles have yet been described. As a matter of fact, the exact structure of the three cited species is still in doubt inasmuch as the location of the N-bonded proton has not yet been established.

III Monomeric Pyrazol-1-ylboranes

III.A Species Containing Trigonal Boron

The first reported monomeric pyrazol-1-ylborane, i.e., a species not dimerized to a pyrazabole structure (see Sect. IV.), was 3,5-bis(trifluoromethyl)pyrazol-1-ylborane [2]. However, this compound is known only as its 1:1 molar adduct with trimethylamine, i.e., it contains four-coordinate boron. The first monomeric pyrazol-1-ylborane containing trigonal boron was reported in 1980 [3]: 1,3-dimethyl-2-(1-pyrazolyl)-1,3,2-diazaboracyclohexane, *4*, was obtained on condensation of pyrazole with 1,3-dimethyl-1,3,2-diazaboracyclohexane:

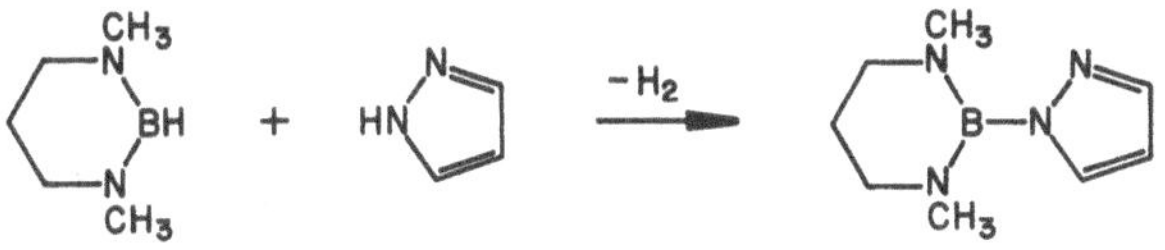

4

Subsequently, several additional monomeric pyrazol-1-ylboranes containing trigonal boron have been described. They are, however, all species in which the boron is incorporated into a 1,3,2-diazaboracycloalkane ring as shown in *5*. This feature is believed to provide for a sufficient electronic saturation of the boron by annular π-backbonding from the two adjacent nitrogen atoms to reduce the Lewis acidity of the boron and thus prevent dimerization to a pyrazabole. The known species of type *5* are surveyed in Table 1.

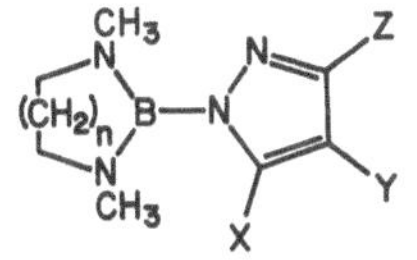

5

Table 1. Monomeric Pyrazol-1-ylboranes of Type *5*

Nr.	n	X	Y	Z	Ref.
1	2	H	H	H	4, 5)
2	2	H	H	CH_3	6)
3	2	CH_3	H	CH_3	6)
4	3	H	H	H	3, 4, 5)
5	3	H	Cl	H	7)
6	3	Br	Br	Br	7)
7	3	H	H	CH_3	5)
8	3	CH_3	H	CH_3	5)

Two principal syntheses of monomeric pyrazol-1-ylboranes containing trigonal boron have been developed. One is the condensation reaction as described above in Eq. (1) (method *A*). The other process involves the symmetrical cleavage of a pyrazabole skeleton upon reaction with the appropriate aliphatic α,ω-diamine (Eq. (2); method *B*).

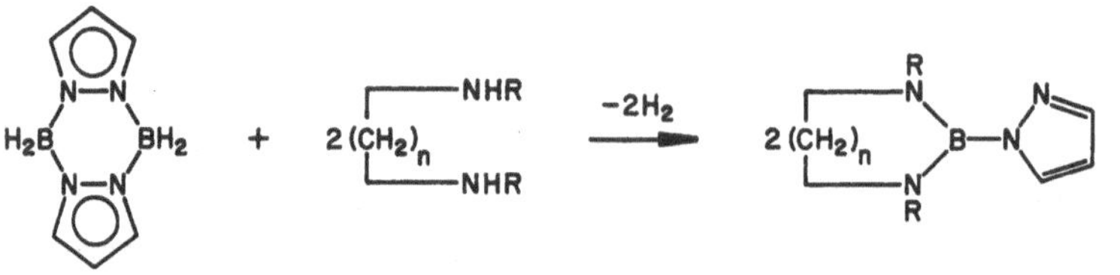

(2)

Although the known monomeric pyrazol-1-ylboranes of type *5* do not dimerize with the formation of pyrazaboles, their boron atom has residual Lewis acidity. Hence, they have been found to form 1:1 molar complexes with various nitrogen bases (see Sect. III.B). In addition, an unusual dimerization has been observed for 1,3-dimethyl-2-(1-pyrazolyl)-1,3,2-diazaboracyclopentane (compound Nr. 1 in Table 1) at low temperatures: On the basis of NMR data, structure *6* was assigned to this dimer [4)].

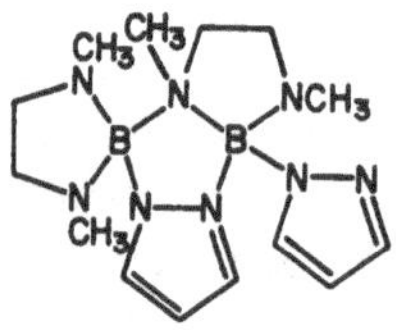

6

Indeed, recent CNDO calculations have shown that dimerization of species of type *5* to yield pyrazabole structures is energetically not favored [203)]. In addition, a sterie effect of the N-methyl groups could be demonstrated.

Most of the known monomeric pyrazol-1-ylboranes containing trigonal boron are stereochemically non-rigid: A sigmatropic boryl group migration from one pyrazole-nitrogen site to the other can readily be observed by NMR spectroscopy [4–6)]. Such dynamic behavior has also been observed for N-bonded silicon [8,9)] and germanium [9)] derivatives of pyrazoles and was discussed in terms of an intramolecular 1,2-shift of the N-substituents. In contrast, the boryl group migration of pyrazol-1-ylboranes involves in intermolecular process which is second order and may well proceed via an intermediate such as *6*[10)].

All monomeric pyrazol-1-ylboranes containing trigonal boron are extremely sensitive to moisture. In addition, *5* with n = 2 and X=Y=Z=H was found to undergo a reversible (by simple redistillation) rearrangement on standing at room temperature [4)]. The rearranged material has not yet been studied but there is evidence for it to contain the species *7* [6)].

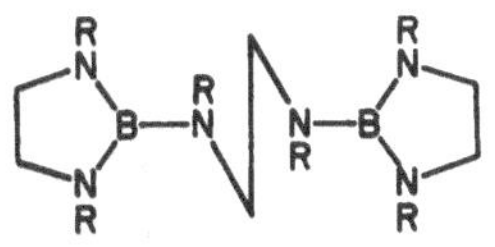

7

$R = CH_3$

The interaction of species of type *5* with various nitrogen bases including pyrazoles is described in Section III.B (below). On interaction with (dimethylamino)dialkylboranes, a simple ligand exchange was observed leading to the formation of 2-dimethylamino-1,3,2-diazaboracycloalkanes and 4,4,8,8-tetraalkylpyrazaboles (see Sect. IV).

III.B Species Containing Four-Coordinate Boron

As noted above, the first monomeric pyrazol-1-ylborane was isolated as its trimethylamine adduct: The compound $(CH_3)_3N{-}BH_2[pz\text{-}3,5\text{-}(CF_3)_2]$ (Hpz = pyrazole) was obtained as a distillable material on reaction of $(CH_3)_3N{-}BH_3$ with 3,5-bis(trifluoromethyl)pyrazole = $H[pz\text{-}3,5\text{-}(CF_3)_2]$ = Hpz*. Surprisingly, when THF$-BH_3$ (THF = tetrahydrofuran) was employed as reagent, the dimeric species $H_2B(\mu\text{-}pz^*)_2BH_2$, a pyrazabole (see Sect. IV.), was obtained without difficulty [2]. Apparently, the two-coordinate nitrogen of the pyrazolyl group in THF$-BH_2[pz\text{-}3,5\text{-}(CF_3)_2]$ is sufficiently basic to displace THF but this base displacement cannot occur in the corresponding $(CH_3)_3N{-}BH_2[pz\text{-}3,5\text{-}(CF_3)_2]$.

Subsequently, several Lewis base adduct of monomeric pyrazol-1-ylboranes have been described. For example, the species $(CH_3)_2HN{-}B(pz)_3$ is a major product of the interaction of tris(dimethylamino)borane, $B[N(CH_3)_2]_3$, with pyrazole; and $(CH_3)_2HN{-}BC_6H_5(pz)_2$ has been obtained from $C_6H_5B[N(CH_3)_2]_2$ in an analogous reaction [11]. However, the nature of these two species has not yet been explored. At ambient temperature, the N-bonded proton seems to be labile and not localized; only at low temperatures it appears to be bound to the nitrogen of the dimethylamine group. This observation suggests a relationship to poly(1-pyrazolyl)borate anions or the corresponding acids, respectively (see Sect. VI.). Indeed, it is possible to synthesize metal derivatives of the ion $[(CH_3)_2NB(pz)_3)]$ [184] and the species $[(CH_3)_2NBC_6H_5\text{-}(pz)_2]^-$ may undergo analogous reactions.

In this context it is of interest to note that the mass spectral fragmentation data on $(CH_3)_2HN{-}B(pz)_3$ suggest its fragmentation to Hpz and $(CH_3)_2NB(pz)_2$, which seems to contradict the low-temperature NMR data; on the other hand, this observation is quite in consonance with a delocalized proton at higher temperatures.

In an unusual reaction, the 3,5-dimethylpyrazole (= Hpz*) adduct of tris(3,5-dimethylpyrazol-1-yl)borane was obtained in 20% yield when $(C_5H_5)_2TiCl_2$ was reacted with $K[HB(pz^*)_3]$. The room temperature spectrum of the compound, $Hpz^*{-}B(pz^*)_3$, showed scrambling of the NH proton; however, the latter is localized at -53 °C and also in the solid state as shown by an X-ray diffraction study [30].

Originating from isolated monomeric pyrazol-1-ylboranes containing trigonal boron (see Sect. III.A), a number of 1:1 molar adducts with nitrogen donor molecules have been identified. Since all of the former are derivatives of a 1,3,2-diazaboracycloalkane, most such studies were performed with species of the type *5* with n = 2: In these compounds the N—B—N bond angle of the diazaboracycloalkane ring is presumably closer to that of tetrahedral boron (as compared to derivatives with n = 3) and, hence, they appear to be the better coordinating agents.

The formation of such adducts seems to be a function of the basicity of the donor molecule: Neither triethylamine nor pyridine formed an adduct with compound 1 (Table 1), whereas pyrrole or 2,6-dimethylpyridine yielded adducts in equilibrium with the individual components. With diethylamine, a chemical transformation occurred; but with pyrazoles, imidazole or 1,2,4-triazole, adduct formation with compound 1 (Table 1) is clearly the favored product [5]. Several other such 1:1 molar adducts have been identified and are surveyed in Table 2. However, it should be noted that, based on NMR data, several of these adducts in solution are in equilibrium with the individual components, probably due to steric effects.

Table 2. 1:1 Molar Adducts of Pyrazol-1-ylboranes with Nitrogen Donor Molecules (compounds with an asterisk exist in solution in equilibrium with the individual components; compounds 6 and 12 are identical; compound 11 is stable only at low temperatures)

Nr.	Pyrazol-1-ylborane	Donor Molecule	Ref.
1	$B(pz)_3$	dimethylamine	11)
2	$B[pz\text{-}3,5\text{-}(CH_3)_2]_3$	3,5-dimethylpyrazole	30)
3	$C_6H_5B(pz)_2$	dimethylamine	11)
4	$H_2B[pz\text{-}3,5\text{-}(CF_3)_2]$	trimethylamine	2)
5	Nr. 1, Table 1	pyrazole	4)
6	Nr. 1, Table 1	3-methylpyrazole	6)
7*	Nr. 1, Table 1	3,5-dimethylpyrazole	6)
8	Nr. 1, Table 1	1,2,4-triazole	5)
9*	Nr. 1, Table 1	2,6-dimethylpyridine	5)
10*	Nr. 1, Table 1	pyrrole	5)
11	Nr. 1, Table 1	(dimethylamino)diethylborane	6)
12	Nr. 2, Table 1	pyrazole	6)
13*	Nr. 2, Table 1	3-methylpyrazole	6)
14*	Nr. 2, Table 1	3,5-dimethylpyrazole	6)
15*	Nr. 3, Table 1	3-methylpyrazole	6)
16*	Nr. 3, Table 1	3,5-dimethylpyrazole	6)
17*	Nr. 4, Table 1	pyrazole	5)

As has been mentioned above, the nature of these adducts involving a secondary amine has not yet been clearly established. One may view these also as acids of type *8*.

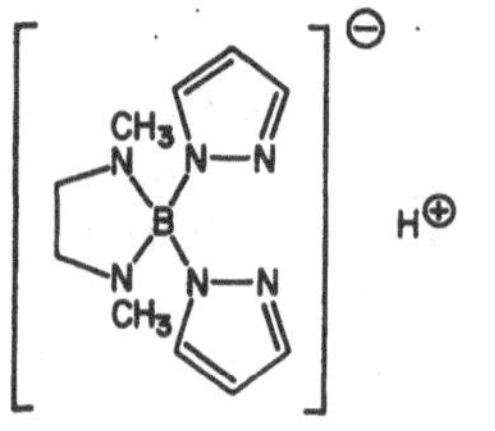

8

This latter formulation may give access to metal derivatives in which such a di(1-pyrazolyl)borate unit serves as a chiral ligand, i.e., by employing a C-substituted pyrazole as donor molecule. There exists only one report in the literature on forming poly(1-pyrazolyl)borate ions containing two different pyrazole groups, which were obtained in quite laborious manner [12].

The lability of the proton is clearly demonstrated by the fact that, for example, compound 6 (Table 2) can be prepared by either reaction of 1,3-dimethyl-2-(1-pyrazol-yl)-1,3,2-diazaboracyclopentane with 3-methylpyrazole or by reaction of 1,3-di-methyl-2-(methylpyrazol-1-yl)-1,3,2-diazaboracyclopentane with pyrazole [6]; see also Section VI.6.

IV Pyrazaboles

IV.A General Remarks

The condensation of pyrazole (=Hpz) or C-substituted derivatives thereof with a trigonal borane (which may be employed as its adduct with a Lewis base) proceeds readily to yield N-borylated pyrazoles. However, the resultant species generally exist in the dimeric "pyrazabole" structure, i.e., $R_2B(\mu\text{-pz})_2BR_2$ = *9* [2,13].

Note: Henceforth, when written between B and B or between B and M (= metal), the group (pz) is meant to be bridging, e.g., $B(pz)_2B = B(\mu\text{-pz})_2B$.

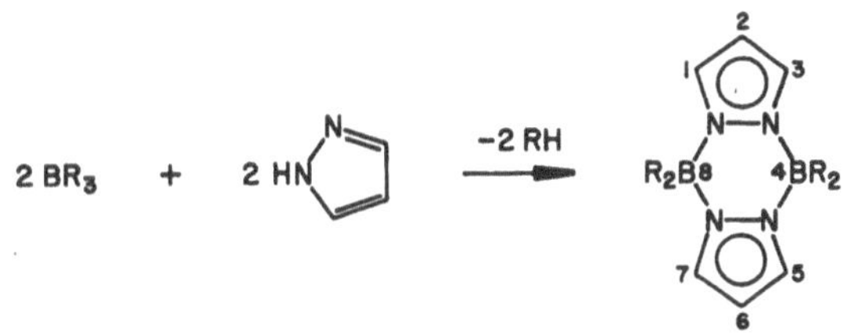

(2)

9

The existence of boron in four-coordinate environment and the presence of the heteroaromatic system both appear to contribute to the extreme chemical stability of the pyrazaboles. This latter feature is documented by the fact that organic substitutions at the pyrazole-carbon sites are readily effected without destruction of the central B_2N_4 skeleton [14]. Also, electrophilic displacement of boron-bonded hydrogen of pyrazaboles proceeds smoothly without apparent attack of the B_2N_4 ring [15]. Most pyrazaboles are unaffected by air or water; they are stable toward aqueous alkali but slowly decompose in boiling hydrochloric acid. These chemical characteristics as well as their physical properties differentiate the pyrazaboles from the structurally related hydrazinoboranes, which also contain the B—N—N arrangement.

The pyrazaboles constitute the bulk of the presently known neutral boron derivatives of pyrazoles; more than 70 different species have been described.

IV.B Preparative Aspects

The preparation of the symmetrical pyrazaboles of the type $R_2B(pz^*)_2BR_2$ is normally based on the thermal condensation of pyrazole or one of its C-substituted derivatives (= Hpz*) with trimethylamine-borane [2], triorganylborane [2], (dialkylamino)diorganylborane [5,6] or tris(organylthio)borane [7] (method *A*). Alternatively, boron-bonded hydrogen of a preformed B-hydropyrazabole, $H_2B(pz^*)_2BH_2$, can be replaced by reaction with elemental halogen [14] or boron trihalides [15,16] at low temperatures (method *B*). However, this electrophilic-induced substitution proceeds stepwise and unsymmetrically boron-substituted pyrazaboles have been obtained from this latter reaction by careful adjustment of the stoichiometry of the reactants [16] (method *C*).

It is noteworthy that in this latter reaction no exchange of boron atoms between $H_2B(pz)_2BH_2$ and the BX_3 occurs, as was demonstrated by employing boron-10 labelled boron tribromide [15]. Hence, the central B_2N_4 ring of the pyrazabole apparently is not opened during the course of this low-temperature substitution process.

Similar reactions of B-hydropyrazaboles with active hydrogen compounds such as pyrocatechol [14], phenol [14], thiols [7] or additional pyrazole [14,32] are equally facile but require high temperatures (method *D*). Therefore, the reaction of pyrazabole with o-phenylenediamine may proceed via the expected substituted pyrazabole as an intermediate. However, the latter is unstable under the reaction conditions and condenses further to yield the borazine derivative *10* with the elimination of pyrazole [14].

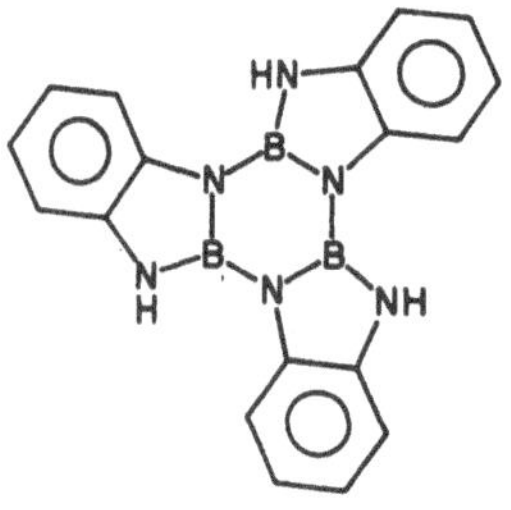

10

Similarly, the reaction of pyrazaboles with α,ω-diaminoalkanes proceeds with symmetrical cleavage of the pyrazabole skeleton. This latter process was used for the synthesis of monomeric pyrazol-1-ylboranes [7] (see Sect. III.A).

The reaction of $H_2B(pz)_2BH_2$ with pyrazole has been stopped at the intermediates $H(pz)B(pz)_2BH_2$ [15] and $H(pz)B(pz)_2BH(pz)$ [14], respectively, by appropriate adjustment of the stoichiometry of the reactants (method *E*).

It is noteworthy that when $H_2B(pz)_2BH_2$ was reacted with a C-substituted pyrazole, Hpz*, a mixture of products was obtained containing all possible products of the composition $B_2(pz^*)_{6-n}(pz)_n$ rather than the expected $(pz^*)_2B(pz)_2B(pz^*)_2$ [17]. This observation clearly suggests that the high-temperature condensation proceeds via intermediate opening of the central B_2N_4 ring or, perhaps, even by complete dissociation into monomeric species, i.e., pyrazol-1-ylboranes containing trigonal boron (see Sect. III.A).

Table 3. Survey of Pyrazaboles

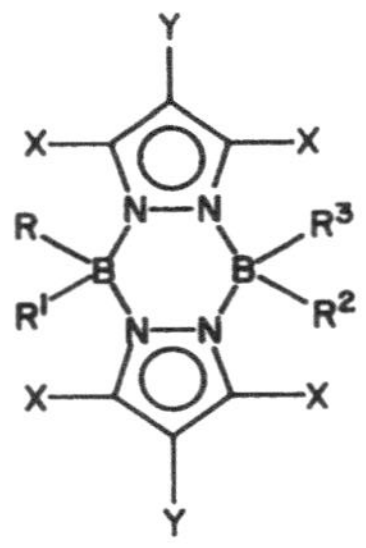

Nr.	R	R^1	R^2	R^3	X	Y	Method	Ref.
1	H	H	H	H	H	H	A	2, 13, 19)
							H	17)
2	H	H	H	H	H	Cl	A	2)
3	H	H	H	H	H	Br	A	2)
4	H	H	H	H	H	CN	A	2)
5	H	H	H	H	H	NO_2	A	2)
6	H	H	H	H	H	CH_3	A	2)
7	H	H	H	H	H	i-C_3H_7	A	14)
8	H	H	H	H	H	i-C_3F_7	A	2)
9	H	H	H	H	CH_3	H	A	2)
10	H	H	H	H	CF_3	H	A	2)
11	H	H	H	H	C_6H_5	H	A	2)
12	H	H	H	H	Br	Br	A	2)
13	H	H	H	H	CH_3	CH_3	A	2)
14	H	H	H	H	CH_3	n-C_4H_9	A	14)
15	Br	H	H	H	H	H	C	16)
16	pz	H	H	H	H	H	E	15)
							F	17)
17	F	F	H	H	H	H	F	18)

Table 3. (continued)

Nr.	R	R^1	R^2	R^3	X	Y	Method	Ref.
18	pz	pz	H	H	H	H	F	[17]
19	C_2H_5	C_2H_5	H	H	H	H	F	[18]
20	C_2H_5	C_2H_5	H	H	CH_3	H	F	[18]
21	n-C_4H_9	n-C_4H_9	H	H	H	H	F	[18, 20]
22	Cl	H	Cl	H	H	i-C_3F_7	C	[14]
23	Br	H	Br	H	H	H	C	[16]
24	Br	H	Br	H	CH_3	H	B	[32]
25	pz	H	pz	H	H	H	E	[14]
26	pz-3,5-$(CH_3)_2$	H	pz-3,5-$(CH_3)_2$	H	CH_3	H	E	[17]
27	C_6H_5	H	C_6H_5	H	H	H	A	[14]
28	Br	Br	Br	H	H	H	C	[16]
29	F	F	F	F	H	H	A B	[14] [21]
30	F	F	F	F	H	NH_2	G	[21]
31	F	F	F	F	H	NO_2	G	[21]
32	Cl	Cl	Cl	Cl	H	H	B	[13, 14]
33	Cl	Cl	Cl	Cl	H	i-C_3H_7	B	[14]
34	Cl	Cl	Cl	Cl	CH_3	n-C_4H_9	B	[14]
35	Br	Br	Br	Br	H	H	B	[14, 16]
36	Br	Br	Br	Br	CH_3	H	B	[32]
37	I	I	I	I	H	H	B	[14]
38	C_6H_4-1,2-$(O)_2$		C_6H_4-1,2-$(O)_2$		H	H	A	[13, 14]
39	C_6H_5O	C_6H_5O	C_6H_5O	C_6H_5O	H	H	D	[14]
40	$S(CH_2)_2S$		$S(CH_2)_2S$		H	H	D	[7]
41	SCH_3	SCH_3	SCH_3	SCH_3	H	H	A	[7]
42	$S(CH_2)_3S$		$S(CH_2)_3S$		H	H	D	[7]
43	SC_2H_5	SC_2H_5	SC_2H_5	SC_2H_5	H	H	A	[7]
44	pz	pz	pz	pz	H	H	D	[14]
45	pz	pz	pz	pz	CH_3	H	D	[32]
46	pz	pz	pz	$N(CH_3)_2$	H	H	A	[11]
47	pz	$N(CH_3)_2$	pz	$N(CH_3)_2$	H	H	A	[11]

Table 3. (continued)

Nr.	R	R^1	R^2	R^3	X	Y	Method	Ref.
48	pz	$N(CH_3)_2$	pz	C_6H_5	H	H	A	11)
49	Cl	Cl	C_1H_5	C_2H_5	H	H	F	18)
50	Br	Br	C_2H_5	C_2H_5	CH_3	H	B	32)
51	Br	Br	C_6H_5	C_6H_5	H	H	B	32)
52	pz	pz	C_2H_5	C_2H_5	H	H	A	11)
53	pz	C_6H_5	pz	C_6H_5	H	H	A	14)
54	$N(CH_3)_2$	C_6H_5	$N(CH_3)_2$	C_6H_5	H	H	A	11)
55	C_6H_4-1,2-$(O)_2$		C_2H_5	C_2H_5	H	H	D	18)
56	C_2H_5	C_2H_5	C_2H_5	C_2H_5	H	H	A	2, 5)
57	C_2H_5	C_2H_5	C_2H_5	C_2H_5	H	Cl	A	2)
58	C_2H_5	C_2H_5	C_2H_5	C_2H_5	H	Br	A	2, 13)
59	C_2H_5	C_2H_5	C_2H_5	C_2H_5	H	CN	A	2, 13)
60	C_2H_5	C_2H_5	C_2H_5	C_2H_5	H	NH_2	A	2)
							G	13)
61	C_2H_5	C_2H_5	C_2H_5	C_2H_5	H	NO_2	A	2, 13)
62	C_2H_5	C_2H_5	C_2H_5	C_2H_5	H	$N(COCH_3)_2$	G	13)
63	C_2H_5	C_2H_5	C_2H_5	C_2H_5	H	CHO	H	2, 13)
64	C_2H_5	C_2H_5	C_2H_5	C_2H_5	H	COOH	G	2, 13)
65	C_2H_5	C_2H_5	C_2H_5	C_2H_5	H	i-C_3H_7	A	13)
66	C_2H_5	C_2H_5	C_2H_5	C_2H_5	H	Li	G	14)
67	C_2H_5	C_2H_5	C_2H_5	C_2H_5	CH_3	H	A	2)
68	C_2H_5	C_2H_5	C_2H_5	C_2H_5	CH_3	CH_3	A	2)
69	n-C_3H_7	n-C_3H_7	n-C_3H_7	n-C_3H_7	H	H	A	5, 6)
70	n-C_4H_9	n-C_4H_9	n-C_4H_9	n-C_4H_9	H	H	A	2)
71	C_6H_5	C_6H_5	C_6H_5	C_6H_5	H	H	A	2, 13)

Another route which can be used for the preparation of unsymmetrically boron-substituted pyrazaboles of the type $RR^1B(pz)_2BR^2_2$ involves the interaction of poly(1-pyrazolyl)borate anions, $[RR^1B(pz)_2]^-$, with a borane containing a readily leaving group [17, 18] (method *F*). Also, substitutions at the 2 and 6 carbon atoms of a pyrazabole skeleton (*9*) by typical organic reactions have been found possible [14] (method *G*). Finally, reaction of a B-halogenated pyrazabole with potassium pyrazolate was found to yield the desired B-pyrazolylpyrazabole [32] (method *H*).

The known pyrazaboles containing symmetrical bridging pyrazolyl groups are surveyed in Table 3. The knowledge on pyrazaboles of type *9* but where the 1,3,5,7-positions in the bridging pyrazolyl groups are not equivalent is extremely limited. Only a few such derivatives have been described: The structures of the species *11* and *12*, respectively, were elucidated by ^1H NMR studies. In addition, 4,4,8,8-tetra-bromo-, tetrahydro- and tetrakis(3-methylpyrazol-1-yl)-1,5(7)-dimethylpyrazabole have been described [189].

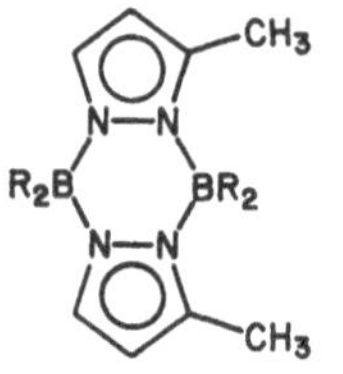

11
R = CH_3 [31]

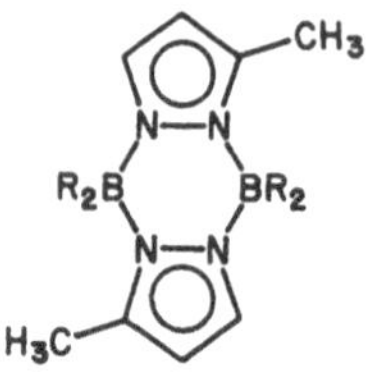

12
R = CH_3 [31]
R = C_2H_5 [5]

IV.C Physical Data

Most pyrazaboles are well crystallized solids, a notable exception being the liquid 4,4,8,8-tetra(n-butyl)pyrazabole [2].

The first X-ray crystal and molecular structure study of a pyrazabole was reported for the species $H(pz')B(pz')_2BH(pz')$ (pz = 3,5-dimethylpyrazolyl) [22]. The compound was found to exist in chair conformation of the central B_2N_4 ring with the two terminal pyrazolyl groups being in trans arrangement. This particular structure was thought to be a consequence of the bulkiness of the pyrazolyl groups, since the structurally related $D_2Ga(pz)_2GaD_2$ exists in the boat arrangement of the Ga_2N_4 ring [23]. This argument was supported by the finding that the B_2N_4 ring of $(C_2H_5)_2B(pz^*)_2B(C_2H_5)_2$ (Hpz* = 4-bromopyrazole) also has a boat conformation [24]. It is, however, no longer acceptable since pyrazaboles were later [15] found to exist in either boat, chair or planar conformation of the B_2N_4 ring and no steric influence of the terminal substituents could be confirmed. Rather, individual structures of pyrazaboles seem to result from crystal packing effects. Structures of pyrazaboles of the type $RR'B(pz)_2BRR'$ are surveyed in Table 4.

The structure of $(pz)_2B(pz)_2BH_2$ has also been determined [25]. As expected, the central B_2N_4 ring exists in boat conformation and the two fragments $(pz)_2B(\mu\text{-}pz)_2$ and $(\mu\text{-}pz)_2BH_2$, respectively, are quite similar to those of the corresponding symmet-

Table 4. X-Ray Diffraction Studies on Pyrazaboles of the Type $RR'B(pz)_2BRR'$

R	R'	Bridging pz Groups	B_2N_4 Ring Conformation	Ref.
H	H	pz	boat	[16]
H	H	pz-3,5-$(CH_3)_2$	boat	[25]
H	Br	pz-4-Cl	boat	[15]
Cl	Cl	pz	planar	[15]
Br	Br	pz	planar	[16]
$S(CH_2)_2S$		pz	boat	[15]
SCH_3	SCH_3	pz	boat	[15]
H	pz-3,5-$(CH_3)_2$	pz-3,5-$(CH_3)_2$	chair	[22]
pz	C_6H_5	pz	chair	[15]
pz	pz	pz	boat	[25]
C_2H_5	C_2H_5	pz-4-Br	boat	[24]

rically substituted pyrazaboles. An important finding of the X-ray diffraction studies is the fact that axial and equatorial B—H distances differ significantly, thus providing for noticeable differences in chemical reactivity.

Many NMR spectral data on individual pyrazaboles have been reported. Most interesting appears to be an extensive study including 2D-NMR experiments that is primarily concerned with pyrazolylpyrazaboles [26]. It was possible to assign $\delta(^1H)$ to specific $\delta(^{13}C)$ signals and also to assign $\delta(^1H)/\delta(^{13}C)$ pairs to individual pyrazole groups and locations. The ^{11}B NMR signals of pyrazaboles were found to cover a range of less than 20 ppm from approximately +8 to −9 ppm.

The mass spectra of B-hydropyrazaboles exhibit three major features [28]. Symmetrical cleavage of the pyrazabole skeleton followed by further breakdown is common to all compounds. In addition, those pyrazaboles containing H or CH_3 at the C atoms of the pyrazole rings undergo an electron impact-induced rearrangement which appears to result in the formation of a species containing a B_2N_3 ring as a structural entity; subsequent breakdown leads to a B_2N_2 ring system. The mass spectra of C-halogenated pyrazaboles and the corresponding pseudohalogen derivatives evidence the ready loss of hydrogen halide as a predominant feature; no rearrangement ions as cited above were observed [28]. Boron-alkylated pyrazaboles feature the ready loss of hydrocarbon moieties under electron impact [29]. It appears that B-alkylation enhances the stability of the B_2N_4 pyrazabole ring toward electron impact.

IV.D Chemical Behavior

Most studies of the chemical behavior of pyrazaboles have been devoted to transformations leading to derivatives as discussed in Section IV.B. It seems apparent that low-temperature electrophilic induced substitutions at boron sites of pyrazaboles occur without rupture of the central B_2N_4 ring. This has been demonstrated by employing $^{10}BBr_3$ as brominating agent [15]. However, in other processes a cleavage of the B_2N_4 ring does occur during the substitution process. This was observed for the condensation of the parent pyrazabole with C-substituted pyrazoles [17] but also for the reaction of 4,4,8,8-tetrabromopyrazabole with potassium pyrazolide [32]. The limited available data on such processes suggest a symmetrical cleavage of the pyrazabole skeleton to form a trigonal boron derivative which may undergo scrambling and subsequent dimerization.

Studies other than preparative-type transformations have so far been limited to B-pyrazolylpyrazaboles. For example, the latter compounds have been found to react with R_2BX (where X is a readily leaving group) to form polynuclear boron spiro-cations (see Sect. V.A). It has also been shown that the pyrazolylpyrazaboles offer themselves as a class of neutral bidentate chelating ligands [35]. For example, the complexes $H_2B(pz)_2B(pz)_2ZnCl_2$ and $Cl_2Zn(pz)_2B(pz)_2B(pz)_2ZnCl_2$ were readily accessible by combination of the two reactants in an appropriate solvent. Obviously, a more detailed study of this feature is mandated.

IV.E Polymeric Pyrazaboles

Three representatives of polymeric pyrazaboles have been described [36]. The first two were obtained when 4,4′-methylenedipyrazole was reacted with triethylborane or trimethylamine-borane, respectively, to yield species of type *13*.

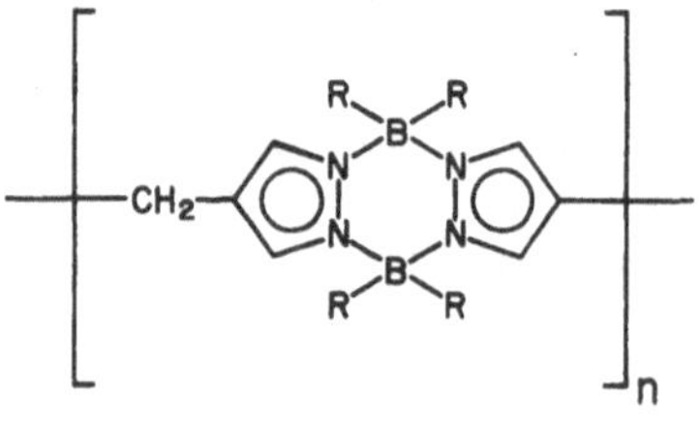

13
R = H, C_2H_5

Both polymers (of undetermined chain length) were found to be thermally stable. An analogous species, *14*, was synthesized from 3,5,3′,5′-tetramethyl-1,4-xylene-4,4′-dipyrazole. This polymer softens above 360 °C.

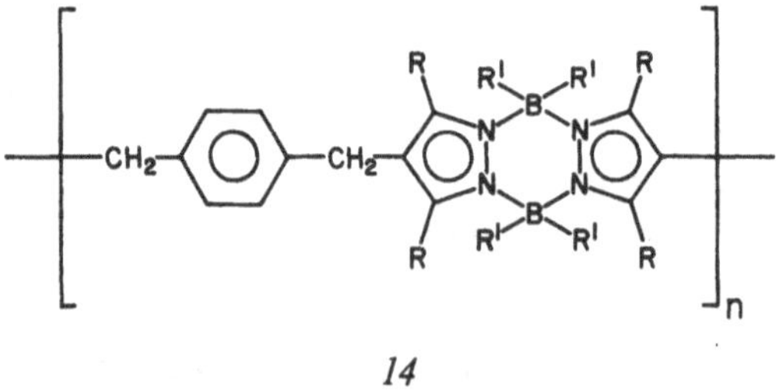

14
R = CH_3, R′ = C_2H_5

V Pyrazolylboronium Cations

V.A Polynuclear Boron Spiro-Cations

There is but one brief report on the formation of boron spiro-cations in which three or more boron atoms are bridged by pyrazolyl groups. The first such species was obtained on interaction of the $[B(pz)_4]^-$ ion with two molar equivalents of diethylboryl tosylate, $(C_2H_5)_2B(OSO_2C_6H_4\text{-}4\text{-}CH_3)$ [27]:

$$K[B(pz)_4] + 2\,(C_2H_5)_2B(OSO_2C_6H_4\text{-}4\text{-}CH_3) \rightarrow K(OSO_2C_6H_4\text{-}4\text{-}CH_3) + [(C_2H_5)_2B(\mu\text{-}pz)_2B(\mu\text{-}pz)_2B(C_2H_5)_2]^+(OSO_2C_6H_4\text{-}4\text{-}CH_3)^- \quad (4)$$

Similarly, the pyrazabole $(pz)_2B(pz)_2B(pz)_2$ was found to react with $(C_2H_5)_2$-$B(OSO_2C_6H_4\text{-}4\text{-}CH_3)$ to yield the di-cation $[(C_2H_5)_2B(pz)_2B(pz)_2B(pz)_2B(C_2H_5)_2]^{+2}$.

The two chelating pyrazaboles were also found to react with π-allylpalladium chloride dimer to yield the species $[(\eta^3\text{-}CH_2CRCH_2)Pd(pz)_2B(pz)_2Pd(\eta^3\text{-}CH_2CRCH_2)_2]^+$ and $[(\eta^3\text{-}CH_2CRCH_2)Pd(pz)_2B(pz)_2B(pz)_2Pd(\eta^3\text{-}CH_2CRCH_2]^{2+}$, respectively. These cations were characterized by their ^1H NMR spectra but no additional data are yet known [27].

V.B Additional Boron Cations

Closely related to the spiro-cations described above are a few examples of ions containing three bridging pyrazolyl groups between the same two boron atoms. Such species were first obtained on interaction of pyrazolide anion with ethylboryl ditosylate, $C_2H_5B(OSO_2C_6H_4\text{-}4\text{-}CH_3)_2$ [33]:

$$3\,Mpz + 2\,C_2H_5B(OSO_2C_6H_4\text{-}4\text{-}CH_3)_2 \rightarrow 3\,M(OSO_2C_6H_4\text{-}4\text{-}CH_3)\; [C_2H_5B(pz)_3BC_2H_5]^+(OSO_2C_6H_4\text{-}4\text{-}CH_3)^- \quad (5)$$

The X-ray crystal and molecular structure of the salt $[C_2H_5B(pz)_3BC_2H_5]PF_6$ has been described [34]. Also, the salts $[HB(pz^*)_3BH]MCl_6$ (Hpz^* = 3,5-dimethylpyrazole; M = Nb, Ta) were obtained in an unusual process on interaction of $K[HB(pz^*)_3]$ with MCl_5. Their structure was again confirmed by X-ray diffraction and also NMR data [30]

Another series of interesting cyclic boron cations of type *15* has been reported [36].

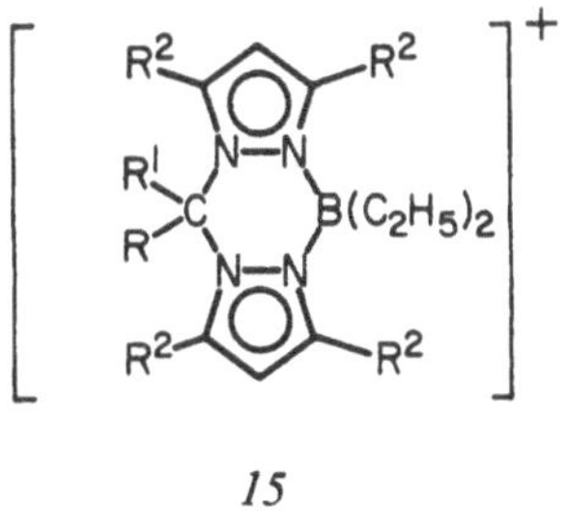

15

These were obtained from the reaction of a geminal dipyrazolylalkane with a trigonal borane containing a ready leaving group, R_2BX. However, little is known about

Table 5. Survey of Ions of Type *15* [36]

R	R^1	R^2	Anion
H	H	H	PF_6^- or 1/2 $B_{12}H_{12}^{2-}$
CH_3	H	H	PF_6^-
CH_3	H	CH_3	PF_6^-
CH_3	CH_3	H	PF_6^- or 1/2 $B_{12}H_{12}^{2-}$
$(CH_2)_4$		H	PF_6^-
pz	H	H	PF_6^-
pz	pz	H	PF_6^-

these species other than their ^{1}H NMR spectra. The known compounds of type *15* are listed in Table 5.

VI Poly(1-pyrazolyl)borates

VI.A General Remarks

The largest class of pyrazole derivatives of boron are the poly(1-pyrazolyl)borates. These are species of the general formula $M[R_nB(pz^*)_{4-n}]_m$, where M is a cation of effective charge m (as it may also contain other ligands), R is a substituent other than pz* (usually H, alkyl or aryl), pz* is a 1-pyrazolyl moiety which may contain various substituents at the carbon sites, and n can be 0, 1 or 2.

The importance of poly(1-pyrazolyl)borates resides in their electronic and steric features, which make the anion $[R_nB(pz^*)_{4-n}]^-$ ideally suited for coordination to transition metals, thus establishing poly(1-pyrazolyl)borates as an important and versatile family of ligands. Coordination to metal ions occurs through the N2 atom of the pyrazole rings, and the uninegativity of the ion makes $M[R_nB(pz^*)_{4-n}]_2$ derivatives of divalent transition metals uncharged and soluble in organic solvents. Despite their seemingly exotic nature and cumbersome nomenclature, poly(1-pyrazolyl)borates possess a number of very attractive features:

1. Their alkali metal salts are relatively easy to prepare, and they are stable to storage.
2. Their isolable free acids, obtained upon acidification of the anion, can be used to prepare salts with cations not obtainable by the direct route.
3. The three hydrogens of the unsubstituted pyrazole ring provide a convenient NMR probe for molecular symmetry of complexes.
4. Methods exist for placing up to ten substituents on the basic $HB(pz)_3$ skeleton, altering in predetermined fashion the steric and electronic features of the ligand but retaining its original C_{3v} symmetry.
5. For each $(R_nB(pz^*)_{4-n}]^-$ anion there esists an isosteric and isoelectronic, but neutral, $R_nC(pz^*)_{4-n}$ ligand, which produces identical complexes, except that they bear a higher charge (+1 per ligand) than their boron analogs. This is a truly unique feature which permits the study, by analogy, of complexes which cannot be made in the poly(1-pyrazolyl)borate system. For example, many $[H_2B(pz)_2]^-$ derivatives of Pd(II) or Au(I) decompose as they are formed by reduction to the metal; their $H_2C(pz)_2$ derivatives, however, are chemically stable.

It is understandable, therefore, that since its introduction in 1966 this ligand system has been explored to a considerable extent, and by 1984 poly(1-pyrazolyl)borate derivatives of most transition metals (except for Sc, Hf and Os) up to Np have been described and a few brief reviews of this subject matter have appeared [37-39]. Since the present emphasis is on the boron-pyrazole system, and much of poly(1-pyrazolyl)-borate chemistry is concerned with the chemistry of transition metals, only the more interesting aspects of the coordination chemistry of poly(1-pyrazolyl)borate derivatives will be highlighted here.

In most cases poly(1-pyrazolyl)borates ligands were found to impart higher chemical stability to the various complexes so that certain types of compounds, otherwise inacessible or of very low stability, could be isolated and studied as their poly(1-pyrazolyl)borate = L* derivatives. Some of these include stable L*CuCO species, five-coordinate Pt(II) derivatives, complexes containing $MoNH_2$ and $MoNHNH_2$ moieties, various monomeric low-valent Mo complexes, etc. The poly(1-pyrazolyl)borates were also employed in constructing model compound approximating the active site in various biologically active molecules such as the blue copper proteins [181, 182], the ethylene receptor sites in plants [186], hemerythrin [117, 118], and oxo-type enzymes such as xanthine oxidase, sulfite oxidase and nitrate reductase [187].

In terms of their coordinating ability, the main poly(1-pyrazolyl)borate ligand types are:

1. Bidentate ligands derived from $[R_2B(pz^*)_2]^-$, which coordinate to transition metal ions forming structures such as *16* (= $R_2B(pz)_2M(pz)_2BR_2$) or *17* (= $R_2B(pz)_2$-MXY).

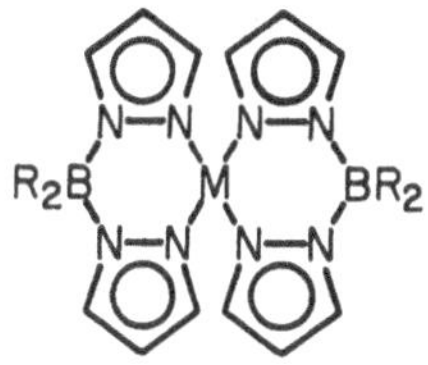

16

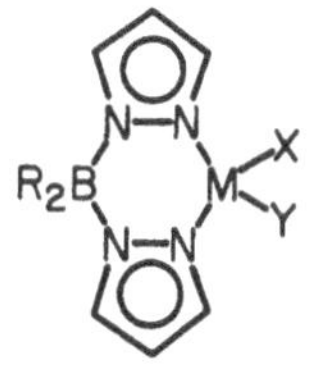

17

2. Tridentate ligands derived from $[RB(pz^*)_3]^-$ and which usually form very stable complexes exemplified by *18* (= $RB(pz)_3M(pz)_3BR$) and *19* (= $RB(pz)_3MXYZ$), although in some cases they may act as bidentate ligands.

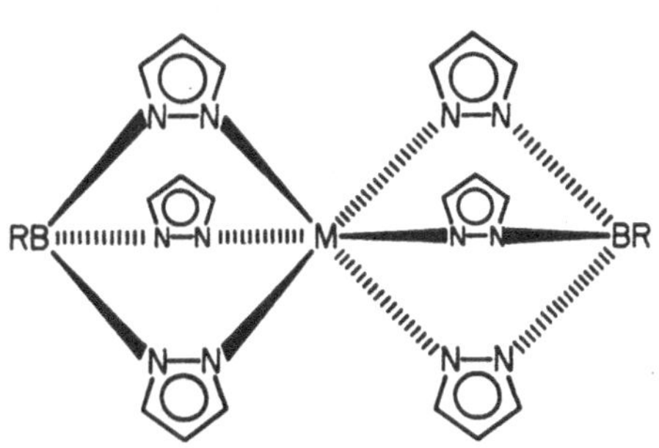

18

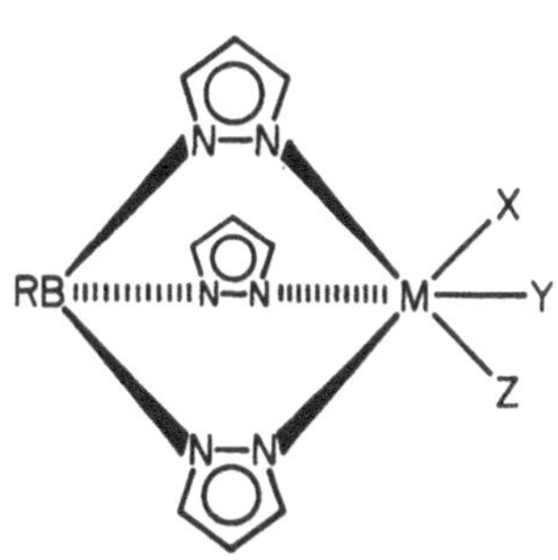

19

In $HB(pz)_3Cu(C_2H_4)CuCl$, the $HB(pz)_3$ moiety is bidentate, the third pyrazolyl group coordinating to CuCl [186]. A similar structure was found in the analogous cyclohexene complex [188].

3. The ligand $[B(pz)_4]^-$ acts most commonly in tridentate fashion although it, too, occasionally forms bidentate complexes. When this happens, the uncoordinated pyrazolyl groups retain their chelating ability, and it is possible to prepare dinuclear species of the type $M(pz)_2B(pz)_2M$ (where M may be identical or even dissimilar metal moieties) in which $[B(pz)_4]^-$ acts as a bis-bidentate ligand.

VI.B Preparation of Poly(1-pyrazolyl)borates

The parent ligand system, $[H_nB(pz)_{4-n}]^-$, has been synthsized by heating $K[BH_4]$ in molten pyrazole. Through careful temperature control it is possible to stop the reaction at the di- or tri-substitution stage [40,41]:

$$K[BH_4] + Hpz(excess) \xrightarrow{\approx 90\,°C} K[H_2B(pz)_2] \xrightarrow{\approx 170\,°C} K[HB(pz)_3]$$
$$\xrightarrow{> 200\,°C} K[B(pz)_4]$$

These salts are water soluble, and their hydrolytic stability increases with increasing substitution of boron-bonded hydrogen by pyrazolyl groups, $K[B(pz)_4]$ being the most stable. Acidification gives rise to the free acids $H[H_nB(pz)_{4-n}]$ containing a chelated proton. The latter can be converted to other poly(1-pyrazolyl)borate salts by titration with metal or tetraalkylammonium hydroxides. Pyrolysis of the free acids results in elimination of pyrazole, and the resulting trigonal $H_nB(pz)_{3-n}$ instantaneously dimerizes to a pyrazabole (see Sect. IV.).

It should be noted that the reaction of $K[BH_4]$ with pyrazole cannot be stopped at the $K[H_3B(pz)]$ stage; from an incomplete reaction only $K[BH_4]$ and $K[H_2B(pz)_2]$ could be isolated. It has been possible, though, to synthesize the ion $[H_3B\{pz\text{-}3,5\text{-}(CH_3)_2\}]^-$ by reaction of the borane adduct of 3,5-dimethylpyrazole with sodium hydride, and from it by a carefully controlled reaction with pyrazole in N,N-dimethylacetamide to prepare the first example of an asymmetric poly(1-pyrazolyl)borate ligand, i.e., $Na[H_2B(pz)\{pz\text{-}3,5\text{-}(CH_3)_2\}]$ [12].

The above general synthesis scheme is also applicable to various C-substituted pyrazoles, Hpz*, provided they contain no functionality which would react with the $[BH_4]^-$ ion [43]. Steric effects are very important. Thus, the reaction of $K[BH_4]$ with 3-methylpyrazole gives exclusively the $[H_2B(pz\text{-}3\text{-}CH_3)_2]^-$ ion [56], $[HB(pz\text{-}3\text{-}CH_3)_3]^-$ [57,189] and $[B(pz\text{-}3\text{-}CH_3)_4]^-$ [189] with no evidence for the formation of any $B(pz\text{-}5\text{-}CH_3)$-type derivatives. This is consistent with the attachment of boron to the sterically least encumbered pyrazole nitrogen. Steric reasons are also responsible for the reaction of $K[BH_4]$ with 3,5-dimethylpyrazole stopping at the trisubstitution state. However, the free acid $H[B\{pz\text{-}3,5\text{-}(CH_3)_2\}_4]$ has been obtained adventitiously and its structure was proven by X-ray crystallography [30]. This implies that, even though the species $[B\{pz\text{-}3,5\text{-}(CH_3)_2\}_4]^-$ is capable of existence, there is a steric barrier for the transition state necessary for the reaction according to Eq. (6).

$$[HB\{pz\text{-}3,5\text{-}(CH_3)_2\}_3]^- + H[pz\text{-}3,5\text{-}(CH_3)_2] \rightarrow H_2 + [B\{pz\text{-}3,5\text{-}(CH_3)_2\}_4]^- \quad (6)$$

When $K[BH_4]$ is heated in an even more sterically hindered pyrazole, i.e., 3,5-bis(t-butyl)pyrazole, up to its boiling point, no reaction occurs at all.

In some instances complexes were obtained derived from poly(1-pyrazolyl)borate ligands which never existed by themselves. This was the case with halogenation of the 4-position in $HB[pz\text{-}3,5\text{-}(CH_3)_2]_3Mo(CO)_2NO$ to produce derivatives of the type $HB[pz\text{-}3,5\text{-}(CH_3)_2\text{-}4\text{-}X]_3Mo(NO)X_2$ [44,45] and in $HB[pz\text{-}3,5\text{-}(CH_3)_2]_3Re(CO)_3$ which, on bromination, yielded the species $HB[pz\text{-}3,5\text{-}(CH_3)_2\text{-}4\text{-}Br]_3Re(CO)_3$ [46]. Another example is the formation of *20* by the reaction depicted in Eq. (7).

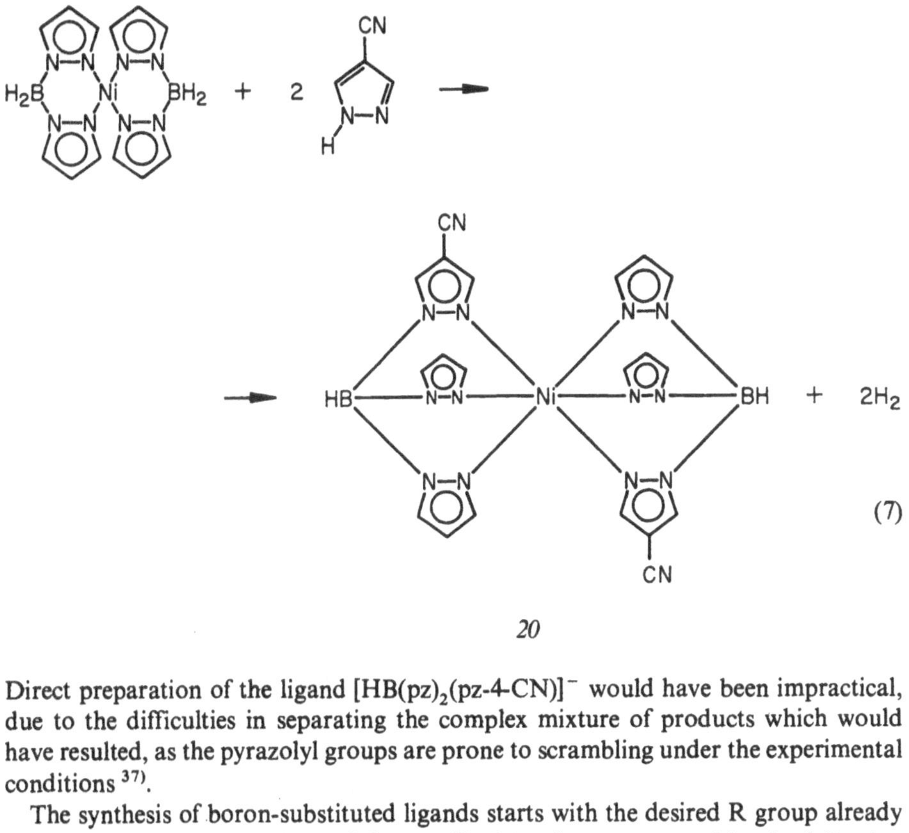

(7)

20

Direct preparation of the ligand $[HB(pz)_2(pz\text{-}4\text{-}CN)]^-$ would have been impractical, due to the difficulties in separating the complex mixture of products which would have resulted, as the pyrazolyl groups are prone to scrambling under the experimental conditions [37].

The synthesis of boron-substituted ligands starts with the desired R group already attached to the boron. Anions of the type $[R_2B(pz)_2]^-$ are prepared by the following reactions [32,43]:

$$BR_3 + Napz + Hpz(excess) \rightarrow Na[R_2B(pz)_2] + RH \quad (8)$$

$$Na[B(C_6H_5)_4] + Hpz(excess) \rightarrow Na[(C_6H_5)_2B(pz)_2] + 2\,C_6H_6 \quad (9)$$

$$H_3N{-}B(C_6H_5)_3 + Kpz + Hpz(excess) \rightarrow K[(C_6H_5)_2B(pz)_2] + NH_3 + C_6H_6 \quad (10)$$

In all these reactions an excess of pyrazole is used, and the reaction stops cleanly at the $[R_2B(pz)_2]^-$ stage, even after prolonged reflux in boiling pyrazole. The presence of one molar equivalent of the pz^- anion is necessary; otherwise pyrazaboles are formed.

The $[RB(pz)_3]^-$ ligands are synthesized from monoorganylboron precursors as follows [43,47–50]:

$$RB(OH)_2 + Napz + Hpz(excess) \rightarrow Na[RB(pz)_3] + 2\,H_2O \quad (11)$$

$$RBCl_2 + 3\,Napz \rightarrow Na[RB(pz)_3] + 2\,NaCl \quad (12)$$

$$RBH_2 + Napz + Hpz(excess) \rightarrow Na[RB(pz)_3] + 2\,H_2 \quad (13)$$

The bis-chelate L_2^*Co derived from $[(C_6H_4\text{-}4\text{-}Br)B(pz)_3]^-$ was prepared and converted to the 4-Li, 4-D, 4-C_4H_9, 4-COOH and 4-$COOCH_3$ derivatives, which may have some promise as intermediates for covalently bound NMR shift reagents [50]

Known poly(1-pyrazolyl)borates of the general structure *21* and of the type $M[R_2B(pz^*)_2]$ (n = 2) are listed in Table 6, those of the type $M[RB(pz^*)_3]$ (n = 1) are compiled in Table 7.

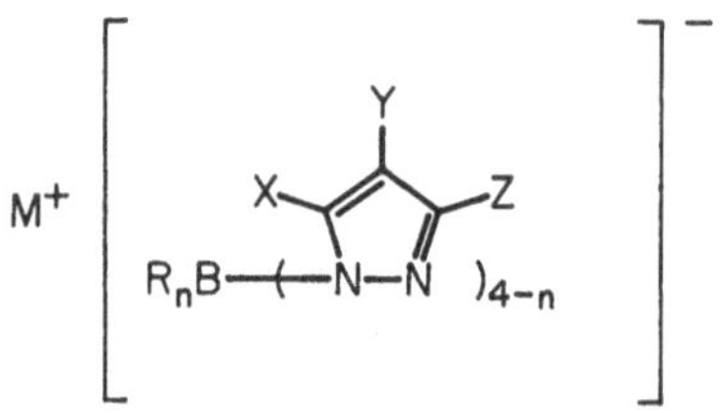

21

Table 6. Poly(1-pyrazolyl)borates of Structure *21* with n = 2

R	X	Y	Z	M	Ref.
H	H	H	H	K	40,41)
H	H	H	H	$N(CH_3)_4$	40)
H	H	H	CH_3	K	56,189)
H	CH_3	H	CH_3	K	80)
H	C_2H_5	H	C_2H_5	K	80*)
H	C_6H_5	H	C_6H_5	K	80*)
H	CH_3	CH_3	CH_3	K	43*)
CH_3	H	H	H	Na	81)
C_2H_5	H	H	H	Na	43)
n-C_4H_9	H	H	H	Na	43)
C_6H_5	H	H	H	Na	43)
C_6H_5	H	H	H	K	32)
$N(CH_3)CH_2^-$	H	H	H	H	4)
CH_3	CH_3	H	CH_3	Na	81)
F	CH_3	H	CH_3	Na	43*)

Table 7. Poly(1-pyrazolyl)borates of Structure *21* with n = 1

R	X	Y	Z	M	Ref.
H	H	H	H	Li	40)
H	H	H	H	Na	40)
H	H	H	H	K	40, 41)
H	H	H	H	$N(CH_3)_4$	40)
H	H	H	CH_3	K	189)
H	H	H	CH_3	$N(C_2H_5)_4$	57)
H	H	i-C_3H_7	H	K	43*)
H	H	Cl	H	K	43*)
H	H	Br	H	K	83)
H	CH_3	H	CH_3	Li	43)
H	CH_3	H	CH_3	K	43)
H	C_2H_5	H	C_2H_5	K	82)
H	CH_3	CH_3	CH_3	K	43)
H	CH_3	n-C_4H_9	CH_3	K	43*)
H	CH_3	Cl	CH_3	$Mo(CO)_2NO$	44*)
H	CH_3	I	CH_3	$Mo(CO)_2NO$	44*)
H	CH_3	I	CH_3	$Re(CO)_3$	46*)
i-C_3H_7	H	H	H	Na	47)
n-C_4H_9	H	H	H	Na	43, 47)
C_6H_5	H	H	H	H_2pz	43)
C_6D_5	H	H	H	1/2 Co	48)
C_6H_4-4-D	H	H	H	1/2 Co	50)
C_6H_4-3-CH_3	H	H	H	1/2 Co	49)
C_6H_4-4-CH_3	H	H	H	1/2 Co	49)
C_6H_4-4-C_4H_9	H	H	H	1/2 Co	50)
C_6H_4-4-COOH	H	H	H	1/2 Co	50)
C_6H_4-4-$COOCH_3$	H	H	H	1/2 Co	50)
C_6H_4-4-Br	H	H	H	1/2 Co	50)
C_6H_4-4-Li	H	H	H	NI	50)
C_6H_4-4-Li	H	H	H	1/2 Co	50)
pz	H	H	H	Li	40)
pz	H	H	H	Na	40)
pz	H	H	H	K	40, 41)
pz	H	H	H	Cs	40)
pz	H	H	H	$N(CH_3)_4$	40)
$N(CH_3)_2$	H	H	H	H	11)
pz-3-CH_3	H	H	CH_3	K	189)
pz-3,5-$(CH_3)_2$	CH_3	H	CH_3	H	30)

Included are ligands which have not been isolated in the free state but were either prepared in situ or were converted without isolation to complex species; relevant references are marked with an asterisk. Not included are the various indazolylborates. The structures of the latter have not yet been convincingly established and may be in error, since it is claimed that the boron is bonded to the most sterically hindered nitrogen atom [51–55].

VI.C Coordination Chemistry of Poly(1-pyrazolyl)borates

VI.C.1 Complexes of $[R_2B(pz^*)_2]^-$ Ligands

The $[R_2B(pz^*)_2]^-$ ligand resembles a beta-diketonate ion in its geometry and charge, and it readily forms chelates of the type *16* (see p. 17) with divalent transition metal ions; these complexes may be square planar (Ni, Cu) or tetrahedral (Mn, Fe, Co, Zn). The are two main features distinguishing $[R_2B(pz^*)_2]^-$ ligands from diketonates:

1) The steric crowding around the metal due to the 3-(C)H of the pyrazole ring, which is even more pronounced if there is a substituent in the 3-position. Thus, unlike the case of diketonates, all known complexes of type *16* are monomeric.
2) The $B(pz^*)_2M$ ring is not planar but puckered in the boat form as shown in *22*, which places the pseudoaxial R group close to M.

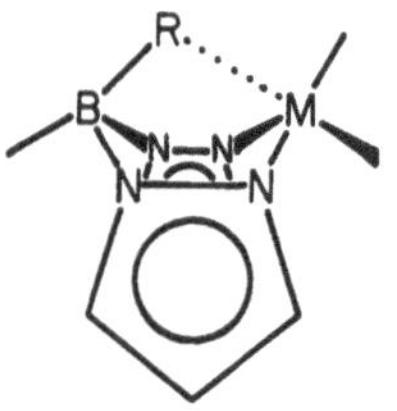

22

The effect of this proximity ranges from blocking access to the metal by other ligands to outright bond formation between the boron substituent R and M. The first case may be illustrated by comparing the reactivity of the orange, square planar nickel chelates $[H_2B(pz)_2]_2Ni$ and $[(C_2H_5)_2B(pz)_2]_2Ni$. The former quite readily forms blue-violet octahedral complexes with donor molecules such as pyridine or pyrazole; the latter is completely unreactive. Such steric protection by the pseudoaxial ethyl groups also accounts for the remarkable stability of the Cr(II) complex $[(C_2H_5)_2B(pz)_2]_2Cr$ [69].

An example of the second case is $H_2B(pz\text{-}3,5\text{-}(CH_3)_2]_2Mo(CO)_2(\pi\text{-allyl})$, where the presence of a B—H—Mo bond (H—Mo distance = 2.14 Å) was established by X-ray crystallography [70]; similar B—H—M bridging was also encountered in $L^*Mo(CO)_2(\eta^3\text{-}C_7H_7)$ [71], in $L^*Ta(CH_3)_3Cl$ [72] and in $L^*Pt(CH_3)_3$ [73]. The B—H—Pt bond is broken in the last compound by other ligands L such as phosphites, carbon monoxide or pyrazole with the formation of $L^*Pt(CH_3)_3L$ [73].

Even more surprising was the finding that the pseudoaxial methylene group in $(C_2H_5)_2B(pz)_2MXY$ interacts with the metal beyond just providing steric protection. It has been noted that in complexes such as L_2^*Ni [43] or $L^*Mo(CO)_2(\eta^3\text{-}CH_2CRCH_2)$ [80] protons of one methylene group are shifted considerably in the NMR spectrum, implying an interaction with the metal. More detailed studies, including X-ray crystallographic investigations of $L^*Mo(CO)_2(\eta^3\text{-}CH_2CC_6H_5CH_2)$ [74] and of $L^*Mo(CO)_2(\eta^3\text{-}C_7H_7)$ [75] revealed that one of the methylenic protons forms a three-center C—H—M bond. This is particularly intriguing since in the C_7H_7 complex, which is η^3, the Mo could have achieved an 18-electron configuration by becoming $\eta^5\text{-}C_7H_7$ instead of opting for the C—H—Mo bond. At room temperature there is a rapid oscillatory exchange between the methylenic protons, while at −60 °C the structure found in the crystal is also present in solution. The first exchange process has an activation energy of about 14 kcal/mole; while breaking the C—H—Mo bond, which is accompanied by inversion of the $(C_2H_5)_2B(pz)_2Mo$ ring and occurs at higher temperatures, has E_a in the 17–20 kcal/mole range [76].

The three-center bond formation, while being helped by the geometry of the $(C_2H_5)_2B(pz)_2M$ ring, is primarily a result of the electron deficiency of $(C_2H_5)_2B(pz)_2Mo(CO)(\eta^3\text{-}CH_2CHCH_2)$, which has a 16-electron configuration, and in the absence of other electron donors seeks electron density even from an aliphatic methylene group. In the presence of a donor molecule, e.g., pyrazole, the C—H—Mo bond is broken by a trans attack of the base to yield the electronically saturated 1:1 molar adduct with the donor molecule. In this latter, the $B(pz)_2Mo$ ring is, for steric reasons, in a seldom encountered chair conformation [77].

While the ligands $[H_2B(pz^*)_2]^-$ and $[(C_2H_5)_2B(pz)_2]^-$ give rise to bonding interactions of their pseudoaxial substituent with the metal, such interactions are totally absent in compounds derived from $[(C_6H_5)_2B(pz)_2]^-$, as has been established by X-ray crystallography for $[(C_6H_5)_2B(pz)_2]_2Ni$ [78] and also for $(C_6H_5)_2B(pz)_2Mo(CO)(\eta^3\text{-}CH_2CHCH_2)$ [79]. The reason for this lack is steric in nature, as non-bonding repulsions force the pseudoaxial phenyl group to be at right angles to the B—Mo axis and thus completely out of range for any bonding.

The $[H_2B(pz)_2]^-$ ligand, apart from forming L_2^*M complexes of structure *16*, also produces octahedral $[L_3^*M]^-$ anions in solution which are, however, easily converted back to L_2^*M and $[L^*]^-$ [61], although tetraethylammonium salts of $[L_3^*M]^-$ with M = Co or Ni are stable [42]. Stable species with more than two L* are obtained with large ions, e.g., L_4^*U, $L_3^*UCl_2$ [58, 59] and L_4^*Np [60]. The structure of distorted octahedral $K[L_3^*V]$ was established by X-ray crystallography [195].

The boron-bonded hydrogens in L_2^*M chelates (M = Co, Ni, Cu, Zn, Cd) still retain reducing power toward organic carbonyl groups; in a study using cyclohexanone and cyclohex-3-enone as substrates, fairly good selectivity was found for the Ni and Co derivatives [119].

VI.C.2 Complexes of $[RB(pz^*)_3]^-$ Ligands

The $[RB(pz^*)_3]^-$ ions represent an interesting class of ligands which form a host of complexes with transition and main group metal ions. Being uninegative and usually occupying three coordination sites, these anions resemble in a way the cyclopentadienide ion. Hence, they form many complexes similar in their stoichiometry to the cyclopentadienide counterparts, but they also show different, and at times unique, behavior. The $[RB(pz^*)_3]^-$ ligands display a tremendous affinity for metal ions and they will often displace such strong ligands as cyclopentadienide or phosphines.

All $[RB(pz^*)_3]^-$ ligands are strong six-electron donors of C_{3v} (or local C_{3v}) symmetry. They promote the formation of six-coordinate octahedral compounds with some trigonal distortion (e.g., *18* and *19*, p. 17). One of the distinguishing features of $[RB(pz^*)_3]^-$ ligands is the extensive screening of the coordinated metal ion by the 3-(C)H or 3-(C)R substituents. This can be appreciated best from the values of cone angles, i.e., 100° for $[C_5H_5]^-$, 180° for $[HB(pz)_3]^-$ and 225° for $[HB\{pz\text{-}3,5\text{-}(CH_3)_2\}_3]^-$ [62], with even larger cone angles to be expected for $[RB(pz^*)_3]^-$ ligands containing substituents larger than methyl groups in the 3-position of the pyrazole ring.

The known octahedral L_2^*M complexes are very compact, unusually stable and often sublimable under vacuum. The approximate D_{3d} symmetry was established by X-ray crystallography for $[HB(pz)_3]_2Co$ [63], $[HB(pz)_3]_2Co^+$ [64], $[HB(pz)_3]_2Fe$ [65] and $[HB\{pz\text{-}3,5\text{-}(CH_3)_2\}_3]_2Fe$ [65].

While $[RB(pz)_3]_2M$ compounds can be roughly compared to the traditional sandwich complexes and there are also numerous examples of $RB(pz)_3MXYZ$ half-sandwiches, only a few instances of mixed sandwiches, i.e., compounds with one $[RB(pz)_3]^-$ ligand and one carbocyclic moiety, have been reported. These are $[RB(pz)_3Co(\eta^5\text{-}C_5H_5)]PF_6$, $RB(pz)_3Co[C_4(C_6H_5)_4]$, $[HB(pz)_3Rh\{\eta^5\text{-}C_5(CH_3)_5\}]PF_6$ and $[B(pz)_4Ru(\eta^6\text{-}C_6H_6)]PF_6$. Structures of the last two compounds were confirmed by X-ray crystallography [193,194].

When $[RB(pz)_3]^-$ ligands react with first-row transition metal ions in the M(III) state, i.e., Fe(III), Cr(III) or Co(III), the usual products are the cations $(L_2^*M)^+$ rather than neutral L_3^*M complexes. On the other hand, a whole series of L_3^*M complexes derived from lanthanides (M = La, Ce, Pr, Sm, Gd, Er, Y) has been obtained [190]. The structure of $[HB(pz)_3]_3Yb$, as determined by X-ray crystallography, shows two tridentate and one bidentate ligand and contains eight-coordinate Yb [191]. What is most surprising, however, is the finding that, on the basis of detailed NMR spectroscopic studies of $[HB(pz)_3]_3Yb$ and the corresponding Lu complex, the species are stereochemically rigid in solution, retaining the structure found in the crystal [192].

The only reported example of two different $RB(pz^*)_3$ groups bonded to the same metal was obtained by the reaction of Co(II) with a 1/1 mixture of $[(C_6H_4\text{-}4\text{-}Br)B(pz)_3]^-$ and $[C_6H_5B(pz)_3]^-$ and separating from the resultant mixture of all three possible products the desired $(C_6H_4\text{-}4\text{-}Br)B(pz)_3Co(pz)_3BC_6H_5$ by means of HPLC [50]. The structure of the analogously prepared complex $(C_6H_4\text{-}4\text{-}Br)B(pz)_3\text{-}Co[pz\text{-}3,5\text{-}(CH_3)_2]_3BH$ was confirmed by X-ray crystallography [196].

By far the greatest usefulness of $[RB(pz^*)_3]^-$ ligands has been established in the area of half-sandwich complexes, i.e., species of the type $RB(pz)_3MXYZ$, where stabilization of otherwise unattainable or highly unstable MXYZ structures can be achieved, thus permitting their study. An excellent example is provided by the synthesis of remarkably stable Cu(I) carbonyl complexes L*CuCO with L* = $HB(pz)_3$, $B(pz)_4$ and $HB[pz\text{-}3,5\text{-}(CH_3)_2]_3$, of which the last complex, containing the ligand with the deepest protective pocket for Cu, was found to be the most stable [66,67]. The structure of $HB(pz)_3CuCO$ was determined by X-ray crystallography [68].

Other interesting structure types stabilized by the $[HB(pz)_3]^-$ ligand are five-coordinate Pt complexes of the general composition $HB(pz)_3Pt(CH_3)L$. These include examples where L is a diversely substituted olefin, allene, substituted acetylene, carbon monoxide, isonitrile or phosphite [85,86]. While all the olefin complexes are stereochemically rigid at room temperature, the carbon monoxide complex is fluxional, retaining five-coordination in solution (^{195}Pt coupling to all 3-H and 4-H protons was observed) but being four-coordinate with bidentate $[HB(pz)_3]^-$ in the crystal [87,88]. Detailed studies of dynamic behavior of $HB(pz)_3Pt(CH_3)L$ and of $B(pz)_4Pt(CH_3)L$ species were carried out using ^1H and ^{13}C NMR spectroscopy [89]. Packing forces seem to determine the coordination number in the crystal: for $HB(pz)_3Pt(CH_3)(CF_3C{\equiv}CCF_3)$ the structure showed trigonal-bipyramidal coordination and tridentate $HB(pz)_3$ [90]; but in $HB(pz)_3Pt(CH_3)(CN\text{-}t\text{-}C_4H_9)$ coordination was square planar and $[HB(pz)_3]^-$ bidentate [91].

Bidentate coordination of $[RB(pz^*)_3]^-$ ligands is also observed in several other complexes. For example, in $RB(pz)_3Mo(CO)_2(C_5H_5)$ the 18-electron configuration is achieved by way of η^5-C_5H_5 and bidentate $RB(pz)_3$ rather than through η^3-C_5H_5 and tridentate $RB(pz)_3$, as was proven by X-ray crystal structure studies [104, 105]. On the other hand, in $C_6H_5B(pz)_3Mo(CO)_2(C_7H_7)$, the structure contains η^3-C_7H_7 and tridentate $C_6H_5B(pz)_3$ [106]. In related $RB(pz)_3RhL_2$ compounds, $RB(pz)_3$ is bidentate in the crystal when L_2 is 1,5-cyclooctadiene or norbornadiene, but tridentate (trigonal bipyramidal geometry) when L_2 is duroquinone [107]. The duroquinone and 1,5-cyclooctadiene complexes are both five-coordinate in solution, as is indicated by ^1H and ^{103}Rh NMR studies [108].

Altogether, the denticity of $[RB(pz)_3]^-$ ligands is often determined by subtle interplay of steric and electronic factors pertaining to the transition metal as well as to the other ligands present. Generally, changing the R group in $[RB(pz)_3]^-$ does not alter its coordinative behavior or the chemistry of its complexes, since the substituent R is far removed from the coordinated metal. However, the solubility in organic solvents is increased when R is a long-chain alkyl group. On the other hand, the introduction of substituents in the 3-position of the pyrazolyl moieties as in $[HB\{pz\text{-}3,5\text{-}(CH_3)_2\}_3]^-$ alters substantially the accessibility of the metal in $HB(pz\text{-}3,5\text{-}Me_2)_3MXYZ$ complexes to potential reactants and also hides the BH in a protective pocket of three methyl groups.

The use of $[HB\{pz\text{-}3,5\text{-}(CH_3)_2\}_3]^-$ as a stabilizing ligand permitted the synthesis of many new unusual Mo and W derivatives. Starting with $HB[pz\text{-}3,5\text{-}(CH_3)_2]_3Mo(NO)I_2$ it was possible to replace both iodine atoms concurrently or sequentially with a wide variety of nucleophiles. Some of the products obtained are:

$L^* = HB[pz\text{-}3,5\text{-}(CH_3)_2]_3$: L*Mo(NO)I(OR) [45, 126]
L*Mo(NO)X(SR) [127]
L*Mo(NO)I(NHR) [45, 126, 128]
L*Mo(NO)(OR)(OR') [129, 130]
L*Mo(NO)(OR)(SR) [127]
L*Mo(NO)(OR)(NHR') [130]
L*Mo(NO)(SR)(NHR') [127]
L*Mo(NO)I(NHNRR') [129, 131]
$L^*Mo(NO)I(NH_2)$ [129]
$L^*Mo(NO)I(NHNH_2)$ [131]

The last two complexes, in themselves rare examples of stable amido and hydrazido transition metal compounds, react with acetone to yield $HB[pz\text{-}3,5\text{-}(CH_3)_2]_3$-$Mo(NO)I[N{=}C(CH_3)_2]$ [129] and $HB[pz\text{-}3,5\text{-}(CH_3)_2]_3Mo(NO)I[NHN{=}C(CH_3)_2]$ [131], respectively.

Structures of representative complexes from the above mentioned types have been determined by X-ray crystallography. In all of these structures the Mo bond to the ZR group (Z = O, S, NH) is shortened, indicating backbonding to the electron-deficient Mo. No ring formation occurred (presumably for steric reasons) when terminal diols or analogous mercapto or amino reactants were used; only compounds such as $HB[pz\text{-}3,5\text{-}(CH_3)_2]_3Mo(NO)I(OROH)$ were obtained [134, 135].

In contrast to the reaction of $[RB(pz)_3Mo(CO)_3]^-$ (R = H or pz) with aryldiazonium ions, which proceeds cleanly to yield red arylazo derivatives $RB(pz)_3Mo(CO)_2N{=}NAr$, the corresponding reaction of $[HB\{pz\text{-}3,5\text{-}(CH_3)_2\}_3Mo(CO)_3]^-$ produced blue compounds initially thought to have the structure $L^*Mo(CO)_3Ar$ [134] but which have now been identified (by X-ray crystallography) as $L^*Mo(CO)_2(\eta^2\text{-}COAr)$ [135]. This reaction proceeds by a radical pathway, as was shown by the isolation of $L^*Mo(CO)_2(\eta^2\text{-}COC_6H_{11})$ when the reaction was run in cyclohexane [135]. Carrying out the reaction of $[L^*Mo(CO)_3]^-$ with aryldiazonium ion or, even better, aryliodonium ion in methylene chloride leads to the formation of the unusual halocarbyne derivative $L^*Mo(CO)_2{\equiv}CCl$, which contains a reactive chlorine and could be converted to numerous derivatives. Reactions with RS^-, RSe^- or $C_6H_5^-$ produced $L^*Mo(CO)_2{\equiv}CSR$, $L^*Mo(CO)_2{\equiv}CSeR$ and $L^*Mo(CO)_2{\equiv}CC_6H_5$, respectively, while treatment with E^{2-} (E = S, Se, Te) produced the appropriate $[L^*Mo(CO)_2({\equiv}CE)]^-$ derivatives [136,137].

A η^2-acyl structure was also established in the product obtained from the reaction of $[HB(pz)_3Mo(CO)_3]^-$ with methyl iodide [139], which was originally thought to be the seven-coordinate $HB(pz)_3Mo(CO)_3CH_3$ [138]. This implies that the $RB(pz)_3$ group is a strong enforcer of six-coordination in Mo and W derivatives. The only authentic examples of seven-coordinate species seem to be $RB(pz)_3M(CO)_3H$ (M = Mo, W) [138], the monohalo derivatives $HB(pz)_3Mo(CO)_3X$ (X = Br, I), and $HB(pz)_3W(CO)_2(CS)$ [140].

Oxidation of $[HB(pz)_3Mo(CO)_3]^-$ produces a 17-electron radical which is stable in the absence of oxygen; the structure of the species has been confirmed by X-ray crystallography. On heating, it loses carbon monoxide and forms quantitatively the dinuclear species $HB(pz)_3Mo(CO)_2Mo(CO)_2(pz)_3BH$, which contains a Mo—Mo triple bond [141].

Reversible electrochemical reduction of $HB[pz\text{-}3,5\text{-}(CH_3)_2]Mo(NO)I_2$ produces a paramagnetic anion (g = 2.206), which loses iodide ion forming paramagnetic (g = 1.998), $L^*Mo(NO)I$. This compound is thought to be the intermediate in nucleophilic substitution on $L^*Mo(NO)I_2$ when derivatives of type $L^*Mo(NO)I(ZR)$ are produced (Z = O, S, NH) [197]. Confirmation for the paramagnetic center in $L^*Mo(NO)(CH_3CN)_2^{+\cdot}$ (g = 1.982) being on Mo was obtained by observing, at low temperature, the ^{95}Mo and ^{97}Mo satellites in the correct ratio [198]. Other well-characterized and stable low-valent Mo and W derivatives of L* have been reported [142].

A wealth of interesting results was obtained from the higher-valent oxomolybdenum species $HB[pz\text{-}3,5\text{-}(CH_3)_2]_3MoOX_2$ and L^*MoO_2X, most of the studies being directed toward the chemistry of the Mo centers of oxo-type enzymes such as xanthine oxidase, sulfite oxidase and nitrate reductase, which involve change in the number of oxygens in the substrate and Mo going between the IV, V and VI states. The complex L^*MoOCl_2 served as a versatile starting material in which both Cl atoms could be replaced by NCS, SR, OR or even by chelating ligands such as $-XCH_2CH_2Y-$ and 1,2-phenylene-(X—)(Y—). This is in marked contrast to the inability of X_2 in $L^*Mo(NO)X_2$ to be replaced in chelating fashion by the cited bifunctional ligands [132]. Many of the L^*MoOX_2 complexes has Mo(V)/Mo(IV) reduction potentials in the range of the enzymes. The structure of $L^*MoO(SC_6H_5)_2$ was determined by X-ray crystallography [187].

The reaction of $[N(C_2H_5)_4][L^*M(CO)_3]$ (M = Mo or W but not Cr) with $N_3S_3Cl_2$ yielded $L^*Mo(CO)_2(NS)$ along with the dinuclear complex $(L^*Mo(CO)_2)_2S$ [199] which, in the case of Mo was also obtained by using sulfur [114] instead of $N_3S_3Cl_2$;

^{13}C, ^{14}N and ^{95}Mo NMR spectra of $L^*Mo(CO)_2(NS)$ and $L^*Mo(CO)_2(NO)$ have been studied [200].

The free acid $[HB(pz)_3Mo(CO)_3]H$ reacts with sulfur or selenium yielding dinuclear species $[L^*Mo(CO)_2]_2Z$ (Z = S, Se). Their structures and that of $[HB\{pz\text{-}3,5\text{-}(CH_3)_2\}_3Mo(CO)_2]_2S$ were determined by X-ray crystallography. The Mo—S—Mo bonds are quite short (2.180 and 2.200 Å), as is the Mo—Se—Mo bond (2.323 Å), and all are linear. In $[HB\{pz\text{-}3,5\text{-}(CH_3)_2\}_3Mo(CO)_2]_2S$ the Mo—N distance trans to bridging S atom is significantly shorter than the other two Mo—N distances (2.188 versus 2.254 Å) while in the $HB(pz)_3$ analog they are indistinguishable [114].

The stable Mo(VI) complex $HB[pz\text{-}3,5\text{-}(CH_3)_2]_3MoO_2Cl$ led to a variety of products derived from substitution of Cl by diverse nucleophiles; the rates of these substitutions are very slow [187]. The products L^*MoO_2X (X = Cl, Br, OCH_3, NCS) were studied by ^{95}Mo NMR and an inverse halogen dependence of the ^{95}Mo chemical shift was found [201].

VI.C.3 Physical Studies

Numerous physical studies of poly(1-pyrazolyl)borate complexes have been carried out, some as part of structure determinations, others to study different phenomena inherent in the complexed transition metal.

For example, interesting results were obtained in the octahedral $[RB(pz^*)_3]_2Fe$ chelate family with regard to their magnetic moments, which were found to be strongly dependent on the L* substituents. Thus, while $[pzB(pz)_3]_2Fe$ is diamagnetic at room temperature, $[HB\{pz\text{-}3,5\text{-}(CH_3)_2\}_3]_2Fe$ is fully paramagnetic (μ_{eff} ca. 5.2 BM), and $[HB(pz)_3]_2Fe$ has an intermediate value of 2.7 BM [143]. The last magnetic moment is strongly temperature dependent and is due to a spin equilibrium between the $^5A_{1g}$ and $^1A_{1g}$ states. Besides the above results, obtained in solution, similar behavior was found in the solid state and was studied by Mössbauer and magnetic susceptibility techniques down to 4.2 K. Both the high-spin and low-spin states were clearly observed in the equilibrium. In the solid state, $[HB(pz)_3]_2Fe$ was diamagnetic at room temperature and below, although it became paramagnetic above 300 K, while $[C_6H_5B(pz)_3]_2Fe$ was diamagnetic even above room temperature. The compound $[HB\{pz\text{-}3,5\text{-}(CH_3)_2\}_3]_2Fe$ is paramagnetic at room temperature, but becomes fully diamagnetic at 147 K, while the related compound $[HB\{pz\text{-}3,4,5\text{-}(CH_3)_3\}_3]_2Fe$ is fully paramagnetic down to 4.2 K [144, 145]. The spin equilibrium in $[HB(pz)_3]_2Fe$ was studied by resonance and pulsed ultrasonic techniques [146], and its high-temperature crossover by Mössbauer, far-infrared and variable temperature magnetic susceptibility [147]. A comparison of Fe—N bond lengths in the low-spin $[HB(pz)_3]_2Fe$ and the high-spin $[HB\{pz\text{-}3,5\text{-}(CH_3)_2\}_3]_2Fe$ showed the latter to be longer by 0.199 Å. This is one of the largest observed bond length expansions from low-spin to high-spin state [148].

Various investigations were devoted to stereochemical nonrigidity encountered in $HB(pz)_3$ and $B(pz)_4$ derivatives, as these various dynamic processes were amenable to study by NMR. In ionic complexes such as $K[B(pz)_4]$ or $Zn[B(pz)_4]_2$ (the same holds true for $HB(pz)_3$ analogs) all pz groups are equivalent by NMR, which implies rapid exchange of coordinated and uncoordinated pz groups. It is noteworthy that $Zn[B(pz)_4]_2$ is isomorphous with $Co[B(pz)_4]_2$ and, hence, six-coordinate in the crystal but four-coordinate and with rapid exchange of pz groups in solution. The same type of dynamic equivalence of pz groups has been found in numerous Pd(II) derivatives of the type $B(pz)_4PdL_2$. The mechanism of such exchange involves presumably a five-coordinate transition state, in which the pseudoaxial pz group becomes attached to Pd at the same time as one of the coordinated pz group departs:

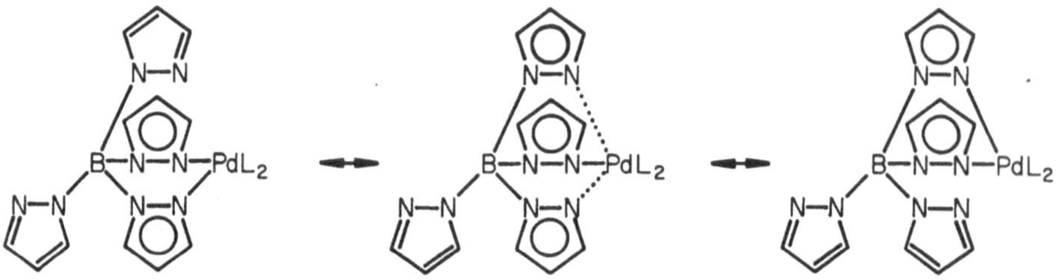

The process is repeated to average the pz environments and also involving inversion of the $B(pz)_2Pd$ boat. Examples of this type of exchange are fairly common and were found in a number of $RB(pz)_3$ complexes of Pd(II) [94-97], Rh(I) [98], Pt(II) [99], Cu(I) [100], Au(III) [101,102], and Hg(II) [102].

In octahedral compounds such as $Co[B(pz)_4]_2$ or $[B(pz)_4Mo(CO)_3]^-$ the fourth, uncoordinated pz group maintains its separate NMR identity. This can be seen particularly well in the Co(II) complex, where the peaks range from −5500 to +6960 Hz (at 60 MHz) because of contact shifts. The fact that the pz groups show up as a 3:1 pattern implies a low barrier to rotation of the uncoordinated pz moiety about the B—N bond, thus making the environments of all 5-protons averaged.

In compounds such as $HB(pz)_3Mo(CO)_2(\eta^3\text{-}CH_2CHCH_2)$ and its various analogs, the room temperature NMR spectrum is indistinct and right between the limiting high- and low-temperature spectra. The high-temperature spectrum (≈60 °C) indicates dynamic C_{3v} symmetry with all three coordinated pyrazolyl groups being identical [83], while the low-temperature spectrum (≈ −60 °C) is in agreement with the static structure found in the crystal [103], which shows two identical and one different pz group. The activation energies for this process, involving rotation about the B—M axis, are in the 13–15 kcal/mole range for a series of related compounds [92]. In the more sterically hindered $HB[pz\text{-}3,5\text{-}(CH_3)_2]_3Mo(CO)_2(\eta^3\text{-}CH_2CCH_3CH_2)$, the room temperature NMR spectrum shows the static structure and only at 165 °C does the dynamic C_{3v} spectrum appear [83]. Altogether, the $RB(pz^*)_3$ class of ligands is exceptionally well suited for the study of their rotation about a transition metal by NMR spectroscopy.

Inversion of the $B(pz^*)_2M$ ring has been studied most thoroughly in bidentate complexes, $R_2B(pz^*)_2MXY$, where it can be followed by means of coalescence of the R signals from the initially non-identical pseudoaxial and pseudo-equatorial groups into a single peak system of an averaged R group (see p. 23).

A large number of poly(1-pyrazolyl) borate complexes had their structure determined by X-ray crystallography; these are listed in Table 8. It should be remembered, however, that in some instances the structure in solution will have a different denticity of the poly(1-pyrazolyl)borate ligand than that found in the crystal.

Table 8. Listing of Poly(1-pyrazolyl)borate Complexes of Known Crystal Structures

L*	Complex	Ref.
$H_2B(pz)_2$	$K[L^*_3V]$	150)
	L^*Co	151)
	L^*_2Ni	149)
$H_2B[pz\text{-}3,5\text{-}(CH_3)_2]_2$	$L^*Ta(CH_3)_3Cl$	154)
	$L^*Mo(CO)_2(C_7H_7)$	152, 153)
	$L^*Mo(CO)_2(\eta^3\text{-}CH_2CHCH_2)$	70)
$(C_2H_5)_2B(pz)_2$	L^*_2Ni	155)
	L^*_2Cr	69)
	$L^*_2Mo_2(OAc)_2$	157)
	$L^*Mo(CO)_2C_7H_7$	75)
	$L^*Mo(CO)_2(\eta^3\text{-}CH_2CHCH_2)$	77)
	$L^*Mo(CO)_2(\eta^3\text{-}CH_2CC_6H_5CH_2)$	74)
	$L^*Mo(\mu\text{-}CH_3COO)_2Mo(\mu\text{-}pz)(\mu\text{-}OH)B(C_2H_5)_2$	122)
	$L^*Pt(CH_3)(C_6H_5C{\equiv}CCH_3)$	156)
$(C_6H_5)_2B(pz)_2$	L^*_2Ni	78)
	$L^*Mo(CO)_2(\eta^3\text{-}CH_2CHCH_2)$	79)
	$L^*Mo(CO)_2(C_7H_7) \cdot Hpz$	158)
$HB(pz)_3$	L^*CuCO	68)
	$L^*Cu(C_2H_4) \cdot CuCl$	167)
	$[L^*Cu]_2$	109)
	L^*CuCl	111)
	$L^*Au(CH_3)_2$	162)
	L^*_2Fe	148)
	$L^*Fe(\mu\text{-}O)(\mu\text{-}HCO_2)_2FeL^*$	119)
	$L^*Fe(\mu\text{-}O)(\mu\text{-}CH_3CO_2)_2FeL^*$	118, 119)
	L^*_2Co	63)
	$[L^*_2Co][Sn_2Co_5Cl_2(CO)_{19}]$	64)
	$L^*Pt(CH_3)CO$	87, 88)
	$L^*Pt(CH_3)(CF_3C{\equiv}CCF_3)$	90)
	$L^*Pt(C{\equiv}N\text{-}t\text{-}C_4H_9)$	91)
	$[L^*Rh\{C_5(CH_3)_5\}]PF_6$	168)
	$L^*Mo(CO)_2NO$	164)
	$[L^*Mo(CO)_3] \cdot$	141)
	$L^*Mo(CO)_3Br$	138)
	$L^*Mo(CO)_2N_2C_6H_5$	169)
	$L^*Mo(CO)_2(\eta^2\text{-}COCH_3)$	138)
	$L^*Mo(CO)_2(\eta^2\text{-}COC_6H_5)$	138)
	$L^*Mo(CO)_2(\eta^3\text{—}CH_2CCH_3CH_2)$	165)
	$(L^*Mo(CO)_2)_2$	141)
	$L^*Mo(N_2C_6H_4\text{-}4\text{-}F)(SC_6H_4\text{-}4\text{-}CH_3)_2$	166)
	24 (p. 31)	117)
	$L^*Mo(\mu\text{-}CH_3COO)_2MoL^*$	115)
	L^*WCl_3 1/2 $[(C_2H_5)_4N,NH_4]$	142)
	$L^*Sn(CH_3)_3$	161
	L^*TcCl_2O	160)
	L^*_3Yb	159)
$HB(pz\text{-}3\text{-}CH_3)_3$	$[N(C_2H_5)_4](L^*Mo(CO)_3]$	57)
$C_2H_5B(pz)_3$	$[L^*BC_2H_5]PF_6$	34)
$C_6H_5B(pz)_3$	$L^*Mo(CO)_2(\eta^3\text{-}C_3H_5)$	106)

Table 8. (continued)

L*	Complex	Ref.
	$L^*Mo(CO)_2(C_7H_7)$	106)
$HB[pz\text{-}3,5\text{-}(CH_3)_2]_3$	$[L^*BH]TaCl_6$	30)
	$(L^*Cu)_2$	110)
	$L^*Cu(C_2H_4)$	167)
	$K[L^*Cu(SC_6H_4\text{-}4\text{-}NO_2)]$	181, 182)
	L^*_2Fe	148)
	$L^*Co(SC_6F_5)$	180)
	L^*MoCl_2O	174)
	$L^*Mo(CO)_2(SC_6H_4\text{-}4\text{-}NO_2)$	137)
	$L^*Mo(CO)_2(SC_6H_4\text{-}4\text{-}Cl)$	178)
	$L^*Mo(NO)(1\text{-pyrrolide})_2$	179)
	$L^*Mo(CO)_2(\eta^2\text{-}COC_6H_5)$	135)
	$L^*Mo(CO)_2(\eta^2\text{-}COC_6H_{11})$	135)
	$(L^*Mo(NO)I)_2O$	112, 113)
	$(L^*Mo(CO)_2)_2S$	114)
	$[L^*Mo(NO)(CH_3CN)_2]PF_6$	177)
	$[(C_2H_5)_4N](L^*Mo(CO)_3]$	175)
	$[(C_2H_5)_4N](L^*MoCl_3]$	176)
$B(pz)_4$	$L^*Cu[C_6H_4\text{-}1,2\text{-}\{As(CH_3)_2\}_2]$	173)
	$L^*Mo(CO)_2(C_5H_5)$	104, 105)
	$L^*Mo(CO)_2(\equiv CCl)$	136)
	$L^*Mo(CO)(NO)[P(C_6H_5)_3]$	170)
	$L^*Rh(CO)I_2$	171)
	L*Rh(norbornadiene)	107)
	L*Rh(1,5-cyclooctadiene)	107)
	L*Rh(duroquinone)	107)
	$[L^*Ru(C_6H_6)]PF_6$	172)

VI.C.4 Dinuclear Complexes Capped by $HB(pz^*)_3$ Groups

Although most poly(1-pyrazolyl)borate complexes are of the structure L^*_nM or $L^*_nMX_m$ (where X represents various other types of ligands), there are also examples of complexes containing the structure $[RB(pz)_3M]_2Z$ (where Z can be any type of linkage between the metal atoms). The simplest example of such a species is $HB(pz)_3Cu(\mu\text{-}Cl)_2Cu(pz)_3BH$ and its Co analog, both containing five-coordinate metal. This type of complex is, however, labile (especially the Co derivative) and disproportionates readily on heating to yield $M[HB(pz)_3]_2$ and MCl_2 [111].

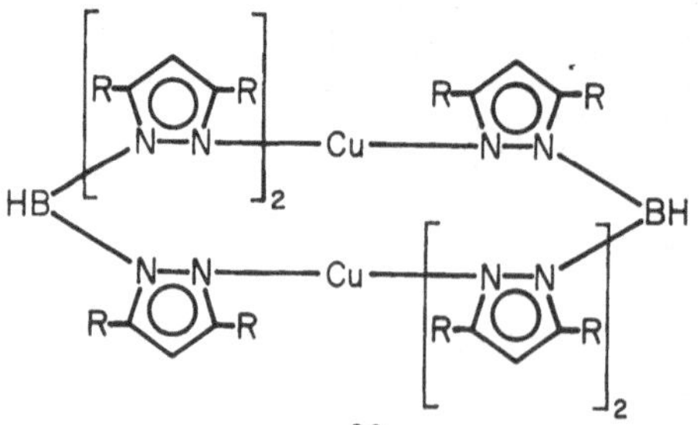

23

R = CH_3

The dimer $[HB(pz)_3Cu]_2$ has an unusual structure which involves a rare example of the pyrazole N2 bonding to two metal atoms. The central core consists of a Cu—N—Cu—N ring; each Cu is further bonded to two pz groups from different L* ligands. At the same time all pz groups are NMR equivalent down to −130 °C [109]. The analogous dimer $[HB\{pz\text{-}3,5\text{-}(CH_3)_2\}_3Cu]_2$ has a different structure, *23*, which contains two trigonal planar copper atoms. This latter dimer is highly dissociated in benzene [110].

Various dinuclear complexes involving Mo have been reported. They include $[(HB\{pz\text{-}3,5\text{-}(CH_3)_2\}_3Mo(NO)I]_2O$, where the Mo—O—Mo bond is bent at a 171° angle [112,113]; and also $[HB(pz)_3Mo(CO)_2]_2S$, which was prepared from $[HB(pz)_3Mo(CO_3]H$ and elemental sulfur. The complex $[HB\{pz\text{-}3,5\text{-}(CH_3)_2\}_3\text{-}Mo(CO)_2]_2S$ was obtained in similar fashion and the two latter species contain a linear Mo—S—Mo unit. A linear Mo—Se—Mo unit is present in $[HB(pz)_3Mo(CO)_2]_2Se$, which was obtained from $[HB(pz)_3Mo(CO)_3]H$ and black selenium [114].

Also known are complexes of the type $[HB\{pz\text{-}3,5\text{-}(CH_3)_2\}_3M(NO)X]_2Z$ where M = Mo or W, X = Cl or I, and Z is a long organic bridge such as $NHC_6H_4OC_6H_4NH$, $NHC_6H_4CH_2C_6H_4NH$, NHC_6H_4NH or OC_6H_4O [115]. In a $HB(pz)_3Mo(\mu\text{-}CH_3COO)_2\text{-}Mo(pz)_3BH$ one of the $HB(pz)_3$ ligands is bidentate; the other is tridentate with a weakly bonded axial pz group, in line with the known resistance of dimolybdenum(II) to attach to axial ligands [116]. Finally, the structure of the rather complicated species *24*, which contains both oxy and methoxy bridges, has been determined by X-ray crystallography [117].

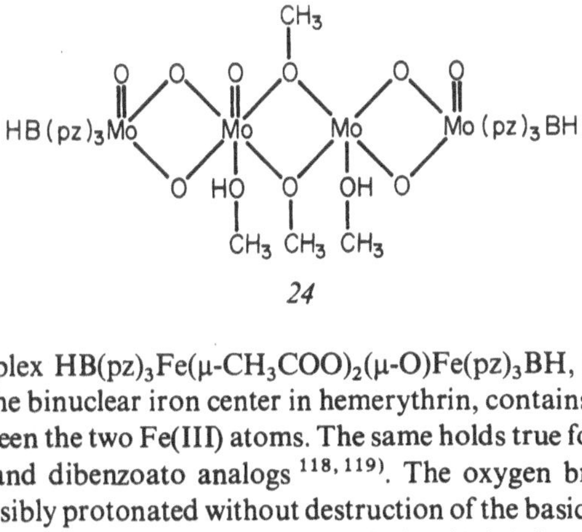

24

The complex $HB(pz)_3Fe(\mu\text{-}CH_3COO)_2(\mu\text{-}O)Fe(pz)_3BH$, prepared in one step as a model for the binuclear iron center in hemerythrin, contains two CH_3COO and one O bridge between the two Fe(III) atoms. The same holds true for the similarly synthesized diformato and dibenzoato analogs [118,119]. The oxygen bridge in these compounds can be reversibly protonated without destruction of the basic structure [124]. A binuclear complex of even higher symmetry, $L^*Rh(CO)_3RhL^*$ (where $L^* = HB(pz)_3$ or $B(pz)_4$) was obtained from the reaction of L*Rh(diolefin) with carbon monoxide [98].

VI.C.5 Complexes of Bis-Bidentate $[B(pz)_4]^-$

Although usually coordinating in tridentate fashion, i.e., as $[pzB(pz)_3]^-$, the ligand $[B(pz)_4]^-$ can also be bis-bidentate when chelating a metal or metalloid which prefers four-coordination. The formation of polyboron spiro cations of the type $[R_2B(pz)_2B(pz)_2BR_2]^+$ from the reaction of $[B(pz)_4]^-$ with two molecules of R_2BX has already been noted (see Sect. V.A.). The same principal structure can also be obtained with various organometallic species. For instance, the cation $[(\eta^3\text{-}CH_2CHCH_2)Pd(pz)_2B(pz)_2Pd(\eta^3\text{-}CH_2CHCH_2)]^+$ can be prepared from $[B(pz)_4]^-$ and π-allylpalladium chloride dimer. Analogs containing methyl or phenyl substituents on the central π-allyl carbon atom are also known. The synthesis proceeds through $(pz)_2B(pz)_2Pd(\eta^3\text{-}CH_2CHCH_2)$, which is a stable compound, and which can be isolated in good yield when the reactants are in the appropriate molar ratio. In addition, the mixed cation $[(C_2H_5)_2B(pz)_2B(pz)_2Pd(\pi\text{-allyl})]^+$ can be prepared by (a) the reaction of the above mentioned intermediate with diethylboryl tosylate, or (b) by treating the pyrazabole $(C_2H_5)_2B(pz)_2B(pz)_2$ with π-allylpalladium chloride dimer. A variety of such mixed bridged spiro-cations can be prepared by the above method [27].

The reaction of $[LPdCl_2]_2$ (L = $P(C_2H_5)_3$ or $P(OC_2H_5)_3$) with $[B(pz)_4]^-$ gave the asymmetric, stereochemically nonrigid compounds $(pz)_2B(pz)_2PdClL$, but when L was $(C_2H_5)_2S$, the binuclear complex $(pz)_2B(pz)_2Pd(pz)_2B(pz)_2PdCl_2$ was obtained; the latter is also available directly from $PdCl_4^-$ [123].

Additional examples of bis-bidentate coordination of the $[B(pz)_4]^-$ include $[(C_5H_5)_2Ti(pz)_2Ti(C_5H_5)_2]^+$ [202], $[B(pz)_4Th(pz)_2B(pz)_2Th(pz)_4B]^{3+}$ [125], and also the rather complex derivative $[\{(C_6H_5)_3P\}(CO)(CH_3CHCN)Ru(pz)_2B(pz)_2RuCl(CO) \cdot (CH_3CHCN)\{P(C_6H_5)_3\}]^+$ [185].

VI.C.6 Poly(1-pyrazolyl)borates Containing Other Donor Groups

This type of ligand, where the coordinating power of the pyrazolyl groups on boron is augmented by some other donor group is currently very poorly represented. One of such ligands, actually a bidentate mono(1-pyrazolyl)borate, i.e., $[(C_2H_5)_2B(pz)(OH)]^-$, was obtained inadvertently through partial hydrolysis of $[(C_2H_5)_2B(pz)_2]^-$ and was isolated in the dinuclear complex $(C_2H_5)_2B(pz)_2Mo(\mu\text{-}CH_3COO)_2Mo(\mu\text{-}pz)(\mu\text{-}OH) \cdot B(C_2H_5)_2$ [122]. The tridentate ligand $[HB\{pz\text{-}3,5\text{-}(CH_3)_2\}_2(SC_6H_4\text{-}4\text{-}CH_3)]^-$ was synthesized intentionally from ArSH and $K[H_2B\{pz\text{-}3,5\text{-}(CH_3)_2\}_2]$. It was converted to L*MSR complexes (M = Cu or Co; SR = O-ethylcysteine, p-nitrobenzenethiolate or pentafluorophenylthiolate). These compounds were studied as synthetic approximations of the proposed active sites in the blue copper proteins (plastocyanin, azurin) [121].

An interesting ligand is $[(CH_3)_2NB(pz)_3]H$ [11]. It was found to coordinate in tridentate fashion through the dimethylamino and two pz groups in the stable complex $pzB(\mu\text{-}pz)_2[\mu\text{-}N(CH_3)_2]Mo(CO)_2(\eta^3\text{-}CH_2CHCH_2)$, as was established by NMR studies. The temperature-dependent spectra are consistent with rotation about the B—Mo axis, but without exchange of the coordinated and uncoordinated pz groups. Two different rotamers in almost statistical distribution can be observed at low temperature [184].

VI.C.7 Concluding Remarks

At the current stage of its development, the coordination chemistry of the parent poly(1-pyrazolyl)borate ligand system has been reasonably well explored, and, to a lesser extent, that derived from 3,5-dimethylpyrazole. This ligand choice was partly based on the commercial availability of the pyrazoles and, lately, of the poly(1-pyrazolyl)borate ligands themselves. By contrast, there are still only limited examples of assessing the relationship between structure and coordinating behavior as the ligand $[R_nB(pz^*)_{4-n}]^-$ is altered through substitution on boron or on pyrazolyl-carbon sites, or even by replacing some pz* groups with alternate donor functionalities.

While early work emphasized high symmetry of this ligand, the current trends may take advantage of asymmetric substitution to adapt the basic poly(1-pyrazolyl)borate framework to various enzyme-like tasks. This could be done by placing appropriate functionalities in such a manner that they will interact, as needed, with a transition metal (or metals). Achievement of these refinements of the ligand system is likely to result from advances in boron chemistry, boron being here the pivotal element.

VII References

1. Matteson, D. S.: J. Org. Chem. *27*, 4293 (1962)
2. Trofimenko, S.: J. Am. Chem. Soc. *89*, 3165 (1967)
3. Niedenzu, K., Weber, W.: J. Organometal. Chem. *195*, 25 (1980)
4. Weber, W., Niedenzu, K.: ibid. *205*, 147 (1981)
5. Alam, F., Niedenzu, K.: ibid *243*, 19 (1983)
6. Alam, F., Niedenzu, K.: ibid. *240*, 107 (1982)
7. Hodgkins, T. G., Niedenzu, K., Niedenzu, K. S., Seelig, S. S.: Inorg. Chem. *20*, 2097 (1981)
8. O'Brien, D. H., Hrung, C. P.: J. Organometal. Chem. *27*, 185 (1971)
9. Cotton, F. A., Ciappenelli, D. J.: Synth. React. Inorg. Metal-Org. Chem. *2*, 197 (1972)
10. Kook, A. M.: Dissertation, University of Kentucky, 1984
11. Niedenzu, K., Seelig, S. S., Weber, W.: Z. Anorg. Allg. Chem. *483*, 51 (1981)
12. Frauendorfer, E., Agrifoglio, G.: Inorg. Chem. *21*, 4122 (1982)
13. Trofimenko, S.: J. Am. Chem. Soc. *88*, 1842 (1966)
14. Trofimenko, S.: ibid. *89*, 4948 (1967)
15. Niedenzu, K., Nöth, H.: Chem. Ber. *116*, 1132 (1983)
16. Hanecker, E., Hodgkins, T. G., Niedenzu, K., Nöth, H.: Inorg. Chem. *24*, 459 (1985)
17. Niedenzu, K., Niedenzu, P. M.: ibid. *23*, 3713 (1984)
18. Trofimenko, S.: ibid. *8*, 1714 (1969)
19. Trofimenko, S.: Inorg. Synth. *12*, 107 (1970)
20. Trofimenko, S.: J. Am. Chem. Soc. *91*, 2139 (1969)
21. Heitsch, C. W.: Abstr. of Papers, 153rd Natl. ACS Meeting, Miami Beach, Florida 1967, p. L109
22. Alcock, N. W., Sawyer, J. F.: Acta Cryst. *B30*, 2899 (1974)
23. Rendle, D. F., Storr, A., Trotter, J.: J. Chem. Soc. Dalton Trans. *1973*, 2252
24. Holt, E. M., Tebben, S. L., Holt, S. L.: Acta Cryst. *B33*, 1986 (1977)
25. Brock, C. P., Niedenzu, K., Hanecker, E., Nöth, H.: ibid. *C00*, 000 (1985)
26. Layton, W. J., Niedenzu, K., Smith, S. L.: Z. Anorg. Allg. Chem. *495*, 52 (1982)
27. Trofimenko, S.: J. Coord. Chem. *2*, 75 (1972)
28. May, C. E., Niedenzu, K., Trofimenko, S.: Z. Naturforsch. *31b*, 1662 (1976)
29. May, C. E., Niedenzu, K., Trofimenko, S.: ibid. *33b*, 220 (1978)
30. Bradley, D. C., Hursthouse, M. B., Newton, J., Walker, N. P. C.: J. Chem. Soc. Chem. Commun. *1984*, 188
31. Peterson, L. K., Thé, K. I.: Can. J. Chem. *57*, 2520 (1979)
32. Layton, W. J., Niedenzu, K., Niedenzu, P. M., Trofimenko, S.: Inorg. Chem. *24*, 1454 (1985)
33. Trofimenko, S.: J. Am. Chem. Soc. *91*, 5410 (1969)
34. Holt, E. M., Holt, S. L., Watson, K. J., Olsen, B.: Cryst. Struct. Comm. *7*, 613 (1978)
35. Hodgkins, T. G.: Dissertation, University of Kentucky, 1984
36. Trofimenko, S.: J. Am. Chem. Soc. *92*, 5118 (1970)
37. Trofimenko, S.: Acc. Chem. Res. *4*, 17 (1979)
38. Trofimenko, S.: Adv. Chem. *150*, 289 (1976)
39. Shaver, A.: J. Organomet. Chem. Lib. *3*, 157 (1977)
40. Trofimenko, S.: J. Am. Chem. Soc. *89*, 3170 (1967)
41. Trofimenko, S.: Inorg. Synth. *12*, 99 (1970)
42. Mani, F.: personal communication
43. Trofimenko, S.: J. Am. Chem. Soc. *89*, 6288 (1967)
44. McCleverty, J. A., Seddon, D., Bailey, N. A., Walker, N. W. J.: J. Chem. Soc. Dalton Trans. *1976*, 898

45. Drane, A. S., McCleverty, J. A.: Polyhedron *2*, 53 (1983)
46. McCleverty, J. A., Wolochowicz, I.: J. Organometal. Chem. *169*, 289 (1979)
47. Reger, D. L., Tarquini, M. E.: Inorg. Chem. *21*, 840 (1982)
48. Domaille, P. J.: J. Am. Chem. Soc. *102*, 5392 (1980)
49. Bothner-By, A. A., Domaille, P. J., Gayathri, C.: ibid. *103*, 5602 (1981)
50. White, D. L., Faller, J. W.: ibid. *104*, 1548 (1982)
51. Zaidi, S. A. A., Neyazi, M. A.: Trans. Met. Chem. *4*, 164 (1979)
52. Siddiqi, Z. A., Khan, S., Zaidi, S. A. A.: Synth. React. Inorg. Metal-Org. Chem. *12*, 433 (1982)
53. Siddiqi, K. S., Neyazi, M. A., Zaidi, S. A. A.: ibid. *11*, 253 (1981)
54. Siddiqi, K. S., Neyazi, M. A., Siddiqi, Z. A., Majid, S. J., Zaidi, S. A. A.: Indian J. Chem. *21A*, 932 (1982)
55. Siddiqi, Z. A., Khan, S., Zaidi, S. A.: Synth. React. Inorg. Metal-Org. Chem. *13*, 425 (1983)
56. McCurdy Jr., W. H.: Inorg. Chem. *14*, 2292 (1975)
57. Desmond, T. J., Lalor, F. J., Ferguson, G., Parvez, M.: J. Organometal. Chem. *277*, 91 (1984)
58. Bagnall, K. W., Edwards, J.: J. Less-Common Met. *48*, 159 (1976)
59. Bagnall, K. W., Edwards, J., du Prees, J. G. H., Warren, R. F.: J. Chem. Soc. Dalton Trans. *1975*, 140
60. Karraker, D. L.: Inorg. Chem. *22*, 503 (1983)
61. Jezorek, J. R., McCurdy Jr., W. H.: ibid. *14*, 1939 (1975)
62. Frauendorfer, E., Brunner, H.: J. Organometal. Chem. *240*, 371 (1982)
63. Churchill, M. R., Gold, K., Maw Jr., C. E.: Inorg. Chem. *9*, 1597 (1970)
64. Curnow, O. J., Nicholson, B. K.: J. Organometal. Chem. *267*, 257 (1984)
65. Oliver, J. D., Mullica, D. F., Hutchinson, B. B., Milligan, W. O.: Inorg. Chem. *19*, 165 (1980)
66. Bruce, M. I., Ostazewski, A. P. P.: J. Chem. Soc. Dalton Trans. *1973*, 2433
67. Abu-Salah, O. M., Bruce, M. I., Hameister, C.: Inorg. Synth. *21*, 107 (1982)
68. Churchill, M. R., DeBoer, B. G., Rotella, F. J., Abu-Salah, O. M., Bruce, M. I.: Inorg. Chem. *14*, 2051 (1975)
69. Cotton, F. A., Mott, G. N.: Inorg. Chem. *22*, 1136 (1983)
70. Kosky, C. A., Ganis, P., Avitabile, G.: Acta Cryst. *B27*, 1859, 2493 (1971)
71. Cotton, F. A., Jeremic, M., Shaver, A.: Inorg. Chim. Acta *6*, 543 (1972);
Calderon, J. L., Cotton, F. A., Shaver, A.: J. Organometal. Chem. *42*, 419 (1972)
72. Reger, D. L., Swift, C. A., Lebioda, L.: J. Am. Chem. Soc. *105*, 5343 (1983)
73. King, R. B., Bond, A.: ibid. *96*, 1338 (1974)
74. Cotton, F. A., LaCour, T., Stanislowski, A. G.: ibid. *96*, 754 (1974)
75. Cotton, F. A., Day, V. W.: J. Chem. Soc. Chem. Commun. *1974*, 415
76. Cotton, F. A., Stanislowski, A. G.: J. Am. Chem. Soc. *96*, 5074 (1974)
77. Cotton, F. A., Frenz, B. A., Stanislowski, A. G.: Inorg. Chim. Acta *7*, 503 (1973)
78. Cotton, F. A., Murillo, C. A.: ibid. *17*, 121 (1976)
79. Cotton, F. A., Frenz, B. A., Murillo, C. A.: J. Am. Chem. Soc. *97*, 2118 (1975)
80. Trofimenko, S.: Inorg. Chem. *9*, 2493 (1970)
81. Breakell, K. R., Patmore, D. J., Storr, A.: J. Chem. Soc. Dalton Trans. *1975*, 749
82. Trofimenko, S.: Inorg. Chem. *10*, 504 (1971)
83. Trofimenko, S.: J. Am. Chem. Soc. *91*, 3183 (1969)
84. Trofimenko, S.: ibid. *90*, 4754 (1968)
85. Clark, H. C., Manzer, L. E.: Inorg. Chem. *13*, 1996 (1974)
86. Clark, H. C., Manzer, L. E.: J. Chem. Soc. Chem. Commun. *1973*, 870
87. Rush, P. E., Oliver, J. D.: ibid. *1974*, 996
88. Oliver, J. D., Rush, P. E.: J. Organometal. Chem. *104*, 117 (1976)
89. Manzer, L. E., Meakin, P. Ẑ.: Inorg. Chem. *15*, 3117 (1976)
90. Davies, B. W., Payne, N. C.: ibid. *13*, 1843 (1974)
91. Oliver, J. D., Rice, N. C.: ibid. *15*, 2741 (1976)
92. Meakin, P., Trofimenko, S., Jesson, J. P.: J. Am. Chem. Soc. *94*, 5677 (1972)
93. Jesson, J. P., Trofimenko, S., Eaton, D. R.: ibid. *89*, 3148 (1967)
94. Onishi, M., Ohama, Y., Sugimura, K., Hiraki, K.: Chem. Letters Chem. Soc. Jpn. *1976*, 955
95. Onishi, M., Hiraki, K., Shironita, M., Yamaguchi, Y., Nakagawa, S.: Bull. Chem. Soc. Jpn. *53*, 961 (1980)
96. Onishi, M., Yamamoto, H., Hiraki, K.: ibid. *53*, 2540 (1980)

97. Onishi, M., Ito, T., Hiraki, K.: J. Organometal. Chem. *209*, 123 (1981)
98. Cocivera, M., Desmond, T. J., Ferguson, G., Kaitner, B., Lalor, F. J., Sullivan, D. J.: Organometallics *1*, 1125 (1982)
99. Gallicano, K. D., Paddock, N. L.: Can. J. Chem. *60*, 521 (1982)
100. Abu-Salah, O. M., Bruce, M. I., Lohmeyer, P. J., Raston, C. L., Shelton, B. W., White, A. H.: J. Chem. Soc. Dalton Trans. *1981*, 962
101. Borkett, N. F., Bruce, M. I.: Inorg. Chim. Acta *12*, L33 (1975)
102. Canty, A. J., Minchin, N. J., Patrick, J. M., White, A. H.: Aust. J. Chem. *36*, 1107 (1983)
103. Holt, E. M., Holt, S. L., Watson, K. J., Olsen, B.: Cryst. Struct. Commun. *7*, 613 (1978)
104. Holt, E. M., Holt, S. L.: J. Chem. Soc. Dalton Trans. *1973*, 1893
105. Calderon, J. L., Cotton, F. A., Shaver, A.: J. Organometal. Chem. *37*, 127 (1972)
106. Cotton, F. A., Murillo, C. A., Stults, B. R.: Inorg. Chim. Acta *22*, 75 (1977)
107. Cocivera, M., Ferguson, G., Kaitner, B., Lalor, F. J., O'Sullivan, D. J., Parvez, M., Ruhl, B.: Organometallics *1*, 1132 (1982)
108. Cocivera, M., Ferguson, G., Lalor, F. J., Szczecinski, P.: ibid. *1*, 1139 (1982)
109. Arcus, C. S., Wilkinson, J. L., Mealli, C., Marks, T. J., Ibers, J. A.: J. Am. Chem. Soc. *96*, 7564 (1974)
110. Mealli, C., Arcus, C. S., Wilkinson, J. L., Marks, T. J., Ibers, J. A.: ibid. *98*, 711 (1976)
111. Roundhill, S. G. N., Roundhill, D. M., Bloomquist, D. R., Landee, C., Willett, R. D., Dooley, D. M., Gray, H. B.: Inorg. Chem. *18*, 831 (1979)
112. Adams, H., Bailey, N. A., Denti, G., McCleverty, J. A., Smith, J. M. A., Wlodarczyk, A.: J. Chem. Soc. Chem. Commun. *1981*, 348
113. Adams, H., Bailey, N. A., Denti, G., McCleverty, J. A., Smith, J. M. A., Wlodarczyk, A.: J. Chem. Soc. Dalton Trans. *1983*, 2287
114. Lincoln, S., Soong, S.-L., Koch, S. A., Sato, M., Enemark, J. H.: Inorg. Chem. *24*, 1355 (1985)
115. Denti, G., Jones, C. J., McCleverty, J. A., Neaves, B. D., Reynolds, S. J.: J. Chem. Soc. Chem. Commun. *1983*, 474
116. Collins, D. M., Cotton, F. A., Murillo, C. A.: Inorg. Chem. *15*, 1861 (1976)
117. Koch, S. A., Lincoln, S.: ibid. *21*, 2904 (1982)
118. Armstrong, W. H., Lippard, S. J.: J. Am. Chem. Soc. *105*, 4837 (1983)
119. Armstrong, W. H., Spool, A., Papaefthymiou, G. C., Frankel, P. B., Lippard, S. J.: ibid. *106*, 3653 (1984)
120. Paolucci, G., Cacchi, S., Cagliotti, L.: J. Chem. Soc. Perkin Trans. *1979*, 1129
121. Thompson, J. S., Zitzman, J. L., Marks, T. J., Ibers, J. A.: Inorg. Chim. Acta *46*, L101 (1980)
122. Cotton, F. A., Kolthammer, B. W. S., Mott, G. N.: Inorg. Chem. *20*, 3890 (1981)
123. Onishi, M., Hiraki, K., Uenoe, A., Yamaguchi, Y., Ohama, Y.: Inorg. Chim. Acta *82*, 121 (1984)
124. Armstrong, W. A., Lippard, S. J.: J. Am. Chem. Soc. *106*, 4632 (1984)
125. Bagnall, K. W., Beheshti, A., Heatley, A.: J. Less-Common Met. *61*, 171 (1978)
126. McCleverty, J. A., Denti, G., Reynolds, S. J., Drane, A. S., Murr, N. E., Rae, A. E., Bailey, N. A., Smith, J. M. A.: J. Chem. Soc. Dalton Trans. *1983*, 81
127. McCleverty, J. A., Drane, A. S., Bailey, N. A., Smith, J. M. A.: ibid. *1983*, 91
128. McCleverty, J. A., Rae, A. E., Wolochowicz, I., Bailey, N. A., Smith, J. M. A.: ibid. *1982*, 429
129. McCleverty, J. A., Rae, A. E., Wolochowicz, I., Bailey, N. A., Smith, J. M. A.: J. Organometal. Chem. *168*, C1 (1979)
131. McCleverty, J. A., Rae, A. E., Wolochowicz, I., Bailey, N. A., Smith, J. M. A.: ibid. *1983*, 71
130. McCleverty, J. A., Rae, A. E., Wolochowicz, I., Bailey, N. A., Smith, J. M. A.: J. Chem. Soc. Dalton. Trans. *1982*, 1
132. Denti, G., McCleverty, J. A., Wlodarczyk, A.: ibid. *1981*, 2021
133. Adams, H., Bailey, N. A., Drane, A. S., McCleverty, J. A.: Polyhedron *2*, 465 (1983)
134. Trofimenko, S.: Inorg. Chem. *10*, 504 (1971)
135. Desmond, T., Lalor, F. J., Ferguson, G., Ruhl, B., Parvez, M.: J. Chem. Soc. Chem. Commun. *1983*, 55
136. Desmond, T., Lalor, F. J., Ferguson, G., Parvez, M.: ibid. *1983*, 457
137. Desmond, T., Lalor, F. J., Ferguson, G., Parvez, M.: ibid. *1984*, 75
138. Trofimenko, S.: J. Am. Chem. Soc. *91*, 588 (1969)
139. Curtis, M. D., Shiu, K.-B., Butler, W. M.: Organometallics *2*, 1425 (1983)
140. Greaves, W. W., Angelici, R. J.: J. Organometal. Chem. *191*, 49 (1980)

141. Shiu, K.-B., Curtis, M. D., Huffman, J. C.: Organometallics *2*, 936 (1983)
142. Millar, M., Lincoln, Sr. S., Koch, S. A.: J. Am. Chem. Soc. *104*, 288 (1982)
143. Jesson, J. P., Trofimenko, S., Eaton, D. R.: ibid. *89*, 3158 (1967)
144. Jesson, J., Weiher, J. F.: J. Chem. Phys. *46*, 1995 (1967)
145. Jesson, J. P., Weiher, J. F., Trofimenko, S.: ibid. *48*, 2058 (1968)
146. Beattie, J. K., Binstead, R. A., West, R. J.: J. Am. Chem. Soc. *100*, 3044 (1978)
147. Hutchinson, B., Daniels, L., Henderson, E., Neill, P., Long, G. J., Becker, L. W.: J. Chem. Soc. Chem. Commun. *1979*, 1003
148. Oliver, J. D., Mullica, D. F., Hutchinson, B., Milligan, W. D.: Inorg. Chem. *19*, 165 (1980)
149. Echols, H. M., Dennis, D.: Acta Cryst. *B32*, 1627 (1976)
150. Dapporto, P., Mani, F., Mealli, C.: Inorg. Chem. *17*, 1323 (1978)
151. Guggenberger, L. J., Prewitt, C. T., Meakin, P., Trofimenko, S., Jesson, J. P.: ibid. *12*, 508 (1973)
152. Cotton, F. A., Jeremic, M., Shaver, A.: Inorg. Chim. Acta *6*, 543 (1972)
153. Calderon, J. L., Cotton, F. A., Shaver, A.: J. Organometal. Chem. *42*, 419 (1972)
154. Reger, D. L., Swift, C. A., Lebioda, L.: J. Am. Chem. Soc. *105*, 5343 (1983)
155. Echols, H. M., Dennis, D.: Acta Cryst. *B30*, 2173 (1974)
156. Davies, B. W., Payne, N. C.: J. Organometal. Chem. *102*, 245 (1975)
157. Collins, D. M., Cotton, F. A., Murillo, C. A.: Inorg. Chem. *15*, 1861 (1976)
158. Cotton, F. A., Murillo, C. A., Stults, B. R.: Inorg. Chim. Acta *22*, 75 (1977)
159. Stainer, M. V. R., Takats, J.: J. Am. Chem. Soc. *105*, 410 (1983)
160. Thomas, R. W., Estes, G. W., Elder, R. C., Deutsch, E.: ibid. *101*, 4581 (1979)
161. Nicholson, B. K.: J. Organometal. Chem. *265*, 153 (1984)
162. Canty, A. J., Minchin, N. J., Patrick, J. M., White, A. H.: Aust. J. Chem. *36*, 1107 (1983)
163. Reger, D. L., Swift, C. A., Lebioda, L.: Inorg. Chem. *23*, 349 (1984)
164. Holt, E. M., Holt, S. L., Cavalito, F., Watson, K. J.: Acta Chem. Scand. *A30*, 225 (1976)
165. Holt, E. M., Holt, S. L., Watson, K. J.: J. Chem. Soc. Dalton Trans. *1973*, 2444
166. Condon, D., Ferguson, G., Lalor, F. J., Parvez, M., Spalding, T.: Inorg. Chem. *21*, 188 (1982)
167. Thompson, J. S., Harlow, R. L., Whitney, J. F.: J. Am. Chem. Soc. *105*, 3522 (1983)
168. Restivo, R. J., Ferguson, G., O'Sullivan, D. J., Lalor, F. J.: Inorg. Chem. *14*, 3046 (1975)
169. Avitabile, G., Ganis, P., Nemiroff, M.: Acta Cryst. *A27*, 725 (1971)
170. de Gil, E. R., Rivera, A. V., Noguera, H.: ibid. *B33*, 2653 (1977)
171. Cocivera, M., Desmond, T. J., Ferguson, G., Kaitner, B., Lalor, F. J., Sullivan, D. J.: Organometallics *1*, 1125 (1982)
172. Restivo, R. J., Ferguson, G.: J. Chem. Soc. Chem. Commun. *1973*, 847
173. Abu-Salah, O. M., Bruce, M. I., Lohmeyer, P. J., Raston, C. L., Skelton, B. W., White, A. H.: J. Chem. Soc. Dalton Trans. *1981*, 962
174. Ferguson, G., Kaitner, B., Lalor, F. J., Roberts, G.: J. Chem. Res. (S) *1982*, 6
175. Marabella, C. P., Enemark, J. H.: J. Organometal. Chem. *226*, 57 (1982)
176. Millar, M., Lincoln, S., Koch, S. A.: J. Am. Chem. Soc. *104*, 288 (1982)
177. Denti, G., Ghedini, M., McCleverty, J. A., Adams, H., Bailey, N. A.: Trans. Met. Chem. *7*, 222 (1982)
178. Begley, T., Condon, D., Ferguson, G., Lalor, F. J., Khan, M. A.: Inorg. Chem. *20*, 3420 (1981)
179. Obaidi, N. A., Brown, K. P., Edwards, A. J., Hollins, S. A., Jones, C. J., McCleverty, J. A., Neaves, B. D.: J. Chem. Soc. Chem. Commun. *1984*, 690
180. Thompson, J. S., Sorrell, T., Marks, T. J., Ibers, J. A.: J. Am. Chem. Soc. *101*, 4193 (1979)
181. Thompson, J. S., Marks, T. J., Ibers, J. A.: Proc. Natl. Acad. Sci. USA *74*, 3114 (1977)
182. Thompson, J. S., Marks, T. J., Ibers, J. A.: J. Am. Chem. Soc. *101*, 4180 (1979)
183. Reger, D. L., Tarquini, M. E., Lebioda, L.: Organometallics *2*, 1763 (1983)
184. Niedenzu, K., Trofimenko, S.: unpublished data
185. Hiraki, K., Ochi, N., Kitamura, T., Sasada, Y., Shinoda, S.: Bull. Chem. Soc. Jpn. *55*, 2356 (1982)
186. Thompson, J. S., Harlow, R. L., Whitney, J. F.: J. Am. Chem. Soc. *105*, 3522 (1983)
187. Enemark, J. E.: personal communication
188. Thompson, J. S., Whitney, J. F.: Acta Cryst. *C40*, 756 (1984)
189. Niedenzu, K., Niedenzu, P. M., Warner, K. R.: Inorg. Chem. *24*, 1604 (1985)
190. Bagnall, K. W., Tempest, A. C., Takats, J., Masino, A. P.: Inorg. Nucle. Chem. Letters *12*, 555 (1976)
191. Stainer, M. V. R., Takats, J.: Inorg. Chem. *21*, 4050 (1982)

192. Stainer, M. V. R., Takats, J.: J. Am. Chem. Soc. *105*, 410 (1983)
193. Restivo, R. J., Ferguson, G.: J. Chem. Soc. Chem. Commun. *1973*, 847
194. Restivo, R. J., Ferguson, G., O'Sullivan, D. J., Lalor, F. J.: Inorg. Chem. *14*, 3046 (1975)
195. Dapporto, P., Mani, F., Mealli, C.: ibid. *17*, 1323 (1978)
196. Faller, J. W.: personal communication
197. McCleverty, J. A., Murr, N. E.: J. Chem. Soc. Chem. Commun. *1981*, 960
198. Atherton, N. M., Denti, G., Ghedini, M., Oliva, C.: J. Magn. Reson. *43*, 167 (1981)
199. Lichtenberg, D. L., Hubbard, J. H.: Inorg. Chem. *23*, 2718 (1984)
200. Minelli, M., Hubbard, J., Lichtenberger, D. L., Enemark, J. H.: ibid. *23*, 2721 (1984)
201. Minelli, M., Yamanouchi, K., Enemark, J. H., Subramanian, P., Kaul, B. B., Spence, J. T.: ibid. *23*, 2554 (1984)
202. Manzer, L. E.: J. Organometal. Chem. *102*, 167 (1975)
203. Companion, A. L., Liu, F., Niedenzu, K.: Inorg. Chem. *24* (1985)

A Survey of Structural Types of Borates and Polyborates

Gert Heller

Institute of Inorganic and Analytical Chemistry, Free University Berlin, Fabeckstraße 34/36, 1000-Berlin 33/FRG

Table of Contents

1 Introduction: Water-Containing Borates

In recent years, interesting new borate structures have been determined by X-ray diffraction. Therefore, it seems appropriate to survey this new development and to update a previous review [1].

The structures of hydrated borates and polyborates (Chapt. 1) are principially different from those of anhydrous species (Chapt. 2), although there are transitions between both the groups. These latter arise if the borates are not fully hydrated, or, for example, if the BO_3 group is associated with isolated OH groups (Chapt. 2.2.2), but also for boracites containing OH groups (Chapt. 2.8).

Silicoborates are not included in the present survey, since most of their structures are governed by the silicate moiety, e.g., axinites; only in a few cases is the structure determined by the borate group, e.g., in howlite.

In general, the crystal chemistry of borates is similar to that of silicates; differences arise from the fact that boron combines with oxygen not only in four-fold (tetrahedral) but also in three-fold (trigonal planar = triangular) coordination. As a result, silicate chemistry is considerably less complicated than borate chemistry.

In crystalline borates, in addition to the mononuclear anions BO_3^{3-} and BO_4^{5-} and their protonated or partially protonated equivalents, there exists an extensive series of polynuclear anions formed by corner-sharing of triangular and/or tetrahedral boron-oxygen moieties. The polyanions, in turn, may polymerize to from infinite chains, sheets or three-dimensional networks. It is not yet clearly understood which factors determine the coordination number of boron in a given boron-oxygen compound. For example, in the α-metaboric acid, α-HBO_2, all the boron atoms have triangular coordination, while both BO_3 and BO_4 groups exist in the monoclinic form of the β-metaboric acid, β-HBO_2; all boron atoms in another form of metaboric acid, γ-HBO_2, have coordination number four.

Several principles have been proposed for the classification of borates. The first one was suggested by C. L. Christ who organized the "crystal chemistry and systematic classification of hydrated borate minerals" in four rules [2]. Based on a more crystallographic point of view, Ch. Tennyson developed "a systematic of borates on crystallochemical basis", considering the analogy of borates with silicates and employing the terms nesoborates, soroborates, inoborates (chains), phylloborates (sheets) and tektoborates (networks) [3]. Edwards and Ross [4] formulated another classification on the basis of the rule that "the ratio of the charge of the tetrahedral boron to the total boron is equal to the ratio of the charge of the cation to the total boron". Unfortunately, this postulate fails for the hexaborates and for partially hydrated polyborates.

In 1970, Heller [1] suggested a classification of borates based on the number of boron atoms in the "fundamental building block". In 1971, J. R. Clark [5] added, in an article on "crystal chemistry of borates", a further principle as the fifth rule, namely that "the boric acid group, $B(OH)_3$, may exist in isolated form in the presence of more complex polyanions, or such insular groups may themselves polymerize and attach to side-chains of more complex polyanions", as first observed in the crystal structures of *veatchite* and *paraveatchite*. In 1977, Christ and Clark [6] reviewed the various principles and classifications in their article on "a crystal-chemical classification of borate structures with emphasis on hydrated borates". In addition to a sixth rule, a shorthand notation using n and the symbols △ = triangle and T = tetrahedron was introduced, where n is the number of atoms characteristic of the fundamental building block. For n = 1, one can distinguish between the isolated[1] triangles 1:△ and isolated[1] tetrahedra 1:T. Each unit may be combined with others of its class to form pairs (dimers: 2:△ or 2:T), chains (∞), sheets (∞_2), or networks (∞_3). The true polyanions in which triangles and tetrahedra may exist together, begin with n = 3, i.e., the triborates.

The best examples for the short-hand notation are the known boric acids. Orthoboric acid H_3BO_3 = mineral *sassolite* crystallizes triclinic in $P\bar{1} = C_i^1$ [2]; it is named 1:△ [7], while metaboric acid crystallizes in three modifications: HBO_2-I = $B_3O_3(OH)_3$ orthorhombic (Pbnm = D_{2h}^{16}) must be formulated as $B_3O_3(OH)_3$ with the notation 3:3△, isolated [18]; HBO_2-II = $HB_3O_4(OH)_2$, monoclinic ($P2_1/a = C_{2h}^5$) is also a triborate with the short-hand notation $3\infty_2$:(2△ + T), chains [9]; and HBO_2-III = mineral *metaborite* is cubic ($P\bar{4}3n = T_d^4$) and consists of a three-dimensional network of tetrahedra, short-hand notation $1\infty_3$:T, network [10].

1.1 Monoborates Containing Tetra-Coordinate Boron

The ion $[B(OH)_4]^-$ with the notation 1:T, isolated (Fig. 1, 1a) has been found in the following species:
orthorhombic (space group Pbca = D_{2h}^{15}) $LiBO_2 \cdot 2\,H_2O = Li[B(OH)_4]$ [11,12]
hexagonal (P3 = C_3^1) $LiBO_2 \cdot 8\,H_2O = Li[B(OH)_4] \cdot 6\,H_2O$ [13,14]
triclinic $NaBO_2 \cdot 4\,H_2O = Na[B(OH)_4] \cdot 2\,H_2O$ [15]
tetragonal ($I\bar{4} = S_4^2$) $CsBO_2 \cdot 4\,H_2O = Cs[B(OH)_4] \cdot 2\,H_2O$ [16]
monoclinic (P2/c = C_{2h}^4) α-$CaB_2O_4 \cdot 4\,H_2O = Ca[B(OH)_4]_2$-I [17,18]
orthorhombic (Pban = D_{2h}^4) α'-$CaB_2O_4 \cdot 4\,H_2O = Ca[B(OH)_4]_2$-II [19]
triclinic β-$CaB_2O_4 \cdot 4\,H_2O = Ca[B(OH)_4]_2$-III = mineral *frolovite* [20], which is isotypic with β-$SrB_2O_4 \cdot 4\,H_2O = Sr[B(OH)_4]_2$-II [21]
monoclinic (P2/c = C_{2h}^4) α-$CaB_2O_4 \cdot 6\,H_2O = Ca[B(OH)_4]_2 \cdot 2\,H_2O$-I = mineral *hexahydroborite* [22,23]
monoclinic (C2/c = C_{2h}^6) β-$CaB_2O_4 \cdot 6\,H_2O = Ca[B(OH)_4]_2 \cdot 2\,H_2O$-II [24]
monoclinic ($P2_1/a = C_{2h}^5$) α-$SrB_2O_4 \cdot 4\,H_2O = Sr[B(OH)_4]_2$-I [25,26], which is isotypic with $BaB_2O_4 \cdot 4\,H_2O = Ba[B(OH)_4]_2$ [27] and also with the substitution product $Sr_{0.8}Ca_{0.2}B_2O_4 \cdot 4\,H_2O = Sr_{0.8}Ca_{0.2}[B(OH)_4]_2$ [28]
monoclinic ($P2_1/c = C_{2h}^5$) $BaB_2O_4 \cdot 5\,H_2O = Ba[B(OH)_4]_2 \cdot H_2O$ [29]
tetragonal (P4/nmm = D_{4h}^7) $Na_2Cl[B(OH)_4]$ = mineral *teepleite* [30,31]
tetragonal (P4/n = C_{4h}^3) $CuCl[B(OH)_4]$ = mineral *bandylithe* [32]
orthorhombic (Pnma = D_{2h}^{16}) $Mg_3(OH)_2(SO_4)[B(OH)_4]_2$ = mineral *sulfoborite* [33,33A]
orthorhombic (Pbnm = D_{2h}^{16}) $Mn_3(OH)_2(PO_4)[B(OH)_4]$ = mineral *seamanite* [34]
tetragonal ($I\bar{4} = S_4^2$) $Ca_2(AsO_4)[B(OH)_4]$ = mineral *cahnite* [35,36] and
monoclinic ($P2_1/n = C_{2h}^6$) $MgCa_2(CO_3)[B(OH)_4]_2 \cdot 4\,H_2O$ = mineral *carboborite* [37].

---

1 Isolated means that the unit is only connected by hydrogen bonds to the next unit, but not by oxygen bridges.

2 All triclinic crystals of borates belong to this space group; if there are doubts, this will be specifically mentioned.

1. Monoborates:

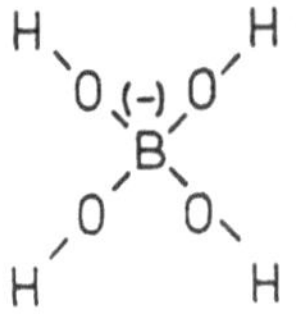

a 1: T, isolated
$[B(OH)_4]^-$

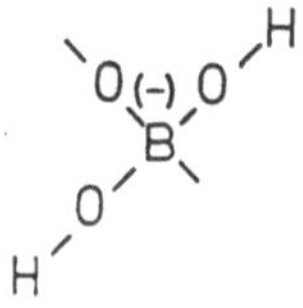

b 1∞: T, chains
$[BO(OH)_2]^-$

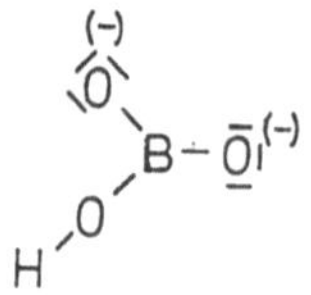

c 1: Δ, partially hydrated
$[BO_2(OH)]^{2-}$

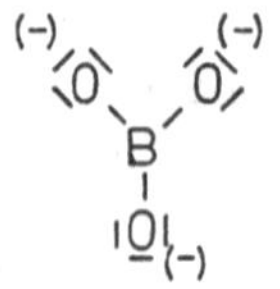

d 1: Δ, waterfree
$[BO_3]^{3-}$

2. Diborates:

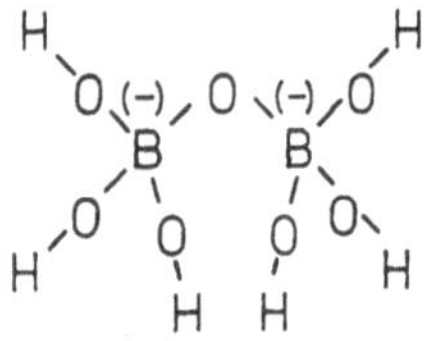

a 2: T, isolated
$[B_2O(OH)_6]^{2-}$

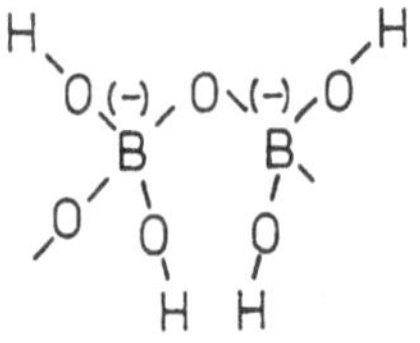

b 2∞: T, chains
$[B_2O_2(OH)_4]^{2-}$

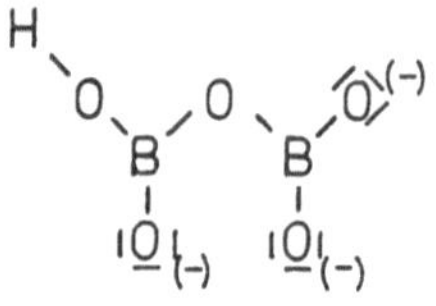

c 2: Δ, isolated, partially hydrated
$[B_2O_4(OH)]^{3-}$

3. Triborates:

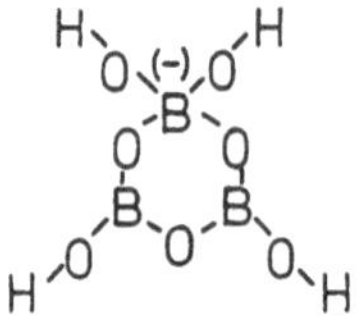

a 3: 2 Δ + T, isolated
$[B_3O_3(OH)_4]^-$

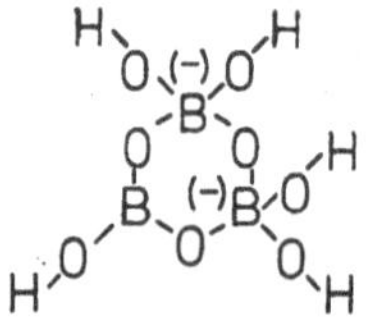

b 3: Δ + 2 T, isolated
$[B_3O_3(OH)_5]^{2-}$

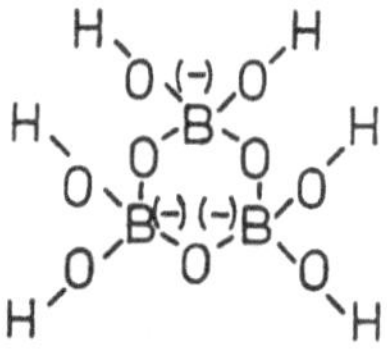

c 3: 3 T, isolated
$[B_3O_3(OH)_6]^{3-}$

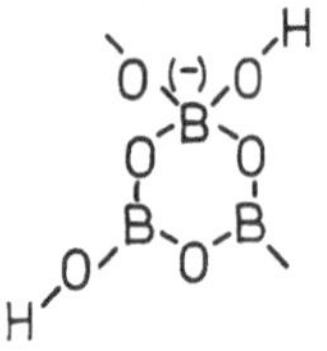

d 3∞: 2 Δ + T, chains
$[B_3O_4(OH)_2]^-$

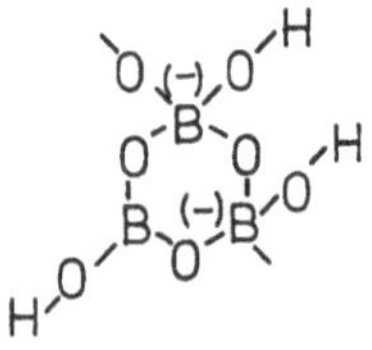

e 3∞: Δ + 2 T, chains
$[B_3O_4(OH)_3]^{2-}$

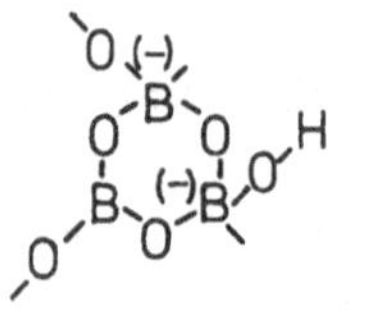

f $3\infty_2$: Δ + 2 T, sheets
$[B_3O_5(OH)]^{2-}$

Fig. 1.

Fig. 1. (continued)

3. Triborates:

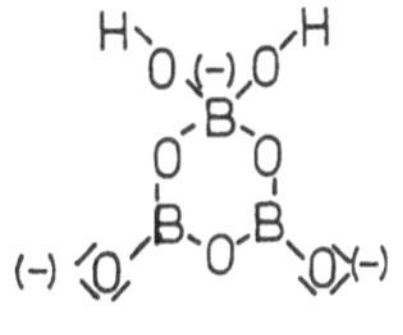

g 3 : (2 Δ + T), isolated, but not fully hydrated $[B_3O_5(OH)_2]^{3-}$

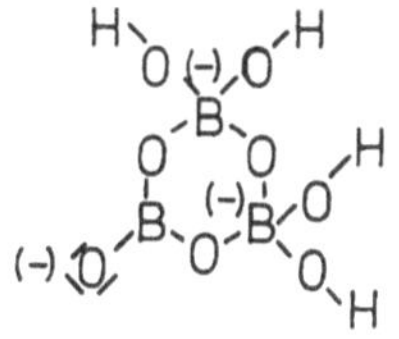

h 3: (Δ + 2 T), isolated, but not fully hydrated $[B_3O_4(OH)_4]^{3-}$

3. Triborates, modified:

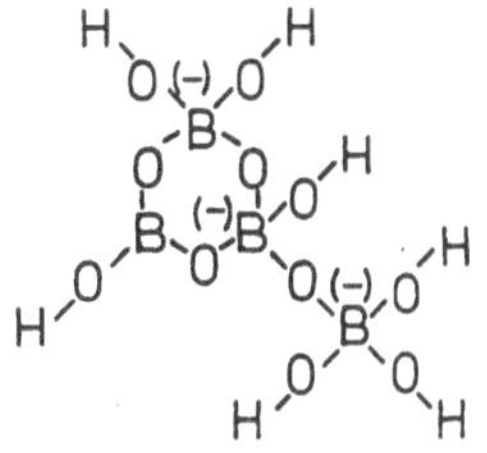

i 3: (Δ + 2 T) + 1: T, modified $[B_4O_4(OH)_7]^{3-}$

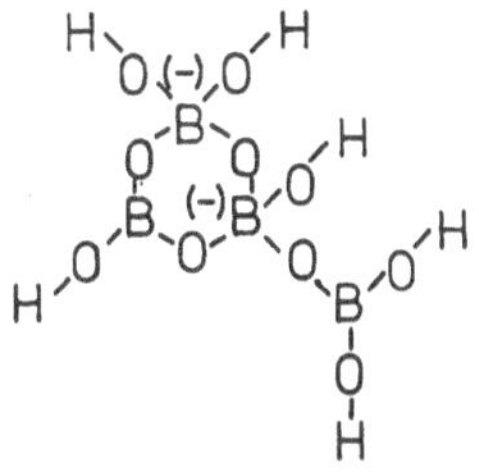

k 3: (Δ + 2 T) + 1: Δ, modified $[B_4O_4(OH)_6]^{2-}$

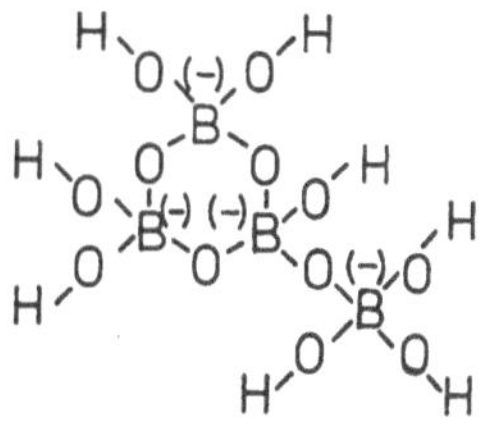

l 3: (3 T) + 1: T, modified $[B_4O_4(OH)_8]^{4-}$

4. Tetraborates:

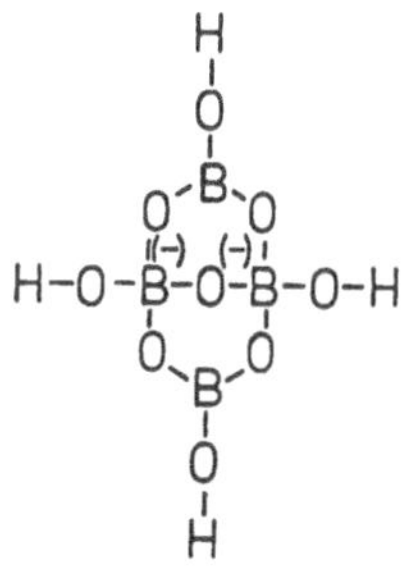

a 4−1: 2 Δ + 2 T, isolated $[B_4O_5(OH)_4]^{2-}$

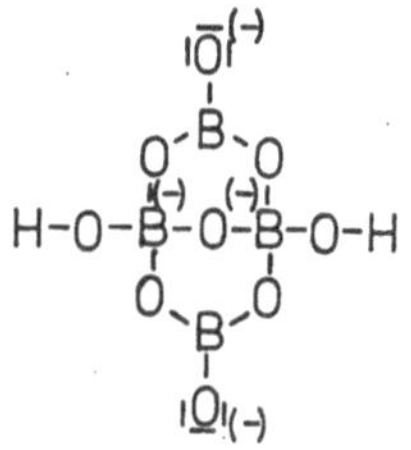

b 4−1: 2 Δ + 2 T, isolated, but only partially hydrated $[B_4O_7(OH)_2]^{4-}$

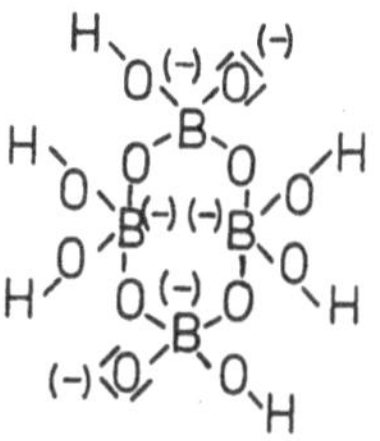

c 4−2: 4 T, isolated, but only partially hydrated $[B_4O_6(OH)_6]^{6-}$

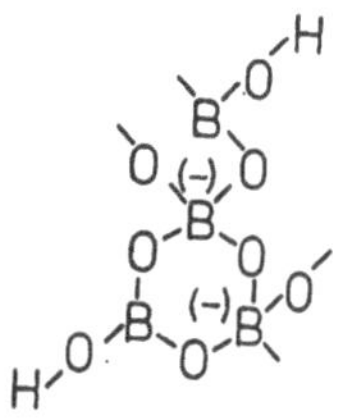

d 4∞: 2 Δ + 2 T, chains $[B_4O_6(OH)_2]^{2-}$

Fig. 1. (continued)

5. Pentaborates:

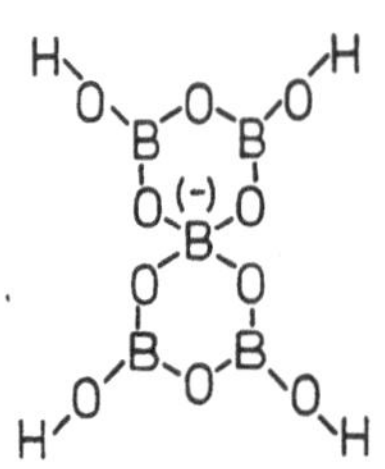

a $5:4\ \Delta + T$, isolated $[B_5O_6(OH)_4]^-$

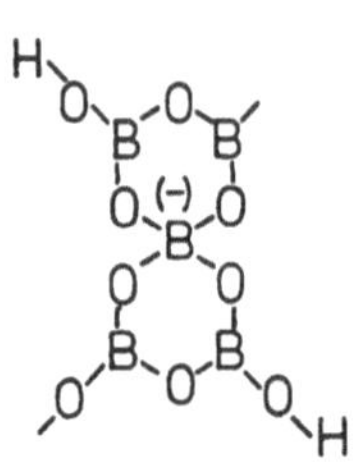

b $5\infty_2: 4\ \Delta + T$, chains $[B_5O_7(OH)_2]^-$

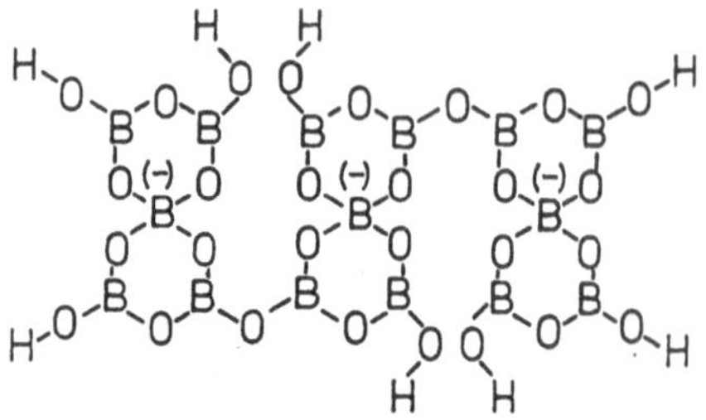

c $5: (4\ \Delta + T)_3$, isolated, trimeric $[B_{15}O_{20}(OH)_8]^{3-}$

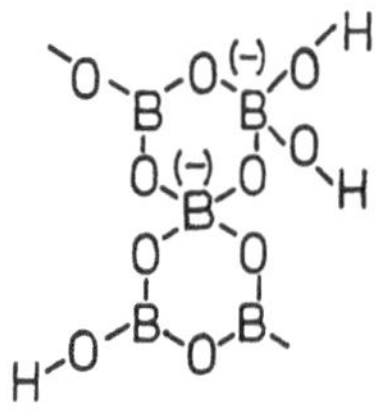

d $5\infty_2: 3\ \Delta + 2\ T$, chains $A \neq B$ $[B_5O_7(OH)_3]^{2-}$

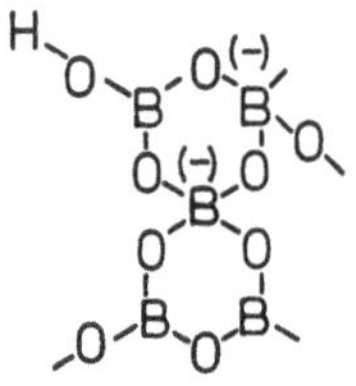

e $5\infty_3: 3\ \Delta + 2\ T$, sheets $A \neq B$ $[B_5O_8(OH)]^{2-}$

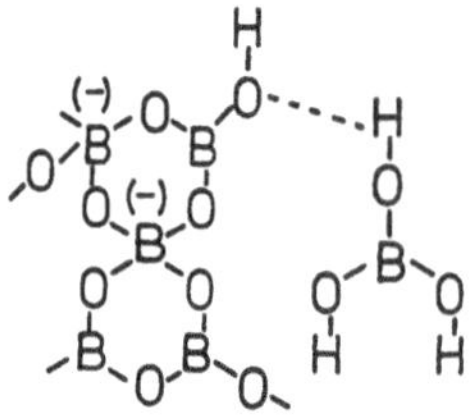

f $5\infty_3: 3\ \dot{\Delta} + 2\ T$, sheets, modified $+ 1\ \Delta$, isolated $B(OH)_3$ $[B_6O_8(OH)_4]^{2-}$

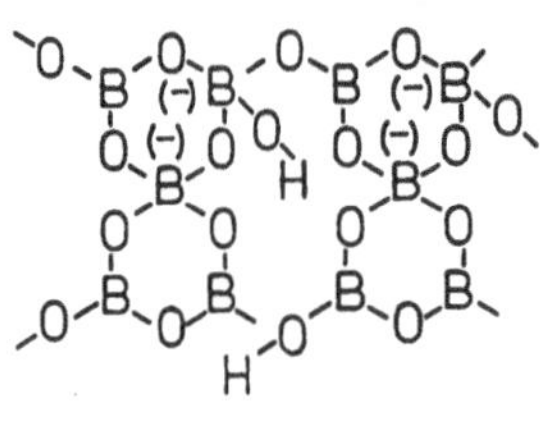

g $5\infty_2: 3\ \Delta + 2\ T$, sheets, dimeric $[B_{10}O_{16}(OH)_2]^{4-}$

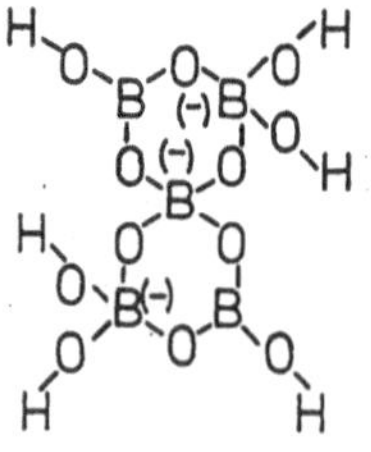

h $5: 2\ \Delta + 3\ T$, $A = B$, isolated $[B_5O_6(OH)_6]^{3-}$

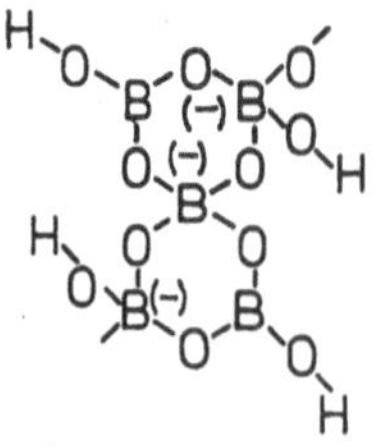

i $5\infty: 2\ \Delta + 3\ T$, $A = B$, chains $[B_5O_7(OH)_4]^{3-}$

Fig. 1. (continued)

5. Pentaborates:

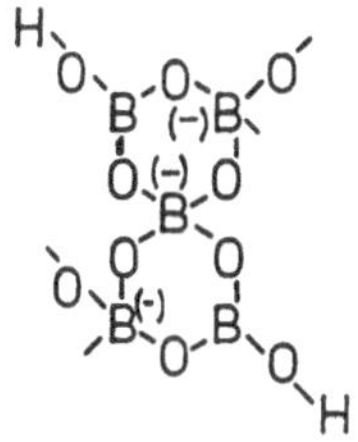

k $5\infty_2$: 2 Δ + 3 T, A=B, sheets $[B_5O_8(OH)_2]^{3-}$

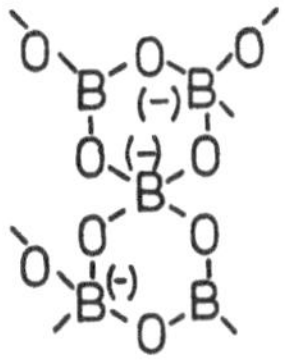

l $5\infty_3$: 2 Δ + 3 T, A=B, three-dimensional $[B_5O_9]^{3-}$

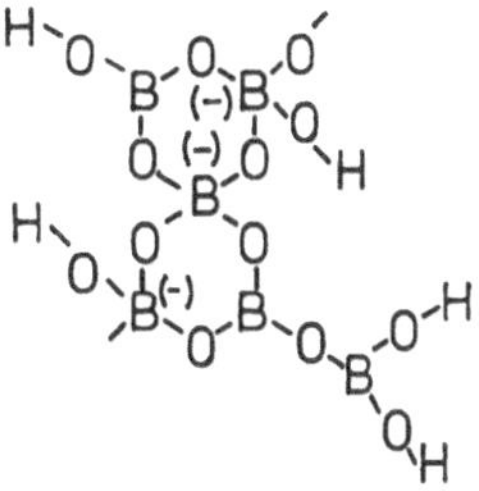

m 5∞: 2 Δ + 3 T, A=B, chains, + 1 Δ, modified $[B_6O_8(OH)_5]^{3-}$

6. Hexaborates:

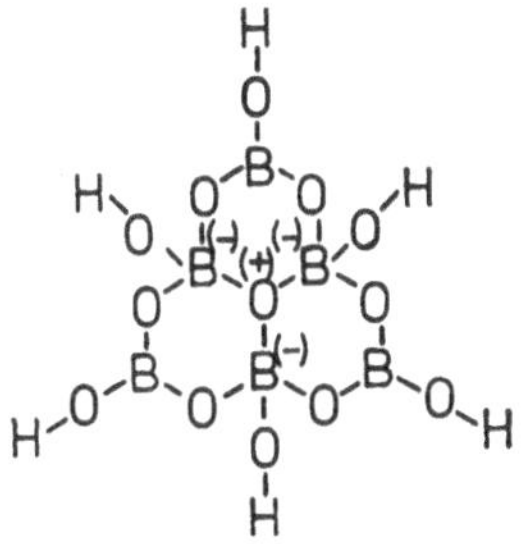

a 6: 3 Δ + 3 T, isolated $[B_6O_7(OH)_6]^{2-}$

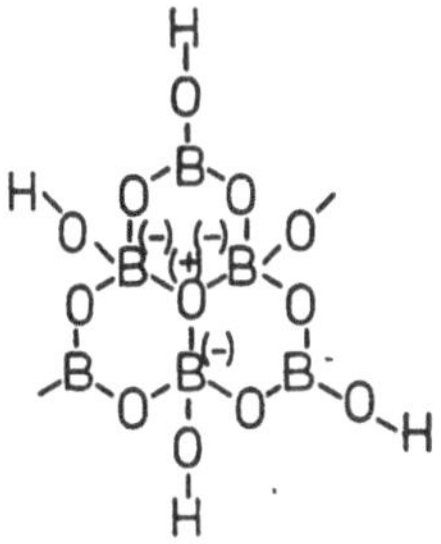

b 6∞: 3 Δ + 3 T, chains $[B_6O_8(OH)_4]^{2-}$

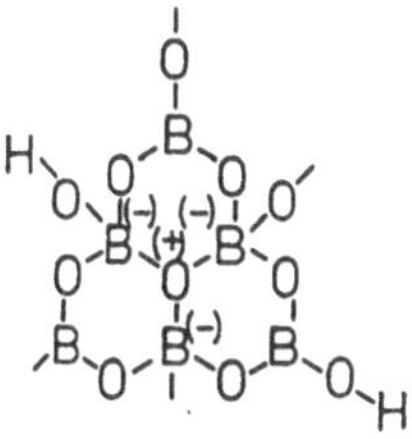

c $6\infty_2$: 3 Δ + 3 T, sheets $[B_6O_9(OH)_2]^{2-}$

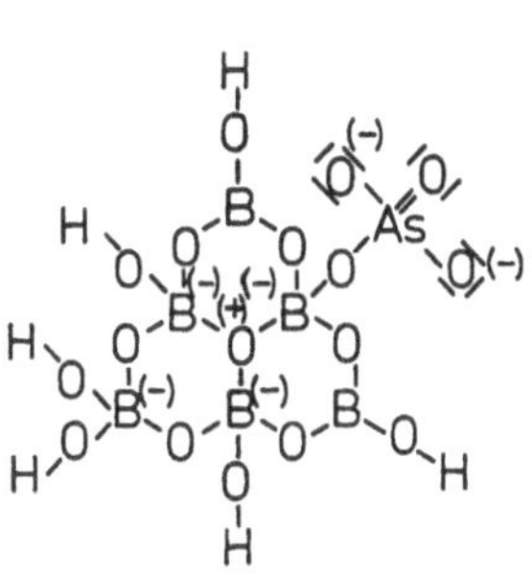

d 6: (2 Δ + 4 T) + As O_4, isolated, modified $[B_6O_7(OH)_6 \cdot OAsO_3]^{5-}$

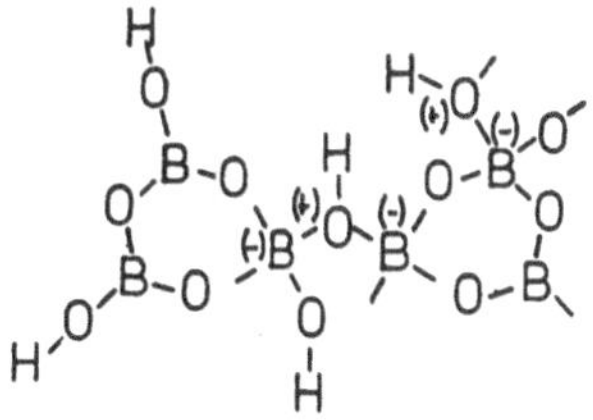

e $6\infty_2$: [(2 Δ + T) + (Δ + 2 T)], layers $[B_6O_7(OH)_5]^{-}$

Fig. 1. (continued)

6 Hexaborates:

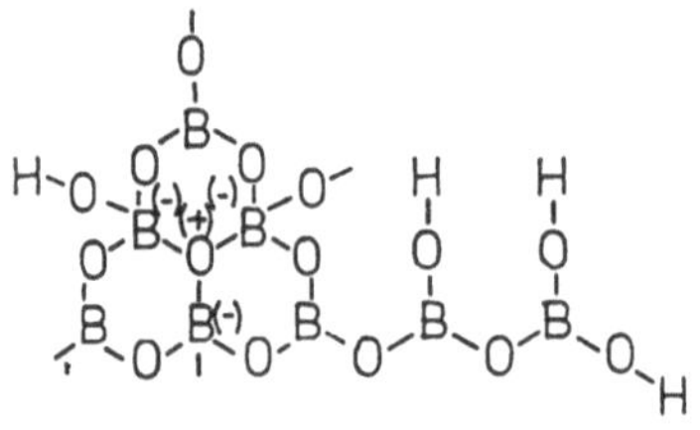

f $6\infty_2$: (3 Δ + 3 T) + 2 Δ, sheets, modified
$[B_8O_{11}(OH)_4]^{2-}$

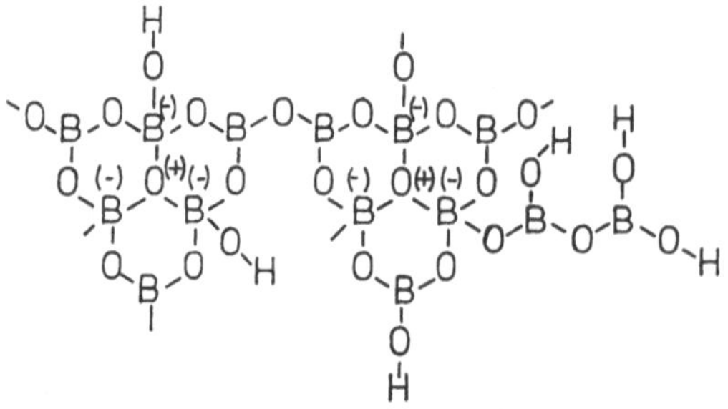

g $6\infty_2$: $(3\,\Delta + 3\,T)_2 + 2\,\Delta$, sheets, modified
$[B_{14}O_{20}(OH)_6]^{4-}$

7. Higher Borates:

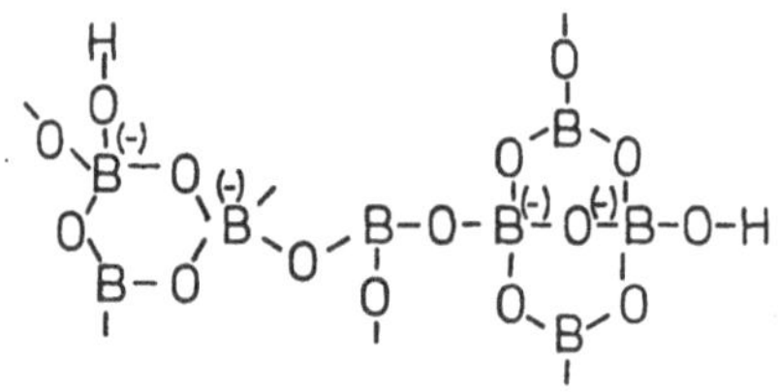

a $8\infty_3$: 3 (Δ + 2 T) + 1Δ + 4 (2 Δ + 2 T), network
$[B_8O_{13}(OH)_2]^{4-}$

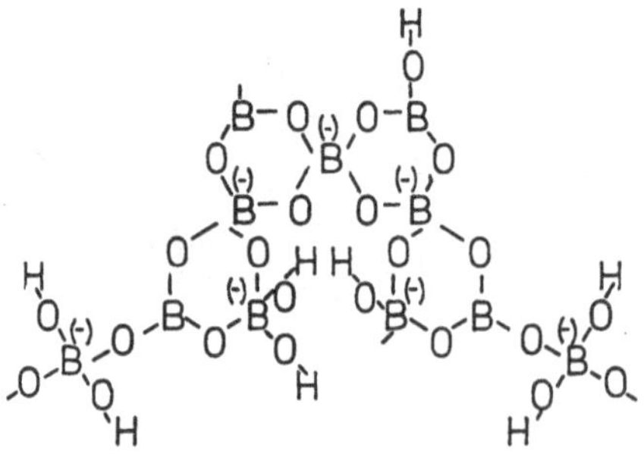

b $11\infty_2$: 9 (4 Δ + 5 T) + 2 T, sheets, modified
$[B_{11}O_{16}(OH)_8]^{7-}$

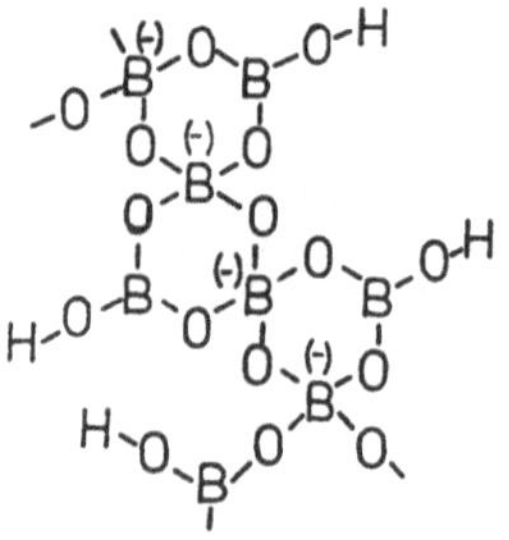

c $8\infty_2$: (4 Δ + 4 T), chains
$[B_8O_{12}(OH)_4]^{4-}$

Fig. 1. (continued)

7. Higher Borates:

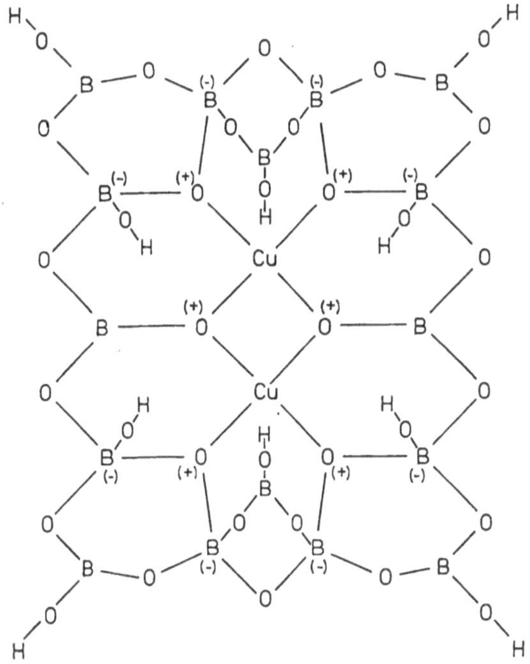

d
Complex
$\{Cu_2O_6[B_{16}O_{18}(OH)_{10}]\}^{6-}$

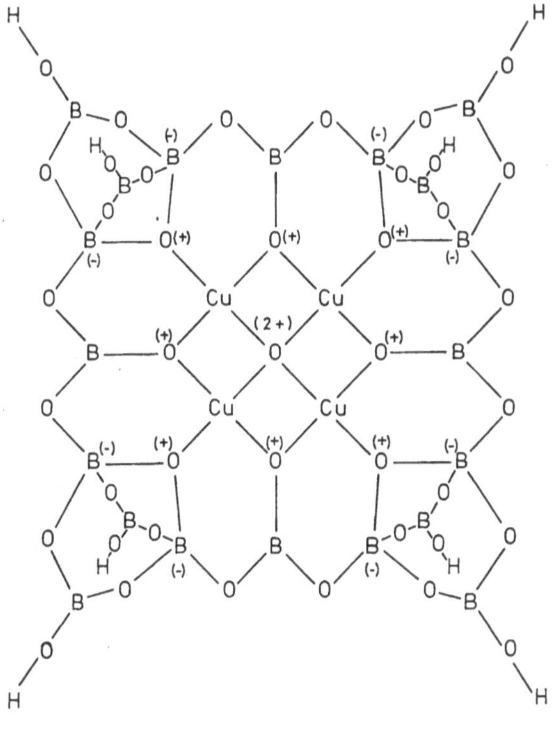

e
Complex
$\{Cu_4O[B_{20}O_{32}(OH)_8]\}^{6-}$

A substituted $[B(OH)_4]^-$ ion is present in the orthorhombic (Pbca = D_{2h}^{15}) $Na[B(OCH_3)_4] \cdot CH_3OH \cdot 1.5\ H_2O$ [38] and in the orthorhombic (Pnma = D_{2h}^{16}) $[C_5H_{12}N][B(OCH_3)_4]$ [39], another one in the hexagonal ($P6_3 = C_6^6$) $Na[B(OH)F_3]$ containing distorted tetrahedrally coordinated boron [40].

1.2 Polymeric Monoborates Containing Three-Coordinate Boron

Partially hydrated monomer anions containing three-coordinate boron (type 1::△, partially hydrated) (Fig. 1, 1c) have been found in the orthorhombic (Pnma = D_{2h}^{16}) $Na_4[B_2O_5] \cdot H_2O = Na_2[BO_2(OH)]$ [41]

1.2.1 Monoborates Containing Three-Coordinate Boron and Other Ions

The isolated $[BO_3]^{3-}$ ion (type 1:△) (Fig. 1, 1d) containing OH groups and other ions has been observed in the
orthorhombic (Pbca = D_{2h}^{15}) $Be_4[B_2O_7] \cdot H_2O = Be_2(OH,F)[BO_3]$ = mineral *hambergite* [42]
trigonal (P321 = D_3^2) $Be_4[B_2O_7] \cdot 3\ H_2O = Be_2(OH,F)[BO_3] \cdot H_2O$ = mineral *berborite* [43,44]
hexagonal ($P6_3/m = C_{6h}^2$) $Mg_3(OH,F)_3[BO_3]$ = *fluoborite* type [45]
monoclinic ($I2/m = C_{2h}^3$) $Mg_{10}[B_2O_{13}] \cdot 9\ H_2O = Mg_5[O(OH)_5(BO_3)] \cdot 2\ H_2O$ = mineral *wightmanite*, a 3 Å wallpaper structure with large, open channels [46], mentioned also as triclinic $Mg_9[(OH)_6(BO_3)]_2 \cdot 2\ H_2O$ [47]
monoclinic $Mg_5[(Cl,OH)_2(OH)_5(BO_3)] \cdot 5\ H_2O$ = mineral *shabynite*, similar to *wightmanite*, but from which it differs in X-ray powder reflections and chemical composition [48]
orthorhombic ($P2_12_12_1 = D_2^4$) $Mg_7[(Cl,OH)_5(BO_3)_3]$ = mineral *karlite* [49]
hexagonal ($P6_3/m = C_{6h}^2$) $Al_{12}[B_{10}O_{33}] \cdot 3\ H_2O = Al_6[(OH)_3(BO_3)_5]$ = mineral *jeremejevite* with a polymeric structure, similar to *fluoborite* and *painite*, with edge-connected BO_3 triangles [50,51] and
cubic ($Fd3 = T_h^4$) $Cd_4[B_2O_7] \cdot H_2O = Cd_4[(OH)_2(BO_3)_2]$ [52].

1.3 Diborates

If two $[B(OH)_4]^-$ ions are condensed, the fully hydrated 2:T, isolated (Fig. 1, 2a) diborate ion $[B_2O(OH)_6]^{2-}$ is formed which exists in
tetragonal ($P4_2 = C_4^3$) $Mg[B_2O_4] \cdot 3\ H_2O = Mg[B_2O(OH)_6]$ = mineral *pinnoite* [53,54] and in
triclinic $Ca[B_2O_4] \cdot 5\ H_2O = Ca[B_2O(OH)_6] \cdot 2\ H_2O$ = mineral *pentahydroborite* [55,56].

Further condensation of this ion leads to the chain 2∞:T, chains (Fig. 1, 1b or 2b), diborate ion $[B_2O_2(OH)_4]^{2-}$ which has been found in
monoclinic ($C2/c = C_{2h}^6$) $Ca[B_2O_4] \cdot 2\ H_2O$-II $= Ca[B_2O_2(OH)_4]$ = mineral *vimsite* [57].

2:△, isolated, but not fully hydrated (Fig. 1, 2c) is the ion $[B_2O_4(OH)]^{3-}$ consisting of one BO_3^{3-} and a $BO_2(OH)$ group which has been found in the monoclinic ($P2_1/a = C_{2h}^5$) members of the szaibelyite group:

$Mg_2[B_2O_5] \cdot H_2O = Mg_2(OH)[B_2O_4(OH)]$ = mineral *szaibelyite* (former *ascharite*) [58, 59]

$Mg_2[B_{2-x}H_{3x}O_5] \cdot H_2O$ ($x \approx 0.18$) = mineral *hydroxyascharite* [60] and
$Mn_2[B_2O_5] \cdot H_2O = Mn_2(OH)(B_2O_4(OH)]$ = mineral *sussexite* [59, 61].

1.4 Triborates

3:2△ + T, isolated (Fig. 1, 3a): The anion $[B_3O_3(OH)_4]^-$ occurs in the monoclinic ($C2/c = C^6_{2h}$) $Na[B_3O_5] \cdot 2\,H_2O = Na[B_3O_3(OH)_4]$ = mineral *ameghinite* [62].

3:△ + 2T, isolated (Fig. 1, 3b): The ion $[B_3O_3(OH)_5]^{2-}$ has been found in monoclinic ($P2_1/a = C^5_{2h}$) $Mg_2[B_6O_{11}] \cdot 15\,H_2O$-I = $Mg[B_3O_3(OH)_5] \cdot 5\,H_2O$-I = mineral *inderite* [63]
triclinic $Mg_2[B_6O_{11}] \cdot 15\,H_2O$-II = $Mg[B_3O_3(OH)_5] \cdot 5\,H_2O$-II = mineral *kurnakovite* [64]
triclinic $Ca_2[B_6O_{11}] \cdot 7\,H_2O = Ca[B_3O_3(OH)_5] \cdot H_2O$ = mineral *meyerhofferite* [65, 66]
triclinic $Ca_2[B_6O_{11}] \cdot 9\,H_2O = Ca[B_3O_3(OH)_5] \cdot 2\,H_2O$ [66, 67]
monoclinic ($P2_1/a = C^5_{2h}$) $Ca_2[B_6O_{11}] \cdot 13\,H_2O = Ca[B_3O_3(OH)_5] \cdot 4\,H_2O$ = mineral *inyoite* [66, 68]
monoclinic ($C2/c = C^6_{2h}$) $MgCa[B_6O_{11}] \cdot 11\,H_2O = MgCa[B_3O_3(OH)_5]_2 \cdot 6\,H_2O$ = mineral *inderborite* [69, 70]
orthorhombic ($Pnma = D^{16}_{2h}$) $Zn_2[B_6O_{11}] \cdot 7\,H_2O = Zn[B_3O_3(OH)_5] \cdot H_2O$ [71].

A partially fluoridized $[B_3O_3(OH)_5]^{2-}$ ion occurs in orthorhombic $K_2[B_3O_3(OH)F_4]$ [72].

3:3T, isolated (Fig. 1, 3c): This ion $[B_3O_3(OH)_6]^{3-}$ is contained in monoclinic

($B2/b = C^6_{2h}$) $Ca[B_2O_4] \cdot \frac{8}{3} H_2O = Ca_3[B_3O_3(OH)_6]_2 \cdot 2\,H_2O$ = mineral *nifontovite* [73].

3∞:(2△ + T), chains (Fig. 1, 3d): The ion $[B_3O_4(OH)_2]^-$ has been found in

orthorhombic ($Pnma = D^{16}_{2h}$) $Tl[B_3O_5] \cdot \frac{3}{2} H_2O = Tl[B_3O_4(OH)_2] \cdot \frac{1}{2} H_2O$ [74].

3∞:(△ + 2T), chains (Fig. 1, 3e): The anion $[B_3O_4(OH)_3]^{2-}$ has been found in monoclinic ($P2_1/a = C^5_{2h}$) high-temperature $Ca_2[B_6O_{11}] \cdot 5\,H_2O$-I = $Ca[B_3O_4(OH)_3] \cdot H_2O$-I = mineral *colemanite* [66, 75, 76]
the monoclinic ($P2_1 = C^2_2$) low-temperature synthetic $Ca_2[B_6O_{11}] \cdot 5\,H_2O$-II = $Ca[B_3O_4(OH)_3] \cdot H_2O$-II [77]
monoclinic ($P2/c = C^4_{2h}$) $MgCa[B_6O_{11}] \cdot 6\,H_2O = MgCa[B_3O_4(OH)_3]_2 \cdot 3\,H_2O$ = mineral *hydroboracite* [78, 79]; and perhaps in $Sr_2[B_6O_{11}] \cdot 5\,H_2O$ [76A].

3∞₂:(△ + 2T), sheets (Fig. 1, 3f): The anion $[B_3O_5(OH)]^{2-}$ has been found in synthetic orthorhombic ($Pbn2_1 = C^9_{2v}$) $Ca_2[B_6O_{11}] \cdot H_2O$-I = $Ca[B_3O_5(OH)]$-I [66, 80, 81], and in
monoclinic ($P2_1/a = C^5_{2h}$) $Ca_2[B_6O_{11}] \cdot H_2O$-II = $Ca[B_3O_5(OH)]$-II = mineral *fabianite* [81, 82].

3:(△ + 2T), isolated, but not fully hydrated (Fig. 1, 3h): The ion $[B_3O_4(OH)_4]^{3-}$ occurs in the isotypic orthorhombic ($Pna2_1 = C_{2v}^9$) triborates $KBO_2 \cdot \frac{4}{3} H_2O = K_3[B_3O_4(OH)_4] \cdot 2 H_2O$ [6,83–85] and

$RbBO_2 \cdot \frac{4}{3} H_2O = Rb_3[B_3O_4(OH)_4] \cdot 2 H_2O$ [86], and in the

monoclinic ($P2_1/b = C_{2h}^5$) $Ca_2Cl[B_3O_4(OH)_4]$ = mineral *solongoite* [87].

3:(2△ + T), isolated, but only partially hydrated (Fig. 1, 3g) is the ion $[B_3O_5(OH)_2]^{3-}$, found in the orthorhombic ($Pnma = D_{2h}^{16}$) $NaBO_2 \cdot \frac{1}{3} H_2O = Na_3[B_3O_5(OH)_2]$ [88,89].

3:(△ + 2T) + 1T, modified (Fig. 1, 3i): The ion $(B_4O_4(OH)_7]^{3-}$ has been found in monoclinic ($I2/a = C_{2h}^4$) $Ca[B_8O_{15}]Cl_2 \cdot 21 H_2O = Ca_2[B_3O_3(OH)_4 \cdot OB(OH)_3]Cl \cdot 7 H_2O$ = mineral *hydrochlorborite* [90].

3:(3T) + 1T, modified (Fig. 1, 3l): The ion $[B_4O_4(OH)_8]^{4-}$ is better formulated as $[B_3O_3(OH)_5 \cdot OB(OH)_3]^{4-}$ and is found in monoclinic ($P2_1/n = C_{2h}^5$) $Ca[B_2O_4] \cdot 2 H_2O$-I $= Ca_2[B_3O_3(OH)_5 \cdot OB(OH)_3]$ = mineral *uralborite* [91,92].

3:(△ + 2T) + 1△, modified (Fig. 1, 3k): The ion $[B_4O_4(OH)_6]^{2-}$ should exist in monoclinic ($P2_1/a = C_{2h}^5$) $Mg[B_4O_7] \cdot 9 H_2O$-I $= Mg[B_3O_3(OH)_4 \cdot OB(OH)_2] \cdot 6 H_2O$ [93].

1.5 Tetraborates

4-1:(2△ + 2T), isolated (Fig. 1, 4a): The anion $[B_4O_5(OH)_4]^{2-}$ has been found in trigonal ($R32 = D_3^7$) $Na_2[B_4O_7] \cdot 5 H_2O = Na_2[B_4O_5(OH)_4] \cdot 3 H_2O$ = mineral *tincalconite* [94]
monoclinic ($C2/c = C_{2h}^6$) $Na_2[B_4O_7] \cdot 10 H_2O = Na_2[B_4O_5(OH)_4] \cdot 8 H_2O$ = mineral *borax* [95,96]
orthorhombic ($P2_12_12_1 = D_2^4$) $K_2[B_4O_7] \cdot 4 H_2O = K_2[B_4O_5(OH)_4] \cdot 2 H_2O$ [97]
monoclinic ($P2_1 = C_{2h}^2$) $(NH_4)_2[B_4O_7 \cdot 4 H_2O = (NH_4)_2[B_4O_5(OH)_4] \cdot 2 H_2O$ [98]
monoclinic (B2/b or $C2/c = C_{2h}^6$) $[NH_3CH_2CH_2NH_3][B_4O_5(OH)_4]$ [99,100]
triclinic $Mg[B_4O_7] \cdot 9 H_2O$-II $= Mg(H_2O)_5[B_4O_5(OH)_4] \cdot 2 H_2O$ = mineral *hungchaoite* [101,102]. The isotypic synthetical orthorhombic ($P2_12_12_1 = D_2^4$) "octaborates" of the $Rb_2Sr[B_4O_5(OH)_4]_2 \cdot 8 H_2O$ type:
$K_2Ca[B_8O_{14}] \cdot 12 H_2O = K_2Ca[B_4O_5(OH)_4]_2 \cdot 8 H_2O$ [103,106]
$Rb_2Ca[B_4O_5(OH)_4]_2 \cdot 8 H_2O$ [107]
$Cs_2Ca[B_4O_5(OH)_4]_2 \cdot 8 H_2O$ [107]
$(NH_4)_2Ca[B_4O_5(OH)_4]_2 \cdot 8 H_2O$ [108]
$K_2Sr[B_4O_5(OH)_4]_2 \cdot 8 H_2O$ [103]
$Rb_2Sr[B_4O_5(OH)_4]_2 \cdot 8 H_2O$ [109]
$Cs_2Sr[B_4O_5(OH)_4]_2 \cdot 8 H_2O$ [107]
$K_2Ba[B_4O_4(OH)_4]_2 \cdot 8 H_2O$ [103] and
$Rb_2Ba[B_4O_5(OH)_4]_2 \cdot 8 H_2O$ [110];
triclinic $Mn[B_4O_7] \cdot 9 H_2O = Mn[B_4O_5(OH)_4] \cdot 7 H_2O$ [111].

4-1:(2△ + 2T) (Fig. 1, 4b), isolated (but only partially hydrated): The anion

$[B_4O_7(OH)_2]^{4-}$ occurs in orthorhombic (Pmcb = D_{2h}^9) $Ca(Mn,Mg)[B_2O_5] \cdot 1.5\ H_2O$ = $Ca_2(Mn,Mg)_2(OH)_4[B_4O_7(OH)_2]$ = mineral *roweite* [112, 113].

4:∞(2△ + 2T), chains (Fig. 1, 4d): The anion $[B_4O_6(OH)_2]^{2-}$ is built not just by polymerization as in the triborate case, but also by breaking of a B—O bond in each polyanion entering the chain (5, 6). It has been found in
monoclinic ($P2_1/c$ = C_{2h}^5) $Na_2[B_4O_7] \cdot 4\ H_2O$ = $Na_2[B_4O_6(OH)_2] \cdot 3\ H_2O$ = mineral *kernite* [114–116]
orthorhombic (Pbca = D_{2h}^{15}) $Na_2[B_4O_7] \cdot H_2O$ = $Na_2[B_4O_6(OH)_2]$ [117, 118]
orthorhombic ($P2_12_12_1$ = D_2^4) $Tl_2[B_4O_7] \cdot 3\ H_2O$ = $Tl_2[B_4O_6(OH)_2] \cdot 2\ H_2O$ [119]
and perhaps
triclinic $Tl_2[B_4O_7] \cdot 1.5\ H_2O$ = $Tl_2[B_4O_6(OH)_6] \cdot \frac{1}{2} H_2O$

which is formulated by the authors as $Tl_4[B_8O_{12}(OH)_{12}] \cdot H_2O$ with the short-hand formulation 7:(3△ + 4T) + △ [120, 120A]; [4]

4-2:(4T), isolated (Fig. 1, 4c), but only partially hydrated: The tetraborate ion $[B_4O_6(OH)_6]^{6-}$ has been reported in monoclinic (A2/m = C_2^3) $Ca_4Mg(CO_3)_2[B_4O_9] \cdot 3\ H_2O$ = $Ca_4Mg(CO_3)_2[B_4O_6(OH)_6]$ = mineral *borcarite* [121].

1.6 Pentaborates

5:4△ + T, isolated (Fig. 1, 5a): The ion $[B_5O_6(OH)_4]^-$ has been found in
monoclinic ($P2_1/c$ = C_{2h}^5) $Na[B_5O_8] \cdot 2\ H_2O$ = $Na[B_5O_6(OH)_4]$ [122]
monoclinic (C2/c = C_{2h}^6) $Na[B_5O_8] \cdot 5\ H_2O$ = $Na[B_5O_6(OH)_4] \cdot 3\ H_2O$ = mineral *sborgite* [123]
orthorhombic (Aba2 = C_{2v}^{17}) $K[B_5O_8] \cdot 4\ H_2O$ = $K[B_5O_6(OH)_4] \cdot 2\ H_2O$ = mineral *santite* [124, 126]
isotypic[3] $NH_4[B_5O_8] \cdot 4\ H_2O$-I = α-$NH_4[B_5O_6(OH)_4] \cdot 2\ H_2O$ [125, 127, 128]
monoclinic (Pn = C_s^1) $NH_4[B_5O_8] \cdot 4\ H_2O$-II = β-$NH_4[B_5O_6(OH)_4] \cdot 2\ H_2O$ [128–130],
probably in
orthorhombic $(n\text{-}C_3H_7)_4N[B_5O_6(OH)_4]$ [131]
monoclinic ($P2_1/a$ = C_{2h}^5) $Cs[B_5O_6(OH)_4] \cdot 2DMSO$ [132] and
monoclinic ($P2_1$ = C_2^2 or $P2_1/m$ = C_{2h}^2 [133], better $P2_1/c$ = C_{2h}^5 [134]) $Tl[B_5O_8] \cdot 4\ H_2O$ = $Tl[B_5O_6(OH)_4] \cdot 2\ H_2O$.

5∞$_2$:(4△ + T), chains (Fig. 1, 5b): The ion $[B_5O_7(OH)_2]^-$ occurs in monoclinic ($P2_1/c$ = C_{2h}^5) $NH_4[B_5O_8] \cdot 2\ H_2O$ = $NH_4[B_5O_7(OH)_2] \cdot H_2O$ = mineral *larderellite* [135].

5:∞$_2$(3△ + 2T), chains (Fig. 1, 5d): The ion $[B_5O_7(OH)_3]^{2-}$ has been found in triclinic $Na_4[B_{10}O_{17}] \cdot 7\ H_2O$ = $Na_2[B_5O_7(OH)_3] \cdot 2\ H_2O$ = mineral *ezcurrite* [136, 137].

5:∞$_3$(3△ + 2T), sheets (Fig. 1, 5e): The ion $[B_5O_8(OH)]^{2-}$ occurs in
orthorhombic ($Pna2_1$ = C_{2v}^9) $Na_4[B_{10}O_{17}] \cdot 5\ H_2O$ = $Na_2[B_5O_8(OH)] \cdot 2\ H_2O$ = mineral *nasinite* [117, 138] and the
isostructural $K_4[B_{10}O_{17}] \cdot 5\ H_2O$ = $K_2[B_5O_8(OH)] \cdot 2\ H_2O$ [139].

3 $Rb[B_5O_8] \cdot 4\ H_2O$ = $Rb[B_5O_6(OH)_4] \cdot 2\ H_2O$ [126A] and
4 +△ should be reserved for side chains

$5:\infty_3(3\triangle + 2T)$, sheets, modified with $B(OH)_3$ (Fig. 1, 5f): The ion $[B_5O_8(OH) \cdot B(OH)_3]^{2-}$ exists in
monoclinic ($P2_1/a = C_{2h}^5$) $Ca[B_6O_{10}] \cdot 5\,H_2O = Ca[B_5O_8(OH) \cdot B(OH)_3] \cdot 3\,H_2O$ = mineral *gowerite* [140]
monoclinic ($Aa = C_s^4$) $Sr_4[B_{22}O_{37}] \cdot 7\,H_2O$-I $= Sr_2\{[B_5O_8(OH)]_2 \cdot B(OH)_3\} \cdot H_2O$-I = mineral *veatchite* [141, 142] and
monoclinic ($P2_1 = C_2^2$) $Sr_4[B_{22}O_{37}] \cdot 7\,H_2O$-II $= Sr_2\{[B_5O_8(OH)]_2 \cdot B(OH)_3\} \cdot H_2O$-II = mineral *paraveatchite* [143].

$5:(2\triangle + 3T)$, A = B, isolated (Fig. 1, 5h): The ion $[B_5O_6(OH)_6]^{3-}$ has been found in triclinic $NaCa[B_5O_9] \cdot 8\,H_2O = NaCa[B_5O_6(OH)_6] \cdot 5\,H_2O$ = mineral *ulexite* [144].

$5:\infty(2\triangle + 3T)$, A = B, chains (Fig. 1, 5i): The ion $[B_5O_7(OH)_4]^{3-}$ occurs in monoclinic ($P2_1/c = C_{2h}^5$) $NaCa[B_5O_9] \cdot 5\,H_2O = NaCa[B_5O_7(OH)_4] \cdot 3\,H_2O$ = mineral *probertite* [145–147].

$5:\infty_2(2\triangle + 3T)$, A = B, sheets (Fig. 1, 5k): The ion $[B_5O_8(OH)_2]^{3-}$ exists in
tetragonal ($P4_12_12_1 = D_4^4$) $Li_3[B_5O_9] \cdot H_2O$-I $= Li_3[B_5O_8(OH)_2]$-I [148, 149]
orthorhombic ($Pn2b = C_{2v}^6$) $Li_3[B_5O_9] \cdot H_2O$-II $= Li_3[B_5O_8(OH)_2]$-II [150]
orthorhombic ($Pbca = D_{2h}^{15}$) $Na_3[B_5O_9] \cdot 2\,H_2O = Na_3(H_2O)[B_5O_8(OH)_2]$ [117, 151] and
monoclinic ($C2/c = C_{2h}^6$) $Na_2Ca_3(SO_4)_2Cl[B_5O_9] \cdot H_2O = Na_2Ca_3(SO_4)_2Cl[B_5O_8(OH)_2]$ = mineral *heidornite* [152].

$5:\infty_3(2\triangle + 3T)$, A = B, network (Fig. 1, 5l): The ion $[B_5O_9]^{3-}$ has been found in
orthorhombic ($Pca2_1 = C_{2v}^5$) $Na_3[B_5O_9] \cdot H_2O$ [153]
monoclinic ($Aa = C_s^4$) $Ca_2[B_5O_9]Cl \cdot H_2O$-I = mineral *hilgardite* [154]
triclinic ($P1 = C_1^1$) $Ca_2[B_5O_9]Cl \cdot H_2O$-II $= [Ca_2(B_5O_9)Cl \cdot H_2O]_3$ = mineral *parahilgardite* [155, 156] and in the
probably isostructural species $(Ca,Sr)_2[B_5O_9](OH,Cl) \cdot H_2O$-III = minerals *strontiohilgardite*, *tyretskite* or *chlortyretskite* (they are a structural pair, depending on the content of Sr or/and OH/Cl) [157–159].

$5:\infty(2\triangle + 3T)$, A = B, chains, modified + $\triangle$ (Fig. 1, 5m): The ion $[B_5O_7(OH)_3 \cdot OB(OH)_2]^{3-}$ has been found in monoclinic ($C2/c = C_{2h}^6$) $K_2Mg_4[B_{24}O_{41}] \cdot 19\,H_2O = KMg_2H[B_5O_7(OH)_3 \cdot OB(OH)_2]_2 \cdot 4\,H_2O$ = mineral *kaliborite* [160].

$5:\infty_2(3\triangle + 2T)$, sheets, modified (Fig. 1, 5g): The ion $[B_5O_8(OH) \cdot OB_5O_7(OH)]^{4-}$ occurs in monoclinic ($P2_1/c = C_{2h}^5$) $Na_4[B_{10}O_{17}] \cdot 3\,H_2O = Na_4[B_{10}O_{16}(OH)_2] \cdot 2\,H_2O = Na_4(H_2O)_2[(HO)B_5O_8 \cdot OB_5O_7(OH)]$ = mineral *biringuccite* [161].

$5:(4\triangle + T)_3$, isolated, trimeric (Fig. 1, 5c): The ion $[B_5O_6(OH)_3 \cdot OB_5O_6(OH)_2\, OB_5O_6(OH)_3]^{3-}$ exists in monoclinic ($C2/c = C_{2h}^6$) $NH_4[B_5O_8] \cdot 2\frac{2}{3}H_2O = (NH_4)_3[B_{15}O_{20}(OH)_8] \cdot 4\,H_2O = (NH_4)_3[B_5O_6(OH)_3 \cdot OB_5O_6(OH)_2 \cdot OB_5O_6(OH)_3] \cdot 4\,H_2O$ = mineral *ammonioborite* [162].

1.7 Hexaborates

$6:3\triangle + 3T$, isolated (Fig. 1, 6a): The ion $[B_6O_7(OH)_6]^{2-}$ has been found in
orthorhombic ($Pbca = D_{2h}^{15}$) $Mg[B_6O_{10}] \cdot 5\,H_2O = Mg[B_6O_7(OH)_6] \cdot 2\,H_2O$ = mineral *aksaite* [163, 164]

monoclinic ($P2_1/c = C_{2h}^5$) $Mg[B_6O_{10}] \cdot 6\ H_2O = Mg[B_6O_7(OH)_6] \cdot 3\ H_2O = Mg(H_2O)_6\{Mg[B_6O_7(OH)_6]_2\}$ [165, 166]
monoclinic ($P2_1/c = C_{2h}^5$) $Mg[B_6O_{10}] \cdot 7\ H_2O = Mg[B_6O_7(OH)_6] \cdot 4\ H_2O$ $Mg(H_2O)_6\{Mg[B_6O_7(OH)_6]_2\} \cdot 2\ H_2O$ = mineral *admontite* [167, 168]
trigonal ($R\bar{3}c = D_3^6$) $Mg[B_6O_{10}] \cdot 7\frac{1}{2}\ H_2O$ $Mg(B_6O_7(OH)_6] \cdot 4\frac{1}{2}\ H_2O$ = $Mg(H_2O)_6\{Mg[B_6O_7(OH)_6]_2\} \cdot 3\ H_2O$ = mineral *macallisterite* [169, 170]
monoclinic ($P2_1/c = C_{2h}^5$) $Na_6Mg[B_{24}O_{40}] \cdot 22\ H_2O = Na_6Mg(H_2O)_8[B_6O_7(OH)_6] \cdot 2\ H_2O$ = mineral *rivadavite* [171]
monoclinic ($P2_1/b = C_{2h}^5$) $Co[B_6O_{10}] \cdot 10\ H_2O = Co(H_2O)_3[B_6O_7(OH)_6] \cdot 4\ H_2O$ [172, 173] and the
isostructural $Ni[B_6O_{10}] \cdot 10\ H_2O = Ni[B_6O_7(OH)_6] \cdot 7\ H_2O$ [172, 173]
triclinic $Ni[B_6O_{10}] \cdot 8\ H_2O = Ni(H_2O)_4[B_6O_7(OH)_6] \cdot H_2O$ [174]
monoclinic ($P2_1/c = C_{2h}^5$) $Co[B_6O_{10}] \cdot 7\ H_2O = Co(H_2O)_6\{Co[B_6O_7(OH)_6]_2\} \cdot 2\ H_2O$ [175]
monoclinic ($P2_1/b = C_{2h}^5$) $Ni(H_2O)_3[B_6O_7(OH)_6] \cdot 3\ H_2O \cdot \frac{1}{2}(CH_3)_2CO$ [176]
monoclinic ($P2_1/b = C_{2h}^5$) $Ni(H_2O)_3[B_6O_7(OH)_6] \cdot \frac{3}{2}\ H_2O \cdot C_2H_5OH$ [177] and
isostructural $Co(H_2O)_3[B_6O_7(OH)_6] \cdot \frac{3}{2}\ H_2O \cdot C_2H_5OH$ [173, 178]
triclinic $K_2Co[B_{12}O_{20}] \cdot 10\ H_2O = K_2(H_2O)_2\{Co[B_6O_7(OH)_6]_2\} \cdot 2\ H_2O$ [179]
monoclinic ($P2_1/c = C_{2h}^5$) $Ni_{0.8}Mn_{0.2}(H_2O)_3[B_6O_7(OH)_6] \cdot C_2H_5OH \cdot 0.42\ H_2O$ [180]
triclinic $Na_6Co_3[B_6O_{10}]_6 \cdot 44\ H_2O = Na_6Co_3[B_6O_7(OH)_6]_6 \cdot 26\ H_2O$ [181] and
triclinic $(NH_4)_2Zn(H_2O)_4[B_6O_7(OH)_6]_2$ and the isotypic $K_2Cd(H_2O)_4[B_6O_7(OH)_6]_2$ [182].
6:∞(3△ + 3T), chains (Fig. 1, 6b): The ion $[B_6O_8(OH)_4]^{2-}$ has been found in monoclinic ($P2_1/a = C_{2h}^5$) $Na_2Mg[B_{12}O_{20}] \cdot 8\ H_2O = Na_2Mg[B_6O_8(OH)_4]_2 \cdot 4\ H_2O$ = mineral *aristarainite* [183, 184].

6:∞_2(3△ + 3T), sheets (Fig. 1, 6c): The ion $[B_6O_9(OH)_2]^{2-}$ occurs in monoclinic ($P2_1/a = C_{2h}^5$) $Sr[B_6O_{10}] \cdot 4\ H_2O = Sr[B_6O_9(OH)_2] \cdot 3\ H_2O$ = mineral *tunellite* [185] and the
isotypic $Ca[B_6O_{10}] \cdot 4\ H_2O = Ca[B_6O_9(OH)_2] \cdot 3\ H_2O$ = mineral *nobleite* [185, 186].

6:∞_2(3△ + 3T) + 2△, sheets, modified (Fig. 1, 6f): The ion $[B_6O_9(OH) \cdot OB(OH) \cdot OB(OH)_2]^{2-}$ occurs in
monoclinic ($P2_1 = C_2^2$) $Sr[B_8O_{13}] \cdot 2\ H_2O = Sr[B_8O_{11}(OH)_4] = Sr[B_6O_9(OH) \cdot OB(OH) \cdot OB(OH)_2]$ = mineral *strontioborite* [187]
isotypic $Ca[B_8O_{13}] \cdot 2\ H_2O = Ca[B_8O_{11}(OH)_4] = Ca[B_6O_9(OH) \cdot OB(OH) \cdot OB(OH)_2$ [188].
6:∞_2[(3△ + 3T)$_2$ + 2△], sheets, modified (Fig. 1, 6g): The ion $[B_6O_9(OH)_2 \cdot OB_6O_8(OH) \cdot OB(OH) \cdot OB(OH)_2]^{4-}$ has been found in
monoclinic ($P2_1/a = C_{2h}^5$) $(Sr,Ca)_2[B_{14}O_{23}] \cdot 8\ H_2O = (Sr,Ca)_2[B_{14}O_{20}(OH)_6] \cdot 5\ H_2O = (Sr,Ca)_2[B_6O_9(OH)_2 \cdot OB_6O_8(OH) \cdot OB(OH) \cdot OB(OH)_2] \cdot 5\ H_2O$ = mineral *volkovite* [189]
the isotypic compounds $Ca_2[B_{14}O_{23}] \cdot 8\ H_2O = Ca_2[B_{14}O_{20}(OH)_6] \cdot 5\ H_2O = Ca_2[B_6O_9(OH)_2 \cdot OB_6O_8(OH) \cdot OB(OH) \cdot OB(OH)_2] \cdot 5\ H_2O$ = mineral *ginorite* [189, 190]

$Sr_2[B_{14}O_{20}(OH)_6] \cdot 5\,H_2O = Sr_2[B_6O_9(OH)_2 \cdot OB_6O_8(OH) \cdot OB(OH) \cdot OB(OH)_2] \cdot 5\,H_2O$ = mineral *strontioginorite* [189].

6:∞_2[(2△ + T) + (△ + 2T)], layers (Fig. 1, 6e): The ion $[B_6O_7(OH)_5]^-$ has been found in monoclinic ($P2_1/c = C^5_{2h}$) $Mg[B_{12}O_{19}] \cdot 5\,H_2O = Mg[B_6O_7(OH)_5]_2$ [191].

6:(2△ + 4T), isolated, modified + AsO_4 (Fig. 1, 6d): The ion $[B_6O_7(OH)_6 \cdot OAsO_3]^{5-}$ has been found in monoclinic ($P2_1/a = C^5_{2h}$) $Ca_4Mg[B_{12}O_{20}](AsO_4)_2 \cdot 20\,H_2O = Ca_4Mg[(B_6O_7(OH)_6) \cdot OAsO_3]_2 \cdot 14\,H_2O$ = mineral *teruggite* [192].

1.8 Higher Borates

8:∞_3[3(△ + 2T) + △ + 4(2△ + 2T)], network (Fig. 1, 7a): The ion $[B_8O_{13}(OH)_2]^{4-}$ has been found in triclinic $Ca[B_4O_7] \cdot \frac{1}{2}\,H_2O = Ca[B_8O_{13}(OH)_2]$ [193, 194].

8:∞_2[4(2△ + 2T) + 4(2△ + 2T)], chains (Fig. 1, 7c): The ion $[B_8O_{12}(OH)_4]^{4-}$ occurs in $Tl_2[B_4O_7] \cdot \frac{3}{2}\,H_2O = Tl_4[B_8O_{12}(OH)_4] \cdot H_2O$; see tetraborates [120, 120A].

6:∞_2[(3△ + 3T) + 2△], sheets: The ion $[B_8O_{11}(OH)_4]^{2-}$ has been found in $Ca[B_8O_{13}] \cdot 2\,H_2O$ and $Sr[B_8O_{13}] \cdot 2\,H_2O$ = mineral *strontioborite*; see hexaborates [187, 188].

5:∞_2(3△ + 2T)$_2$, sheets: The ion $[B_{10}O_{16}(OH)_2]^{4-}$ exists in $Na_4[B_{10}O_{17}] \cdot 3\,H_2O$ = mineral *biringuccite*; see pentaborates [161].

11:∞_2[9(4∞ + 5T) + 2T], sheets, modified (Fig. 1, 7b): The ion $[B_{11}O_{16}(OH)_8]^{7-}$ has been found in orthorhombic ($Pbcn = D^{14}_{2h}$) $Mg_6[B_{22}O_{39}] \cdot 9\,H_2O = Mg_3[B_{11}O_{15}(OH)_9] = HMg_3[B_{11}O_{16}(OH)_8] = HMg_3\{B_9O_{12}(OH)_4 \cdot [O_2B(OH)_2]_2\}$ = mineral *preobratschenskite* [6, 195, 196, 197].

5:∞_2[(2△ + 3T, A = B) + △], chains, modified: The ion $[B_{12}O_{16}(OH)_{10}]^{6-}$ is better formulated as $[B_5O_7(OH)_3 \cdot OB(OH)_2]^{3-}$ and exists in the compound $K_2Mg_4[B_{24}O_{41}] \cdot 19\,H_2O = HKMg_2(H_2O)_4[B_5O_7(OH)_3 \cdot OB(OH)_2]_2$ = mineral *kaliborite* (*paternoite*); see pentaborates [6, 160, 196, 197].

12:(6△ + 6T), isolated: The ion $[B_{12}O_{20}(OH)_4]^{8-}$ exists in monoclinic ($P2_1/c = C^5_{2h}$) $Na_4[B_6O_{11}] \cdot H_2O = Na_8[B_{12}O_{20}(OH)_4]$ [198].

6:∞_2[(3△ + 3T) + 2△], sheets:
The ion $[B_{14}O_{20}(OH)_6]^{4-}$ has been found in $(Sr,Ca)_2[B_{14}O_{23}] \cdot 8\,H_2O$ = mineral *volkovite* [189]
in $Ca_2[B_{14}O_{23}] \cdot 8\,H_2O$ = mineral *ginorite* [190] and in
$Sr_2[B_{14}O_{23}] \cdot 8\,H_2O$ = mineral *strontioginorite* [189, 190]; see hexaborates.

5:(4△ + T)$_3$, isolated, trimeric: The ion $[B_{15}O_{20}(OH)_8]^{3-}$ occurs in $NH_4[B_5O_8] \cdot 2\frac{2}{3}\,H_2O = (NH_4)_3[B_{15}O_{20}(OH)_8] \cdot 4\,H_2O$ = mineral *ammonioborite*; see pentaborates [162].

16:(8 Δ + 8T), complex: The im $[B_{16}O_{24}(OH)_8]^{8-}$ exists in the monoclinic complex $K_6\{UO_2[B_{16}O_{24}(OH)_8]\} \cdot 12\,H_2O$ [201].

16:(6△ + 10T), complex: The ion $[B_{16}O_{24}(OH)_{10}]^{10-}$ exists in the triclinic complex $8\,B_2O_3 \cdot 2\,CuO \cdot 3\,Na_2O \cdot 17\,H_2O = Na_6\{Cu_2[B_{16}O_{24}(OH)_{10}]\} \cdot 12\,H_2O$ [199]; Fig. 1, 7d.

20:(12△ + 8T), complex: The ion $[B_{20}O_{32}(OH)_8]^{12-}$ has been found in tetragonal ($I\bar{4} = S_4^2$) $Na_5H\{Cu_4O[B_{20}O_{32}(OH)_8]\}$. · 32 H_2O; and in triclinic $HK_5\{Cu_4O[B_{20}O_{32}$-$(OH)_8]\}$ · 32 H_2O [200]; Fig. 1, 7e.

1.9 Peroxoborates

Known are the structures of
triclinic $NaBO_3 \cdot 4\,H_2O = Na_2[B_2(O_2)_2(OH)_4] \cdot 6\,H_2O$ [202]
triclinic $NaBO_3 \cdot 3\,H_2O = Na_2[B_2(O_2)_2(OH)_4] \cdot 4\,H_2O$ [202A]
monoclinic ($P2_1/c = C_{2h}^5$) $LiBO_3 \cdot H_2O = Li_2[B_2(O_2)_2(OH)_4]$ [203]. In all three compounds, there exists the ion $[B_2(O_2)_2(OH)_4]^{2-}$ (see Fig. 2).

Unknown are the structures of the triclinic $NaBO_3 \cdot 3\,H_2O$ [204,205]; orthorhombic $Na_2B_2O_5 \cdot 4\,H_2O$ [206]; orthorhombic $KBO_3 \cdot H_2O$ [207,209]; and triclinic $KBO_3 \cdot H_2O_2$ [208,209]; X-ray powder data have been reported for all compounds.

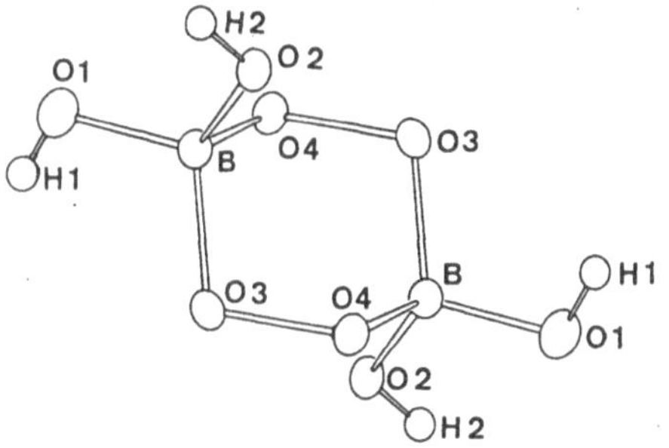

Fig. 2

1.10 Incompletely Characterized Hydrated Borates

The structures of the following hydrated borates and polyborates are unknown but X-ray powder data on the species have been reported:

Trigonal ($C3 = C_3^4$) $B_2O_3 \cdot Li_2O \cdot H_2O = LiBO_2 \cdot \frac{1}{2} H_2O$ and $2\,B_2O_3 \cdot Li_2O \cdot 2\,H_2O$ [210]
$2\,B_2O_3 \cdot Li_2O \cdot 5\,H_2O = Li_2[B_4O_7] \cdot 5\,H_2O = Li_2[B_4O_5(OH)_4] \cdot 3\,H_2O$ (?) [211]
$B_2O_3 \cdot Na_2O \cdot 4\,H_2O = Na[B(OH)_4]$ (?) [212]
triclinic $B_2O_3 \cdot Tl_2O \cdot H_2O = TlBO_2 \cdot \frac{1}{2} H_2O$ and
orthorhombic ($Pbam = D_{2h}^9$ or $Pba2 = C_{2v}^8$) $B_2O_3 \cdot Tl_2O \cdot 2\,H_2O = TlBO_2 \cdot H_2O$ [213];
monoclinic ($P2/c = C_{2h}^4$ or $Pc = C_s^2$) $Mg[B_4O_7] \cdot 3\,H_2O = Mg[B_4O_5(OH)_4] \cdot H_2O$ (?) = mineral *halurgite* [214]
monoclinic $P2/m = C_{2h}^1$ (?) $B_2O_3 \cdot 2\,MgO \cdot 2\,MgCO_3 \cdot H_2O = Mg_4[B_2O_5](CO_3)_2 \cdot H_2O$ = mineral *canavesite* [215]
monoclinic ($C2 = C_2^3$, or $Cm = C_s^3$, or $C2/m = C_{2h}^3$) $B_2O_3 \cdot 3\,MgO \cdot P_2O_5 \cdot 8\,H_2O = Mg_3[B_2O(OH)_4](PO_4)_2 \cdot 6\,H_2O$ (?) = mineral *lüneburgite* [216,217]
$12\,B_2O_3 \cdot MgO \cdot 5\,CaO \cdot 30\,H_2O = Ca_5Mg[B_{24}O_{42}] \cdot 30\,H_2O = Ca_5Mg[B_4O_5(OH)_4]_6 \cdot 18\,H_2O$ (?) = hexagonal (?) mineral *wardsmithite* [218]

cubic ($F4_132 = O^4$) $B_2O_3 \cdot MgO \cdot 2\,CaO \cdot CaCO_3 \cdot 0.36\,H_2O = Ca_3Mg[(BO_3)_2(CO_3)] \cdot 0.36\,H_2O$ = mineral *sakhaite* [219, 220]
orthorhombic ($Pbca = D_{2h}^{15}$) $B_2O_3 \cdot MgO \cdot CaCl_2 \cdot 7\,H_2O = CaMg[B_2O_4]\,Cl_2 \cdot 7\,H_2O$ = mineral *chelkarite* [221]
$B_2O_3 \cdot xMgO \cdot yCaO \cdot zCaCl_2 \cdot 2n\,H_2O = (Ca,Mg)_2(BO_3)Cl \cdot n\,H_2O$ = mineral *aldzhanite* [221, 222]
monoclinic $B_2O_3 \cdot MgCO_3 \cdot 2\,CaO \cdot 10\,H_2O = MgCa_2[B_2O_3(OH)_4](CO_3) \cdot 8\,H_2O$ (?) = mineral *carboborite* [223]
$B_2O_3 \cdot CaO \cdot H_2O = Ca[B_2O_4] \cdot H_2O$ = mineral *korzhinskiite* [90]
monoclinic ($P2_1/m = C_{2h}^2$ or $P2_1 = C_2^2$) $B_2O_3 \cdot 2\,CaO \cdot H_2O = Ca_2[B_2O_5] \cdot H_2O$-I $= Ca[B_2O_4(OH)_2]$-I (?) [224]
orthorhombic $B_2O_3 \cdot 2\,CaO \cdot H_2O = Ca_2[B_2O_5] \cdot H_2O$-II = mineral *sibirskite* [225]
$5\,B_2O_3 \cdot 4\,CaO \cdot 7\,H_2O = Ca_4[B_{10}O_{14}] \cdot 7\,H_2O = Ca_2[B_5O_6(OH)_7]$ or $Ca_2[B_5O_7(OH)_5] \cdot H_2O$ = mineral *priceite (pandermite)* [226]
monoclinic $5\,B_2O_3 \cdot 4\,CaO \cdot 20\,H_2O = Ca_4[B_{10}O_{14}] \cdot 20\,H_2O = Ca_4[B_5O_7(OH)_5] \cdot 15\,H_2O$ (?) = mineral *tertschite* [227]
monoclinic or triclinic $2\,B_2O_3 \cdot 3\,CaO \cdot 9\,H_2O = Ca_3[B_4O_9] \cdot 9\,H_2O = Ca_3[B(OH)_4]_4(OH)_2$ (?) = mineral *olshanskyite* [228]
$2\,B_2O_3 \cdot 2\,(Ca,Sr)O \cdot H_2O = (Ca,Sr)_2[B_4O_7(OH)_2]$ (?) = mineral *kurgantaite* [158]
$3\,B_2O_3 \cdot (Ca,Sr)O \cdot 3\,H_2O = (Ca,Sr)[B_6O_{10}] \cdot 3\,H_2O = Ca[B_3O_4(OH)_2] \cdot H_2O$ (?) = mineral *volkovskite* [229, 230]
hexagonal $2\,B_2O_3 \cdot CaO \cdot CaCl_2 \cdot 3\,H_2O = Ca_2[B_4O_7](Cl, OH)_2$ = mineral *ekaterinite* [231]
monoclinic ($P2/m = C_{2h}^1$ or $P2 = C_2^1$ or $Pm = C_s^1$) $B_2O_3 \cdot xCaO \cdot yCaCl_2 \cdot n\,H_2O$ = mineral *ivanovite* [232]
tetragonal (?) chloride-containing $(Ca,Mg,Mn)_4[B_2O_7] \cdot 2\,H_2O = (Ca,Mg,Mn)_4[B_2O_5(OH,Cl)_4]$ (?) = mineral *wiserite* [61, 225, 233]
monoclinic ($P2_1/c = C_{2h}^5$) $B_2O_3 \cdot BaO \cdot H_2O = Ba[B_2O_4] \cdot H_2O$ [234]
cubic ($P\bar{4}3m = T_d^1$) $11\,B_2O_3 \cdot 4\,Al_2O_3 \cdot 8\,BeO \cdot (Cs,K,Rb)_2O \cdot 2\,H_2O = (Cs,K,Rb)\{Be_4Al_4[B_{11}O_{26}(OH)_2]\}$ = mineral *rhodicite* [235, 236]
orthorhombic ($Pbca = D_{2h}^{15}$) $6\,B_2O_3 \cdot 4\,Al_2O_3 \cdot 4\,NaCl \cdot 2\,KCl \cdot 26\,H_2O = KNa_2Al_4[B_6O_{15}]Cl_3 \cdot 13\,H_2O = KNa_2\{Al_4[B_2O(OH)_6]_3(OH)_6Cl_3\} \cdot H_2O$ (?) = mineral *satimolite* [221, 237, 238]
hexagonal $11\,B_2O_3 \cdot 7\,(Na_2O,CaO) \cdot RE_2O_3 \cdot 7\,H_2O = (Ca,Na_2)_7(RE)_2[B_{22}O_{43}] \cdot 7\,H_2O$ or, more probably, $12\,B_2O_3 \cdot 6\,(Na_2O, CaO) \cdot RE_2O_3 \cdot 6\,H_2O = (Ca,Na_2)_6(RE)_2[B_{24}O_{45}] \cdot 6\,H_2O$ = mineral *braitschite* [239].

1.11 Hydrated Boropolytungstates

The following boron-containing heteropolytungstates have not yet been completely characterized:
hexagonal ($P6_222 = D_6^4$ or $P6_422 = D_6^5$) $K_5[BW_{12}O_{40}] \cdot 18\,H_2O$

cubic ($F\bar{4}3m = T_d^2$ or $Fm3m = O_h^5$) $K_9[BW_{11}O_{39}] \cdot 14\,H_2O$
and tetragonal compounds $K_7[BMn^{2+}W_{11}O_{40}H_2] \cdot 16\,H_2O$ and $Ba_3[BMn^{3+}W_{11}O_{40}H_2] \cdot 26\,H_2O$ [240].

The latter may be isostructural with tetragonal (P4/mnc = D^6_{4h}) $Ba_3[BCo^{3+}(H_2O)W_{11}O_{39}] \cdot 26\ H_2O$, which has been studied recently [240A)].

2 Crystal Chemistry of Anhydrous Borates

The main principles of anhydrous borate systemics are that

a) every boron atom is linked to three or four oxygen atoms;

b) a single structure may contain either triangular or tetrahedral coordination or both;

c) polymerization or formation of chains, layers or framework is brought about by linking the tops of the triangles and tetrahedra;

d) in framework and layered polyanions the triangles and tetrahedra tend to form compact clusters with small negative charge of the type of diborate (2△ + 2T), triborate (2△ + 1T), pentaborate (4△ + 1T), boroxinic (3△), ditriborate (1△ + 2T), dipentaborate (3△ + 2T), and other single and double ring boron-oxygen anions and radicals;

e) complex polyanions tend to "twinning";

f) in the majority of complex polyanions each oxygen atom is bound to two boron atoms; for compounds with the general formula $mB_2O_3 \cdot n\ M_xO_y$ and $m > 1$, the relationship $n = n_\Delta/n_T = m - 1$ is true;

g) a frequent exception for polyanions, including layered and framework structures, is a coordination number of one or three (in cubic boracites and hydrated complex higher polyborates even four) for some oxygen atoms;

h) no triangles and tetrahedra are encountered in anhydrous borates either in isolated forms nor in island polyanions;

i) an increase in the factor $N(B_2O_3)/N(M_xO_y)$ as well as in the cation size leads to a rise in the degree of polymerization of the anion and the number N, while an increased charge causes the opposite [241)].

A classification is proposed for the anhydrous borates, which is based on the composition of anion-forming components, valence of cations, type and size of cations and on the ratio of total numbers of cations to number of boron atoms N. Borates, borosilicates, boroaluminates, boroberyllates, borocarbonates, boromolybdates and borotungstates are arranged into groups where $N = >1$, $N = 1$, $N = 1$ to 0.5, $N = 0.5$ to 0.33, and $N = <0.33$. Based on the nature of univalent elements, orthoborates, fluoroborates, metaborates, and polyborates have been recognized [242)].

The structure of several alkali metal borate crystals has been studied as an aid for interpreting the "boron oxide anomaly". Five kinds of BO_3 units such as $B^\Delta(3\ B_3)$, $B^\Delta(2\ B_3, B_4)$, $B^\Delta(B_3, 2\ B_4)$, $B^\Delta(3\ B_4)$ and B^R, and two kinds of RBO_4 units such as $B^T(4\ B_3)$ and $B^T(3\ B_3, B_4)$ are the basic constituents of the crystals $B_2O_3 \cdot xR_2O$ $(0 \leqq x \leqq 1.0)$. The variation of the fractions of those units with x shows two trends. The number of B_4 units to which a B_3 or a B_4 unit is bonded, increases gradually with x. The fraction of each kind of B^Δ or B^T unit becomes maximal in one of the crystals, in which the unit is the single B_4 or B_3 unit. These tendencies suggest that four kinds of B_4 units bonded with $> 1\ B_4$ units appear in the range $0.5 < x \leqq 1.0$, if the fraction of the B^T units is assumed to increase linearly with x in this range. A relation of the relative stability between the real and hypothetical units is derived from the viewpoint of the energy of the units, defined as, for example, a sum of the bond energies. The B—O bond energies of the RBO_4 units in some crystals have been calculated and were found to vary from 130 to 135 kcal/mol [243].

Based on previous classifications, G. K. Gode [244] and S. García-Blanco [245] have reviewed the crystal chemistry of borates, the latter for hydrated and anhydrous borates.

2.1 Orthoborates Containing the Anion BO_3^{3-}

2.1.1 Orthoborates With M^{3+} Cations, MBO_3

The compounds MBO_3 containing three-coordinate boron are all similar to known types of $CaCO_3$.

2.1.1.1 Orthoborates with Calcite Structure

Trigonal ($R\bar{3}c = D_{3d}^6$) compounds MBO_3 are of the $CaCO_3$-I type with hexagonal dense packing of O atoms and planar BO_3^{3-} groups. To this type belong the high-pressure phases α-$AlBO_3$ [51, 246, 248]; α-$GaBO_3$ [51, 248, 249]; α-$InBO_3$ [250–252]; α-$TlBO_3$ [248]; $BiBO_3$ [251, 253]; α-$ScBO_3$ [251, 254, 255]; the high-temperature phase β-YBO_3-I [256, 257]; the high-temperature/high-pressure phase γ-$YbBO_3$-III [257]; the low-temperature phase β-$LuBO_3$-II [248, 256, 258]; furthermore, $TiBO_3$ [259]; VBO_3 [259, 260]; $CrBO_3$ [259–261]; $FeBO_3$ [262]; $Fe_{0.9}Ga_{0.1}BO_3$ [263]; $Fe_{1-x}Sc_xBO_3$ [264]; $Fe_{1-x}Lu_xBO_3$ [264]; $Fe_{0.9}Ti_{0.1}BO_3$ [263]; $Fe_{0.8}Cr_{0.2}BO_3$ [263]; $RhBO_3$ [248]; and $AmBO_3$ [265].

2.1.1.2 Orthoborates with Vaterite Structure

2.1.1.2.1 Orthoborates with Pseudo-Vaterite Structure. Hexagonal ($P\bar{6}c2 = D_{3h}^2$) compounds MBO_3 with pseudo-*vaterite* structure are of $CaCO_3$-II type. Such compounds exhibiting distorted packing and non-planar BO_3^{3-} groups are called the $YbBO_3$-II type. To this type belong the low-temperature phase α-$YbBO_3$-II [268, 269]; and the high-temperature phase μ-$LuBO_3$-I [268]. Orthoborates similar to the $YbBO_3$-II type (hexagonal, perhaps $P\bar{6}c2 = D_{3h}^2$) are: γ-$CeBO_3$-IV [268]; γ-$PrBO_3$ [268]; and γ-$NdBO_3$ [268].

2.1.1.2.2 Orthoborates with High-Vaterite Structure. Hexagonal ($P6_322 = D_6^6$) compounds MBO_3 of the $CaCO_3$-II′ structure belong to the $GdBO_3$-II type. These are: the high-temperature phases $GdBO_3$-II [270]; $TbBO_3$-II [270]; $DyBO_3$-II [270]; $HoBO_3$-II [270]; $ErBO_3$-II [270]; $TmBO_3$-II [270]; $YbBO_3$-I [270]; and $LuBO_3$-III [270].

2.1.1.2.3 Orthoborates with "Vaterite" Structure. Hexagonal ($P6_3/mmc = D_{6h}^4$) compounds MBO_3 contain BO_4 tetrahedra; these structures are named the YBO_3-II type. To this type belong the low-temperature/low-pressure phases β-$TlBO_3$ [266]; α-YBO_3-II [250, 266, 267]; α-$SmBO_3$-II [250, 267–269]; $EuBO_3$-II [250, 267, 268]; $GdBO_3$-III [250, 267, 268]; $TbBO_3$-III [268]; $DyBO_3$-III [250, 267, 268]; $HoBO_3$-I [250, 267, 268]; $ErBO_3$-I [250, 267, 268]; $TmBO_3$-I [250, 267, 268]; and in older literature α-$YbBO_3$-II [250, 267] as well as the high-temperature phase μ-$LuBO_3$-I [250, 267].

2.1.1.3 Orthoborates with Aragonite Structure

Orthoborates MBO_3 with *aragonite* = $CaCO_3$-III structure are orthorhombic (Pmcn = D_{2h}^{16}) with dolomite-like packing and planar BO_3^{3-} groups; they are called the $LaBO_3$-II type. To this type belong the low-temperature phases α-$LaBO_3$-II [250, 271–273]; α-$CeBO_3$-III [274, 275]; $Ce_{1-x}Tb_xBO_3$ [275]; α-$PrBO_3$-III [276]; α-$NdBO_3$-III [250, 273]; $PmBO_3$ [277]; and the high-pressure phases λ-$SmBO_3$-III [276] and λ-$EuBO_3$-III [276].

2.1.1.4 Orthoborates of the $LaBO_3$-I Type

Orthoborates of the $LaBO_3$-I type are monoclinic ($P2_1/m$ = C_{2h}^2); to this type belong the high-temperature/high-pressure phases β-$LaBO_3$-I [272] and β-$CeBO_3$-II [272].

2.1.1.5 Orthoborates of the High-$NdBO_3$-I Type

The following orthoborates MBO_3 of the high-temperature $NdBO_3$-I type are triclinic (P1 = C_1^1) with isolated BO_3^{3-} groups of nonequilateral triangles:
the high-temperature phases ν-$CeBO_3$-I [272]; ν-$PrBO_3$-I [278]; ν-$NdBO_3$-I [250, 278]; ν-$SmBO_3$-I [278, 279]; and the
high-pressure phases ν-$EuBO_3$-I [278]; ν-$GdBO_3$-I [278]; ν-$TbBO_3$-I [278] and ν-$DyBO_3$-I [278].

2.1.2 Orthoborates With M^{2+} Cations, $M_3(BO_3)_2$

2.1.2.1 Orthoborates with Kotoite Structure

The following orthoborates $M_3(BO_3)_2$ are of orthorhombic (Pnmn = D_{2h}^{12}) $Mg_3(BO_3)_2$ = *kotoite* type with hexagonal dense packing of the O atoms and isolated BO_3^{3-} groups containing non-equilateral B—O bonds: $Mg_3(BO_3)_2$ = mineral *kotoite* [280, 281]; $Mn_3(BO_3)_2$ = mineral *jimboite* [282, 283]; $Co_3(BO_3)_2$ [281, 284]; and $Ni_3(BO_3)_2$ [285, 286].

2.1.2.2 Orthoborates with $Ca_3(BO_3)_2$ Structure

To the trigonal ($R\bar{3}c$ = D_{3d}^6) $Ca_3(BO_3)_2$ type with distorted packing of O atoms, Ca columns and not exactly planar BO_3^{3-} groups belong $Ca_3(BO_3)_2$ [287, 288]; $Sr_3(BO_3)_2$ [289]; $Hg_3(BO_3)_2$ [290]; and $Eu_3(BO_3)_2$ [291].

2.1.2.3 Orthoborates with $Zn_3(BO_3)_2$ Structure

The monoclinic (Bb = C_s^4 or C2/c = C_{2h}^6) $Zn_3(BO_3)_2$ type is so far represented only by $Zn_3(BO_3)_2$ [292, 293], containing four-coordinate boron.

2.1.3 Orthoborates With M^+ Cations, M_3BO_3

α-Li_3BO_3-II is monoclinic ($P2_1/c$ = C_{2h}^5) with nearly planar BO_3^{3-} groups [294], but not isotypic with monoclinic ($P2_1/c$ = C_{2h}^5) Na_3BO_3 [295].

Ag_3BO_3-I is trigonal with R32 = D_3^7 [296], Ag_3BO_3-II with $R\bar{3}c$ = D_{3d}^6 [297]; Tl_3BO_3 is hexagonal ($P6_3/m$ = C_{6h}^2) [298].

β-Li_3BO_3-I is perhaps monoclinic and γ-Li_3BO_3-III perhaps cubic with $a = 614.6$ pm [294].

2.1.4 Orthoborates $M^{1+}M^{2+}[BO_3]$

To the monoclinic ($C2/c = C^6_{2h}$) $LiZn[BO_3]$-II type belong $LiZn[BO_3]$-II [299,300] and $LiMn[BO_3]$ [299,300]; hexagonal ($P\bar{6} = C^1_{3h}$) is $LiCd[BO_3]$-I [301,302]; triclinic ($I\bar{1} = C^1_i$) is $LiCd[BO_3]$-II [301,304,305]. Other hexagonal types are $LiMg[BO_3]$ and $LiZn[BO_3]$-I [306].

2.1.5 Orthoborates $M^{1+}_3M^{3+}[BO_3]_2$

To the monoclinic ($P2_1/c = C^5_{2h}$) $Na_3Nd[BO_3]_2$ type with isolated BO_3^{3-} triangles in a complex structure belong $Na_3La[BO_3]_2$ [307] and $Na_3Nd[BO_3]_2$ [307,308].

2.1.6 Orthoborates $M^{2+}_2M'^{2+}[BO_3]_2$

Tetragonal $Ca_2Mg[BO_3]_2$ is isomorphous with $Na_2Ca[CO_3]_2$; hexagonal $Ba_2Mg[BO_3]_2$ is isomorphous with the mineral *buetschliite*. $(Na,K)_2(Ca,Mg)[CO_3]_2$ [309].

2.1.7 Orthoborates $M^{3+}M'^{3+}[BO_3]_2$

Isomorphous with the mineral *dolomite*, $CaMg[CO_3]_2$, are the trigonal ($R\bar{3} = C^2_{3i}$) compounds with partial cation disorder: $YCr[BO_3]_2$ [310]; $DyCr[BO_3]_2$ [310]; $ErCr[BO_3]_2$ [310]; $YbCr[BO_3]_2$ [310]; $LuCr[BO_3]_2$ [310]; $YFe[BO_3]_2$ [264]; $DyFe[BO_3]_2$ [264]; $HoFe[BO_3]_2$ [264]; $ErFe[BO_3]_2$ [264]; $TmFe[BO_3]_2$ [264]; $YbFe[BO_3]_2$ [264]; and $LuFe[BO_3]_2$ [264].

2.1.8 Orthoborates $M^{2+}M^{4+}[BO_3]_2$

Also isomorphous with *dolomite* are the following trigonal ($R\bar{3} = C^2_{3i}$) compounds with weak rotation at the BO_3^{3-} groups as compared to the ideal position: $MgSn[BO_3]_2$ [311,312]; $CaSn[BO_3]_2$ = mineral *nordenskjöldine* [311–314]; $SrSn[BO_3]_2$ [311–313]; $SrSn_{1-x}Ti_x[BO_3]_2$ [311]; $BaSn[BO_3]_2$ [311–313]; $BaSn_{1-x}Ti_x[BO_3]_2$ [311]; $MnSn[BO_3]_2$ = mineral *tusionite* [314,315]; $FeSn[BO_3]_2$ [314]; $CoSn[BO_3]_2$ [314]; $NiSn[BO_3]_2$ [314]; $CdSn[BO_3]_2$ [314]; $Sr_{1-x}Ba_xTi[BO_3]_2$ [311]; $BaTi[BO_3]_2$ [311,312]; $CaZr[BO_3]_2$ [312,315]; $SrZr[BO_3]_2$ [312,313]; $BaZr[BO_3]_2$ [312,313]; $PbZr[BO_3]_2$ [316]; $CdZr[BO_3]_2$ [312,313]; $CaHf[BO_3]_2$ [313]; $SrHf[BO_3]_2$ [313]; and $BaHf[BO_3]_2$ [313].

2.1.9 Orthoborates $M^{1+}M^{5+}[BO_3]_2$

In monoclinic ($Pn = C^2_s$) $RbNb[BO_3]_2$, octahedra chains of NbO_6 are tilted in the *a*-direction (100); the true formula is $\{RbNb[BO_3]_2\}_5$ [317]. Isotypic are $RbTa[BO_3]_2$ and $TlTa[BO_3]_2$ [317]; orthorhombic ($Pna2_1 = C^9_{2v}$) $TlNb[BO_3]_2$ has a similar, but different structure [303].

2.1.10. Orthoborates $M^{1+}_3M^{3+}_2[BO_3]_3$

Isomorphous with the mineral *shortite*, $Na_2Ca_2[CO_3]_3$, are the orthorhombic ($Amm2 = C^{14}_{2v}$) compounds $Na_3[La_2(BO_3)_3]$ and $Na_3[Nd_2(BO_3)_3]$ [307,308]; while $Li_3[In_2(BO_3)_3]$ belongs to another orthorhombic space group [318]. To the monoclinic ($P2_1/n = C^5_{2h}$) $Li_3[Nd_2(BO_3)_3]$ type belong $Li_3[La_2(BO_3)_3]$ [319], $Li_3[Nd_2(BO_3)_3]$ [247,308,319]; $Li_3[Pr_2(BO_3)_3]$ [320]; and $Li_3[Eu_2(BO_3)_3]$ [321].

2.1.11 Orthoborates $M_6^{1+}M^{3+}[BO_3]_3$

Members of the monoclinic ($P2_1/b = C_{2h}^5$) $Li_6[Yb(BO_3)_3]$ type are: $Li_6[Nd(BO_3)_3]$ [308]; $Li_6[Gd(BO_3)_3]$ [308]; $Li_6[Ho(BO_3)_3]$ [322]; and $Li_6[Yb(BO_3)_3]$ [323].

2.1.12 Orthoborates $M_6^{1+}M_2^{3+}[BO_3]_4$

In the following compounds of the triclinic $Li_3[Ga(BO_3)_2]$ type, $\{M_2(BO_3)_4\}_{\infty}^{6\infty-}$ chains with two different BO_3^{3-} groups have been found: $Li_6[Al_2(BO_3)_4]$ [324] and $Li_6[Ga_2(BO_3)_4]$ [325].

2.1.13 Orthoborates $M_3^{2+}M_2^{3+}[BO_3]_4$

In the following compounds of the orthorhombic ($Pc2_1n = C_{2v}^9$) $Sr_3La_2B_4O_{12}$ type, four different isolated planar BO_3^{3-} groups and chains of distorted La—O and Sr—O polyhedra are found:
$Ca_3[Y_2(BO_3)_4]$ [326, 327]; $Ca_3[La_2(BO_3)_4]$ [327]; $Ca_3(Pr_2(BO_3)_4]$ [327]; $Ca_3[Nd_2(BO_3)_4]$ [327]; $Ca_3[Sm_2(BO_3)_4]$ [327]; $Ca_3[Gd_2(BO_3)_4]$ [327]; $Ca_3[Tb_2(BO_3)_4]$ [327]; $Ca_3[Dy_2(BO_3)_4]$ [327]; $Ca_3[Ho_2(BO_3)_4]$ [327]; $Ca_3[Er_2(BO_3)_4]$ [327]; $Ca_3[Tm_2(BO_3)_4$ [327]; $Ca_3[Yb_2(BO_3)_4]$ [326]; $Ca_3[Lu_2(BO_3)_4]$ [327]; $Sr_3[La_2(BO_3)_4]$ [327–331]; $Sr_3[Pr_2(BO_3)_4]$ [328, 329, 331, 332, 333]; $Sr_3[Nd_2(BO_3)_4]$ [328, 329, 334, 335]; $Sr_3[Sm_2(BO_3)_4]$ [328, 329]; $Sr_3[Eu_2(BO_3)_4]$ [328]; $Sr_3[Gd_2(BO_3)_4]$ [327–329, 331, 333] $Sr_3[Tb_2(BO_3)_4]$ [328, 329]; $Sr_3[Dy_2(BO_3)_4]$ [328, 329]; $Sr_3[Ho_2(BO_3)_4]$ [328, 329]; $Sr_3[Er_2(BO_3)_4]$ [328, 329, 336]; $Sr_3[Tm_2(BO_3)_4]$ [328]; $Sr_3[Yb_2(BO_3)_4$ [327–329]; $Sr_3[Lu_2(BO_3)_4$ [328]; $Ba_3[La_2(BO_3)_4]$ [327, 331, 333]; $Ba_3[Gd_2(BO_3)_4]$ [327]; $Eu_3[Eu_2(BO_3)_4]$ [337]; and $Co_3[Ho_2(BO_3)_4]$ [284].

2.1.14 Orthoborates $M^{3+}M_3'^{3+}[BO_3]_4$

Isomorphous with the mineral *huntite*, $CaMg_3[CO_3]_4$, are the trigonal ($R32 = D_3^7$) compounds of the $Y[Al_3(BO_3)_4]$ type with two different isolated BO_3^{3-} groups and Al—O helices:
$Y[Al_3(BO_3)_4]$ [338–342]; $Y_{1-x}Nd_x[Al_3(BO_3)_4]$ [343, 344] $Pr[Al_3(BO_3)_4]$ [340]; $Nd[Al_3(BO_3)_4]$-I [339, 340, 345]; $Sm[Al_3(BO_3)_4]$ [339, 340, 346]; $Eu[Al_3(BO_3)_4]$ [339, 340]; $Gd[Al_3(BO_3)_4]$ [339, 340]; $Tb[Al_3(BO_3)_4]$ [339, 340] $Dy[Al_3(BO_3)_4]$ [339, 340]; $Ho[Al_3(BO_3)_4]$ [339, 340]; $Er[Al_3(BO_3)_4]$ [338, 339, 340]; $Y_{1-x}Er_x[Al_3(BO_3)_4]$ [338]; $Tm[Al_3(BO_3)_4]$ [340]; $Yb[Al_3(BO_3)_4]$ [339]; $Lu[Al_3(BO_3)_4]$ [340]; $Gd_{1-x}Lu_x[Al_3(BO_3)_4]$ [347]; $Y[Ga_3(BO_3)_4]$ [340, 348–350]; $La[Ga_3(BO_3)_4]$ [340, 349, 350]; $Pr[Ga_3(BO_3)_4]$ [340, 349]; $Nd[Ga_3(BO_3)_4]$ [340, 349, 350, 354]; $Sm[Ga_3(BO_3)_4]$ [340, 348–350]; $Eu[Ga_3(BO_3)_4]$ [340, 348–350]; $Gd[Ga_3(BO_3)_4]$ [340, 348–351]; $Tb[Ga_3(BO_3)_4]$ [340, 348–350];

$Dy[Ga_3(BO_3)_4]$ [348–350]; $Ho[Ga_3(BO_3)_4]$ [349]; $Er[Ga_3(BO_3)_4]$ [340, 349, 350]; $Yb[Ga_3(BO_3)_4]$ [340, 349, 350]; $Sm[Cr_3(BO_3)_4]$ [339]; $Gd[Cr_3(BO_3)_4]$ [339]; $Sc[Fe_3(BO_3)_4]$ [353]; $In[Fe_3(BO_3)_4]$ [353]; $Y[Fe_3(BO_3)_4]$ [340, 350–353]; $La[Fe_3(BO_3)_4]$ [340, 350, 351]; $Pr[Fe_3(BO_3)_4]$ [340, 350, 353]; $Nd[Fe_3(BO_3)_4]$ [340, 350–353]; $(Nd_{0.5}Bi_{0.5})[Fe_3(BO_3)_4]$ [353, 355]; $Er[Fe_3(BO_3)_4]$ [353]; $Sm[Fe_3(BO_3)_4]$ [340, 350–353]; $Eu[Fe_3(BO_3)_4]$ [340, 350, 351, 353]; $Gd[Fe_3(BO_3)_4]$ [340, 350–353]; $Tb[Fe_3(BO_3)_4]$ [340, 350, 351]; $Dy[Fe_3(BO_3)_4]$ [340, 350–352]; $Ho[Fe_3(BO_3)_4]$ [340, 351, 352]; $Er[Fe_3(BO_3)_4]$ [340, 350, 353]; $Tm[Fe_3(BO_3)_4]$ [340]; $Yb[Fe_3(BO_3)_4]$ [340, 350]; $Bi[(Fe_{1.35}Al_{1.65})(BO_3)_4]$ [353].

Isomorphous with the mineral *xanthite* crystallizes the monoclinic ($C2/c = C^6_{2h}$) high-temperature phase $Nd[Al_3(BO_3)_4]$ containing two different isolated BO_3^{3-} groups [356, 358]; the monoclinic high-temperature phase $Gd[Al_3(BO_3)_4]$ crystallizes in $C2 = C^3_2$ or $C2/m = C^3_{2h}$ or $Cm = C^3_s$ [357].

2.1.15 Orthoborates $M_3^{2+}M_3^{3+}(BO_3)_5$

The hexagonal ($P6_3mc = C^4_{6v}$ or $P\bar{6}2c = D^2_{3h}$ or $P6_3/mmc = D^4_{6h}$) $Ca_3[La_3(BO_3)_5]$ [359] crystallizes isomorphous with the mineral *burbankite*.

2.2 BO_3^{3-} Groups Besides Other Ions

2.2.1 BO_3^{3-} Groups Besides O^{2-} Ions

2.2.1.1 Warwickite-Type, $M^{2+}M^{3+}[O(BO_3)]$

The following compounds $M^{2+}M^{3+}[BO_4]$ belong to the orthorhombic ($Pnma = D^{16}_{2h}$) $(Mg,Fe)_3Ti[O(BO_3)]_2$ = mineral *warwickite* type: $Mg(Mg_{0.5}Ti_{0.5})[O(BO_3)]$ = mineral *warwickite* [46];
$MgGa[O(BO_3)]$ [360]; $MgTi^{3+}[O(BO_3)]$ [361]; $MgV^{3+}[O(BO_3)]$ [361]; $MgCr^{3+}[O(BO_3)]$ [360]; $MgFe^{3+}[O(BO_3)]$ [362, 363]; $CaIn[O(BO_3)]$ [360, 362]; $CaSc[O(BO_3)]$ [364]; $CaY[O(BO_3)]$ [270]; $CaGd[O(BO_3)]$ [270]; $CaTb[O(BO_3)]$ [270]; $CaDy[O(BO_3)]$ [270]; $CaHo[O(BO_3)]$ [270] $CaEr[O(BO_3)]$ [270]; $CaTm[O(BO_3)]$ [270]; $CaYb[O(BO_3)]$ [270]; $CaLu[O(BO_3)]$ [364]; $Mn^{2+}Fe^{3+}[O(BO_3)]$ [360]; $Fe^{2+}Fe^{3+}[O(BO_3)]$ [362]; $CoGa[O(BO_3)]$ [360]; $CoSc[O(BO_3)]$ [360]; $CoCr^{3+}[O(BO_3)]$ [360]; $CoFe^{3+}[O(BO_3)]$ [362, 365]; $NiCr^{3+}[O(BO_3)]$ [366]; the low-pressure $NiFe^{3+}[O(BO_3)]$-I [362]; $CdIn[O(BO_3)]$ [360]; $CdSc[O(BO_3)]$ [360].

2.2.1.2 Vonsenite-Ludwigite Type, $M_2^{2+}M^{3+}[O_2(BO_3)]$

The following compounds belong to the orthorhombic ($Pbam = D^9_{2h}$) vonsenite-ludwigite series:
$(Mg,Fe^{2+})_2(Fe^{3+},Al)[O_2(BO_3)]$ = mineral *ludwigite* [362, 367, 368]
$Mg_2(Fe, Ti, Mg)[O_2(BO_3)]$ = mineral *azoproite* [369]
$Fe_2^{2+}Fe^{3+}[O_2(BO_3)]$ = mineral *vonsenite* [362, 367, 370–372]
$Co_2Fe^{3+}[O_2(BO_3)]$ [362]

$Ni_2Fe^{3+}[O_2(BO_3)]$ [367, 372]
$Cu_2Fe^{3+}[O_2(BO_3)]$ [362]
$Fe_2^{2+}V^{3+}[O_2(BO_3)]$ [361], $Co_2V^{3+}[O_2(BO_3)]$ [361]
$Ni_2Al[O_2(BO_3)]$ [373] and
$Co_2^{2+}Co^{3+}[O_2(BO_3)]$ [374].

2.2.1.3 *Hulsite Type, $M_2^{2+}M^{3+}[O_2(BO_3)]$*

To the hulsite type belongs only the monoclinic (P2/m = C_{2h}^1) solid series (Fe^{2+},Mg, Fe^{3+},Sn^{4+})[BO_5], for example $Mg_{0.7}Fe_{1.5}^{2+}Fe_{0.6}^{3+}Sn_{0.2}^{4+}[O_2(BO_3)]$ = mineral *hulsite* [375, 376].

2.2.1.4 *Pinakiolite Type, $M_2^{2+}M^{3+}[O_2(BO_3)]$*

To the pinakiolite type belongs only the monoclinic (C2/m = C_{2h}^3) solid series (Mg,Mn^{2+})$_2$$Mn^{3+}$[$BO_5$], for example $(Mg,Mn)_{1.77}(Mn,Fe,Al)_{1.11}[O_2(BO_3)]$ = mineral *pinakiolite* [46]; or $Mg_{1.9}(Mn,Al,Fe)_{1.0}[O_2(BO_3)]$ = also *pinakiolite* [377].

2.2.1.5 *Takéuchite Type, $M_2^{2+}M^{3+}[O_2(BO_3)]$*

The orthorhombic (Pnnm = D_{2h}^{12}) takéuchite type is represented by the solid series (Mg,Mn^{2+})$_2$(Mn^{3+},Fe^{3+},Ti^{4+})[BO_5], for example, $Mg_{1.59}Mn_{0.42}^{2+}Mn_{0.78}^{3+}Fe_{0.19}^{3+}Ti_{0.01}^{4+}$-[$O_2(BO_3)$] = mineral *takéuchite* [378]; and $(Mg,Mn^{2+},Mn^{3+})_{2.95}[O_2(BO_3)]$ = mineral *orthopinakiolite* [377].

2.2.1.6 *Borates $M_5^{2+}M^{4+}[O_2(BO_3)]_2$*

The orthorhombic (Pbam = D_{2h}^9) borates $Ni_5Ti[B_2O_{10}]$ = $Ni_5Ti[O_2(BO_3)]_2$ and $Co_5Ti[B_2O_{10}]$ = $Co_5Ti[O_2(BO_3)]_2$ are similar to the ludwigite type [379].

2.2.1.7 *Borates $M^{2+}M_2^{3+}[O(BO_3)_2]$*

While α-$CaAl_2[B_2O_7]$ = α-$CaAl_2[O(BO_3)_2]$-II is hexagonal ($P6_322$ = D_6^6) [380], β-$CaAl_2[O(BO_3)_2]$-I is monoclinic [380]; $BaAl_2[O(BO_3)_2]$ is also monoclinic (P2/c = C_{2h}^4 or Pc = C_2^2) [381].

2.2.1.8 *Borates $M_4^{3+}[O_3(BO_3)_2]$*

$Bi_4[B_2O_9]$ = $Bi_4[O_3(BO_3)_2]$ crystallizes monoclinic ($P2_1/c$ = C_{2h}^5) with isolated BO_3^{3-} groups [382].

2.2.1.9 *Borates $M_3^{2+}M_2^{3+}[O_3(BO_3)_2]$*

The $Ca_3Gd_2[B_2O_9]$ = $Ca_3Gd_2[O_3(BO_3)_2]$ type (C2/m = C_{2h}^3 or C2 = C_2^3 or Cm = C_s^3) is represented by $Ca_3La_2[O_3(BO_3)_2]$ [270]; $Ca_3Gd_2[O_3(BO_3)_2]$ [270]; and $Ca_3Yb_2[O_3(BO_3)_2]$ [270].

2.2.1.10 *Borates $M_3^{+}M_3^{5+}[O_6(BO_3)_2]$*

The trigonal (P31m = C_{3v}^2) $K_3Nb_3[B_2O_{12}]$ = $K_3Nb_3[O_6(BO_3)_2]$ and the hexagonal ($P\bar{6}2m$ = D_{3h}^3) $K_3Ta_3[B_2O_{12}]$ = $K_3Ta_3[O_6(BO_3)_2]$ have two nonequivalent non-symmetrical BO_3^{2-} groups with a pentagonal channel-structure like $K_6Nb_6[Si_4O_{26}]$ [383, 384].

2.2.1.11 Borates $M^{3+}M^{6+}[O_3(BO_3)]$

The monoclinic ($P2_1 = C_2^2$) $LaMo[BO_6] = La_2Mo_2[O_3(BO_3)]_2$ type with endless B—O chains of distorted BO_3^{3-} triangles is represented by $La_2Mo_2[O_3(BO_3)]_2$ [385] and $La_2W_2[O_3(BO_3)]_2$ [385], and perhaps also by $CeMo[BO_6]$ [386]. $PrMo[BO_6]$ and $NdMo[BO_6]$ [386] seem to have other structures; considering the B—O distances in the Mo compound (123, 129, and 155 pm), one can classify these compounds to the metaborates $LaBO_2[OMoO_3]$.

2.2.1.12 Borates $M_5^{3+}[O_6(BO_3)]$

The orthorhombic ($Cmc2_1 = C_{2v}^{12}$) $Al_5[BO_9] = Al_5[O_6(BO_3)]$ type is found in the compound $B_2O_3 \cdot 5\,Al_2O_3$ [387] or in $2\,B_2O_3 \cdot 9\,Al_2O_3 = Al_{20-(4/11)}[B_{4+(4/11)}O_{36}]$ [388].

2.2.1.13 Borates $M_{13}^{3+}[O_{12}(BO_3)_5]$

The monoclinic ($P2_1/c = C_{2h}^5$) compound $La_{13}[B_5O_{27}]$ is better formulated as $La_{13}[O_{12}(BO_3)_5]$ [389].

2.2.1.14 Borates $M^{2+}M_9^{3+}M^{4+}[O_{15}(BO_3)]$

The hexagonal ($P6_3 = C_6^6$) $CaAl_9Zr[BO_{18}] = CaZrB[Al_9O_{18}] = CaZrAl_9[O_{15}(BO_3)]$ has been found in the mineral *painite* with BO_3^{3-} triangles in a rigid dense $[Al_9O_{18}]$ octahedral framework topologically identical to those found in *jeremejevite* and *fluoborite* [390].

2.2.1.15 Other Not Completely Characterized Orthoborates with Oxygen Ions

The monoclinic ($P2/m = C_{2h}^1$) $Ba_3Al_4[O_3(BO_3)_4]$ [381] and the monoclinic ($P2 = C_2^1$ or $Pm = C_s^1$) $NaMg_3Mn_3[O_5(BO_3)_2]$ [391] are not yet completely characterized by X-ray structural analysis.

The borates of the type $M_2^{3+}M^{6+}[B_2O_9]$ with M^{6+} = Mo or W and M^{3+} = Pr, Nd, Sm, Eu, Gd, and Tb belong to one type, those with M^{6+} = W and M^{3+} = Sm, Eu, Gd, Tb, and Dy to another [385, 386, 392, 393], while the compounds $M_4^{3+}M^{6+}[B_2O_{12}]$ with M^{6+} = W, M^{3+} = La to Ho [393] and $M_6^{3+}M_3^{6+}[B_4O_{24}]$ with M^{6+} = Mo, M^{3+} = Pr or Nd are isostructural, as based on X-ray diffraction results [386].

2.2.2 BO_3^{3-} Groups besides Halide Ions

2.2.2.1 $Be_2[(BO_3)F]$ Type

The monoclinic ($C2 = C_2^3$) $Be_2[(BO_3)F]$ type is represented only by $B_2O_3 \cdot 3\,BeO \cdot Be(F,OH)_2$ = mineral *hambergite* [42, 394].

2.2.2.2 $KBe_2[(BO_3)F_2]$ Type

The monoclinic ($C2 = C_2^3$) $KBe_2[(BO_3)F_2]$ type is related to the hambergite type; it is found in $B_2O_3 \cdot K_2O \cdot 2\,BeO \cdot 2\,BeF_2 = KBe_2[(BO_3)\,F_2]$ [395]; $RbBe_2[(BO_3)\,F_2]$; $CsBe_2[(BO_3)F_2]$; and probably also $NaBe_2[(BO_3)F_2]$ [396].

2.2.2.3 $Mg_2[(BO_3)F]$-I Type

To this orthorhombic ($Pnma = D_{2h}^{16}$) warwickite type belong the high-temperature phases β-$Mg_2[(BO_3)F]$-I [397] and $Mn_2[(BO_3)F]$ [360].

2.2.2.4 *$Mg_2[(BO_3)F]$-II Type*

To this orthorhombic ($Pna2_1 = C_{2v}^9$) olivine type belong the low-temperature phase α-$Mg_2[(BO_3)F]$-II and the solid series $Mg_2[(BO_3)F_{1-x}(OH)_x]$ and $Mg_2[(BO_3)_{1-y}$-$F_{1-3y}]$ [398].

2.2.2.5 *Fluoborite Type*

The hexagonal ($P6_3/m = C_{6h}^2$) type with BO_3^{3-} groups orientated vertical to the trigonal axes, is represented by γ-$Mg_3[(BO_3)F_3]$ or by $Mg_3[(BO_3)(F,OH)_3]$ = mineral *fluoborite* [45, 390, 397].

2.2.2.6 *$Mg_5[(BO_3)_3F]$ Type*

The orthorhombic ($Pna2_1 = C_{2v}^9$) type is represented only by $Mg_5[(BO_3)_3F]$ [399]; it has a hexagonal dense structure like the olivine-chondrodite group.

2.2.2.7 *$Ca_2[(BO_3)Cl]$ Type*

This monoclinic ($P2_1/c = C_{2h}^5$) type is represented only by the synthetic compound $Ca_2[(BO_3)Cl]$ [400].

2.2.2.8 *$Eu_2[(BO_3)Cl]$ Type*

To this hexagonal ($P6_3mc = C_{6v}^4$) type belong $Eu_2[(BO_3)Cl]$ and $Eu_2[(BO_3)Br]$ [401].

2.2.2.9 *$Mg_7[(BO_3)_3Cl_5]$ Type*

This orthorhombic ($P2_12_12_1 = D_2^4$) type is only represented by $Mg_7[(BO_3)_3(Cl,OH)_5]$ = mineral *karlite* [49].

2.2.2.10 *$Pb_6[(BO_3)_3OCl]$ Type*

This type with nearly planar BO_3^{3-} groups has only one member, the synthetic orthorhombic ($Pcmb = D_{2h}^{11}$) $Pb_6[(BO_3)_3OCl] = Pb_4O[Pb_2(BO_3)_3Cl]$ [402].

2.2.2.11 *$Al_6[(BO_3)_5F_3]$ or Jeremejevite Type*

This hexagonal ($P6_3/m = C_{6h}^2$) type is represented by $Al_6[(BO_3)_5(F,OH)_3]$ = mineral *jeremejevite*; of the two different BO_3^{3-} groups, one is strictly planar whereas the other one deviates significantly from planarity [50, 51, 392, 403, 404].

2.2.3 BO_3^{3-} Groups besides CO_3^{2-} or SO_4^{2-} and/or O^{2-} Ions

The cubic ($F4_132 = O^4$) compound $Ca_3Mg[(BO_3)_2(CO_3)]$ contains $\frac{1}{3}$ mol H_2O as the mineral *sakhaite*, and is therefore mentioned already in chapter 1 [219, 220].

Hexagonal ($P6_3 = C_6^6$) $Ca_4Mn_3^{3+}[O_3(BO_3)_3(CO_3)]$ = mineral *gaudefroyite* contains isolated BO_3^{3-} groups [405].

The synthetic orthorhombic ($Pnma = D_{2h}^{16}$) compound $B_2O_3 \cdot 5\,PbO \cdot PbSO_4$ = $Pb_6[O_2(BO_3)_2(SO_4)]$ also contains isolated distorted BO_3^{3-} groups [406].

2.3 Monoborates Containing Four-Coordinate Boron Only

2.3.1 The Ion BO_4^{5-}

2.3.1.1 Chrysoberyll Type, $M^{2+}M^{3+}[BO_4]$

To the orthorhombic (Pnma = D_{2h}^{16}) $BeAl_2O_4$ = chrysoberyll type with slightly distorted tetrahedral boron belong the following borates: $MgAl[BO_4]$ = mineral *sinhalite* [408]; $Mg_{0.95}Fe^{2+}_{0.05}Al[BO_4]$ [409]; $MgGa[BO_4]$ [366]; $NiCr^{3+}[BO_4]$ [366]; and the high-pressure phase $NiFe^{3+}[BO_4]$-II [407,408].

2.3.1.2 $CaAl[BO_4]$ Type

Only orthorhombic (Ccc2 = C_{2v}^{13}) $CaAl[BO_4]$ represents this type [380,410].

2.3.1.3 $SrAl[BO_4]$ Type

Only orthorhombic (Pccn = D_{2h}^{10}) $SrAl[BO_4]$ represents this type [411].

2.3.1.4 $BaMg[SiO_4]$ Type

To the hexagonal $BaMg[SiO_4]$ type belong $CaLa[BO_4]$ and $SrLa[BO_4]$; $CaY[BO_4]$ is not isomorphous [412].

2.3.1.5 $LiGe[BO_4]$ Type

This orthorhombic (Fmm2 = C_{2v}^{18}) type is represented by $Li_2O \cdot B_2O_3 \cdot 2\,GeO_2$ = $LiGe[BO_4]$ [413].

2.3.1.6 Ta, $BO_4]$ Type

To this tetragonal ($I4_1$/amd = D_{4h}^{19}) zirkone type belong $Nb[BO_4)$ [414]; $Ta[BO_4]$ [414]; and $(Nb,Ta)[BO_4]$ = mineral *behierite* [415].

2.3.2 The Ion BO_4^{5-} Besides O^{2-} Ions

2.3.2.1 Norbergite Type, $M_3^{3+}[BO_6]$

To the orthorhombic (Pnma = D_{2h}^{16}) *norbergite* = $Mg_2SiO_4 \cdot Mg(OH,F)_2$ type with the anion $[O_2(BO_4)]^{9-}$ belong $Al_3[BO_6]$ = $Al_3[O_2(BO_4)]$ [51] and $Fe_3[BO_6]$ = Fe_3-$[O_2(BO_4)]$ [416].

2.3.2.2 $Gd_3[BO_6]$ Type

To the monoclinic (C2/m = C_2^3 or C2 = C_2^3 or Cm = C_s^3) $Gd_3[O_2(BO_4)]$ type belong $Y_3[BO_6]$; the high-temperature phase $Sm_3[BO_6]$-I; $Eu_3[BO_6]$; $Gd_3[BO_6]$; $Tb_3[BO_6]$; $Dy_3[BO_6]$; $Ho_3[BO_6]$; $Er_3[BO_6]$; $Tm_3[BO_6]$; the low-temperature phase $Yb_3[BO_6]$-II [270]; perhaps also $La_3[BO_6]$ [417].

2.3.2.3 $Lu_3[BO_6]$ Type

To the monoclinic (C2/m = C_2^3 or C2 = C_2^3 or Cm = C_s^3) $Lu_3[O_2(BO_4)]$ type belong the high-temperature phase $Yb_3[BO_6]$-I and $Lu_3[BO_6]$ [270].

2.3.2.4 $Ni_2Nb[BO_6]$ Type, $M_2^{2+}M^{5+}[BO_6]$

This orthorhombic (Pnma = D_{2h}^{16}) type is represented by $Ni_2Nb[BO_6]$ = Ni_2Nb-$[O_2BO_4]$ [418].

2.3.2.5 Chondrodite Type, $M^{2+}M_4^{3+}[B_2O_{10}]$

To the monoclinic ($P2_1/c$ = C_{2h}^5) $Mg_5[(OH,F)_2(SiO_4)_2]$ = mineral *chondrodite* type belongs $B_2O_3 \cdot CoO \cdot 2\,Al_2O_3$ = $Al_4Co[O_2(BO_4)_2]$ with BO_4 tetrahedra, connecting chains of Al octahedra [419].

2.3.3 The Ion $[B_3O_7]^{5-}$

All the B atoms are tetrahedral in the ion $[B_3O_7]^{5-}$, found in the orthorhombic (Cmma = D_{2h}^{21}) $CaAl[B_3O_7]$ = mineral *johachidolite* [420, 421]; see also triborates (Chapt. 2.6.2).

2.3.4 Tetracoordinate Boron Besides Halide or Germanate

2.3.4.1 Baryte Type

To the orthorhombic (Pnma = D_{2h}^{16}) baryte = $BaSO_4$-II type belongs the borate $Ba_2[(BO_3)F]$ [422,423].

2.3.4.2 Sulphohalite Type

The cubic ($F\bar{4}3m$ = T_d^2) compound $Ca_3Er_3Ge_2[BO_{13}]$ = $Ca_3Er_3[O(BO_4)(GeO_4)_2]$ is of a *sulphohalite* = $Na_6[ClF(SO_4)_2]$ related type [424].

2.3.5 BO_4^{5-} Ions Besides PO_4^{3-} or AsO_4^{3-}

$Ca[BPO_5]$, $Sr[BPO_5]$, $Ca[BAsO_5]$, and $Sr[BAsO_5]$ are hexagonal (P6cc = C_{6v}^2 or P6/mcc = D_{6h}^2) with tetrahedral boron [425]; $Ba[BPO_5]$ is isomorphous [426].

$Pb[BPO_5]$ and $Pb[AsBO_5]$ are also hexagonal in the stillwellite = $RE[BSiO_5]$ type with tetrahedral boron [427].

α-$Mg_3[BPO_7]$ and α-$Zn_3[BPO_7]$ crystallize in the orthorhombic space group Imm2 = C_{2v}^{20}; β-$Zn_3[BPO_7]$ is hexagonal in $P\bar{6}m2$ = D_{3h}^1 or $P\bar{6}2m$ = D_{3h}^3 [428].

2.3.6 The High-Pressure B_2O_3

The orthorhombic ($Ccm2_1$ = C_{2v}^{12}) high-temperature/high-pressure phase B_2O_3-II contains boroxine rings with distorted tetrahedral boron [429]; the structure of the hexagonal ($P3_1$ = C_3^2) low-pressure B_2O_3 is still unknown [430].

2.4 Metaborates, BO_2^-

2.4.1 Metaborates With M^+ Cations, MBO_2

2.4.1.1 $LiBO_2$-I Type

The $LiBO_2$-I type is represented only by monoclinic ($P2_1/c$ = C_{2h}^5) α-$LiBO_2$-I with the short-hand notation 1:∞(△), chains (Fig. 3a) [13, 431].

Metaborates:

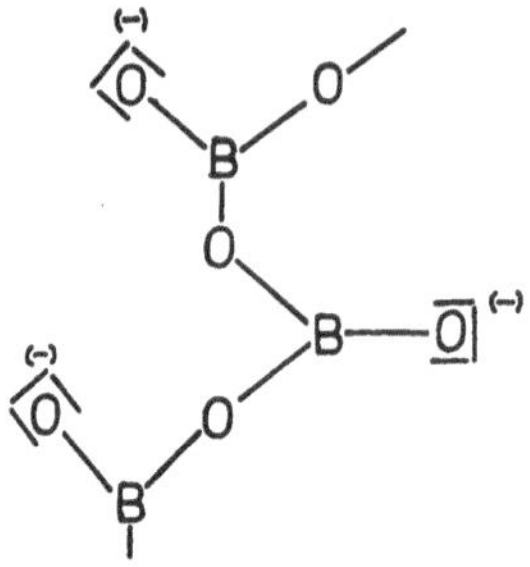

a α-Li B O_2
1 : ∞ (1 Δ), chains

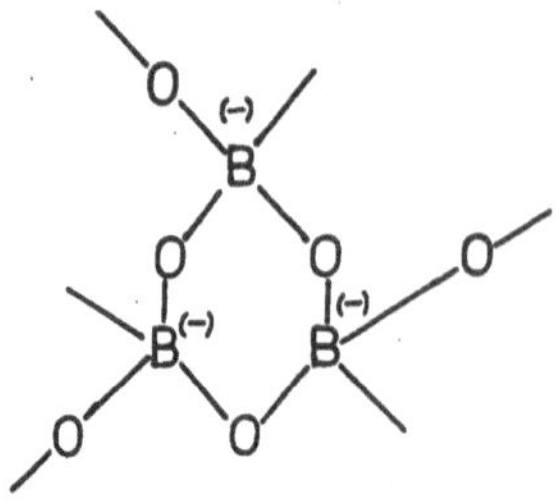

b γ-Li B O_2
3 : ∞_3 (3 T), network

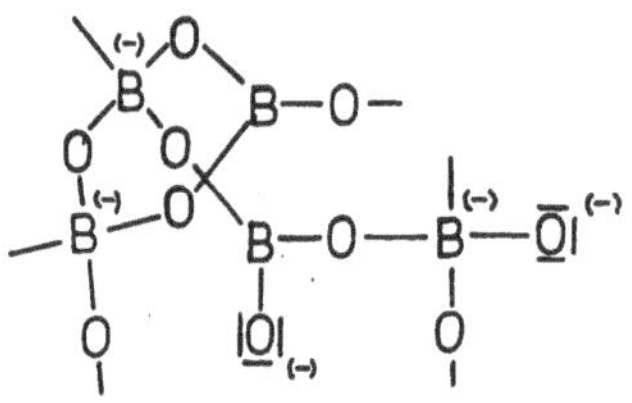

c $B_5 O_{10}^{5-}$
5 : ∞_2 (2 Δ + 3 T), layers

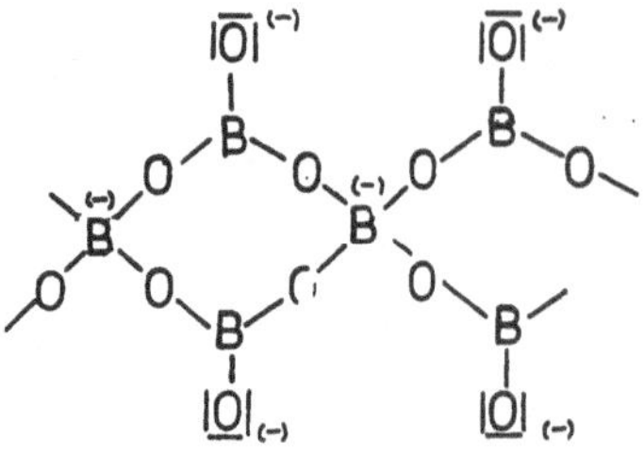

d $B_6 O_{12}^{6-}$
6 : ∞ (4 Δ + 2 T), chains

Diborates:

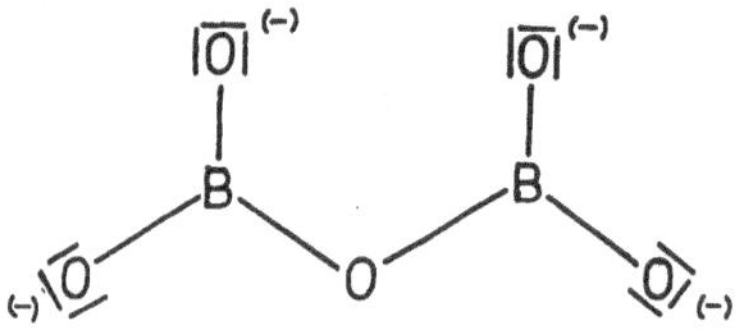

e $B_2 O_5^{4-}$
2 : 2 Δ, isolated

Triborates:

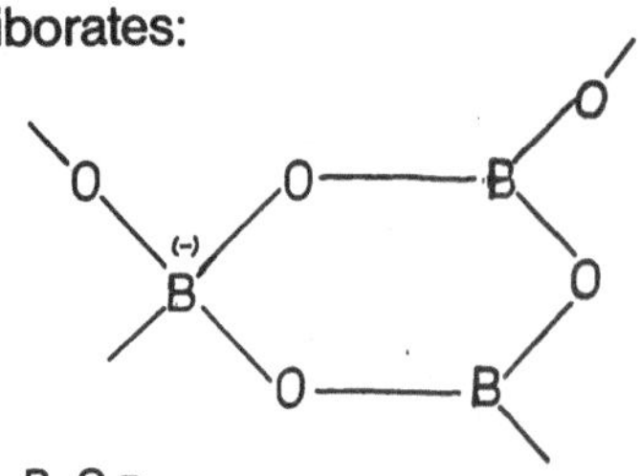

f $B_3 O_5^-$
3 : ∞ (2 Δ + T), chains

Tetraborates:

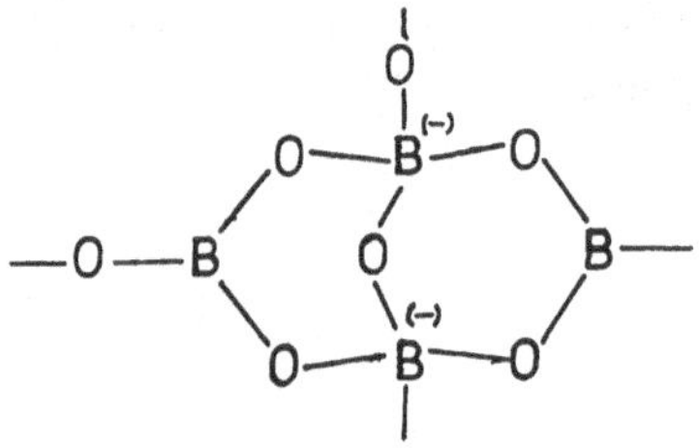

g $B_4 O_7^{2-}$
4 : ∞_3 (2 Δ + 2 T), network

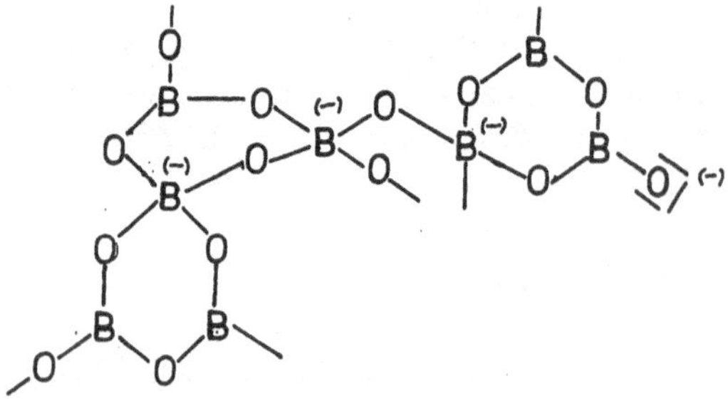

h $Na_4 B_8 O_{14}$
8 : ∞_3 (5 : 3 Δ + 2 T, 3 : 2 Δ + T), network

Fig. 3.

Fig. 3. (continued)

Tetraborates:

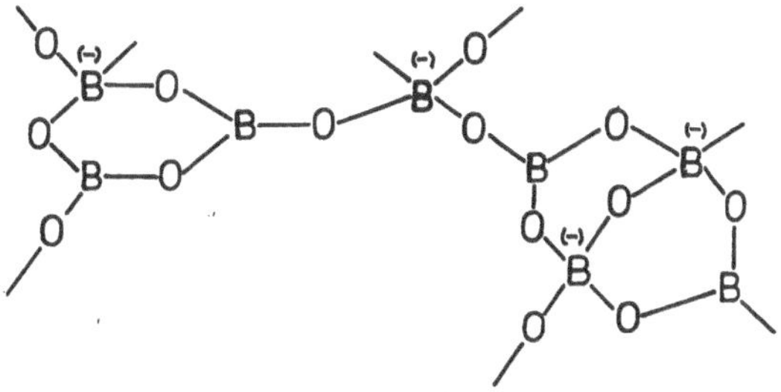

i $Ca_2B_8O_{14}$
8 : ∞_3 (3 : 2 Δ + T, 1 T), 4 : 2 Δ + 2 T)
network

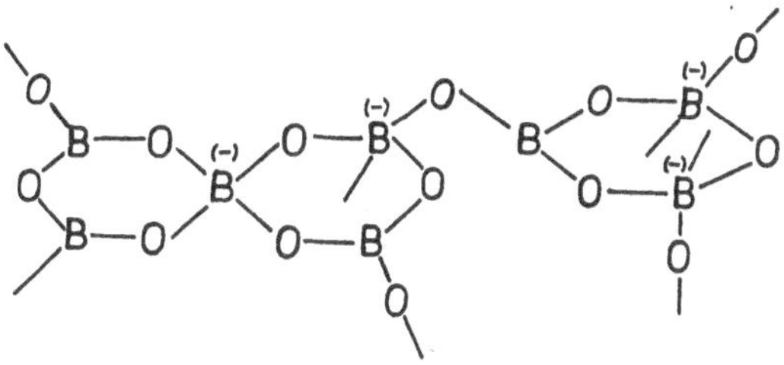

k $Ba_2B_8O_{14}$
8 : ∞_3 (5 : 3 Δ + 2 T, 3 : Δ + 2 T),
network

Pentaborates:

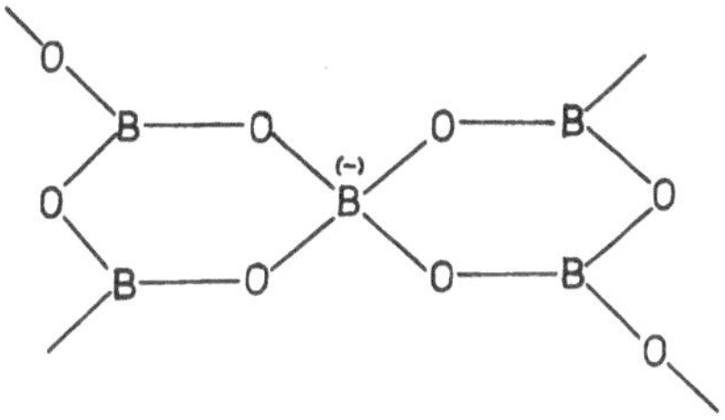

l $B_5O_8^-$
5 : 4 Δ + T,
isolated

Octaborates:

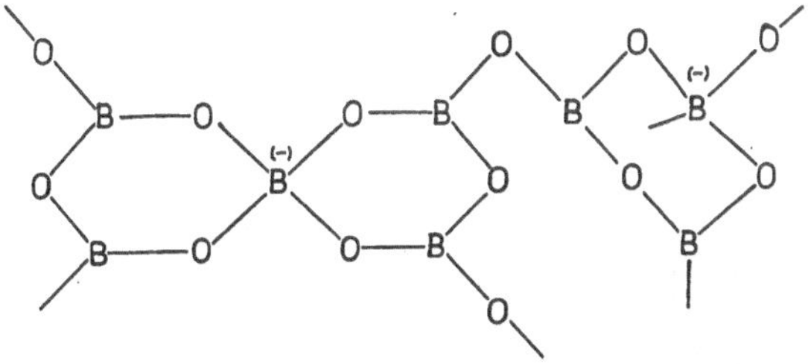

m $B_8O_{13}^{2-}$
8 : ∞_2 (3 : 2 Δ + T, 5 : 4 Δ + T),
layers

Nonaborates:

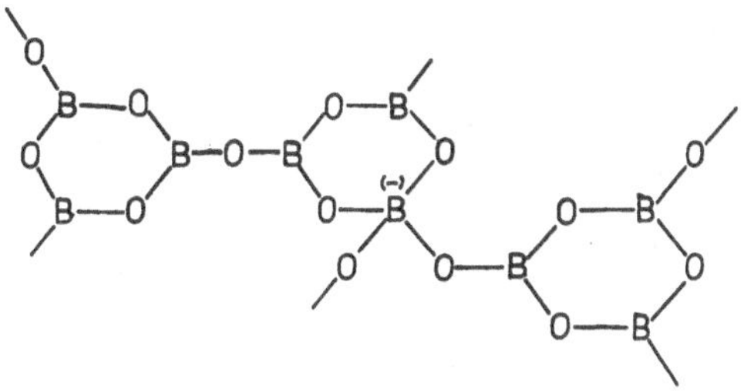

n $B_9O_{14}^-$
9 : ∞_3 (2 x 3 : 3 Δ, 3 : 2 Δ + T),
network

Fig. 3. (continued)

Nonaborates:

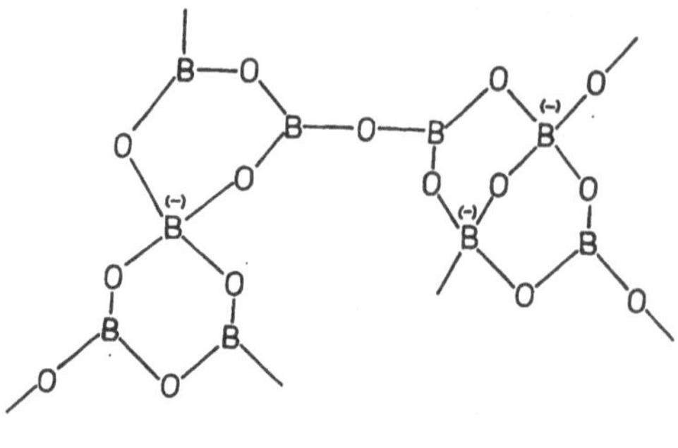

o $B_9O_{15}^{3-}$
9 : ∞_3 (5 : 4 Δ + T, 4 : 2 Δ + 2 T),
network

Higher Borates:

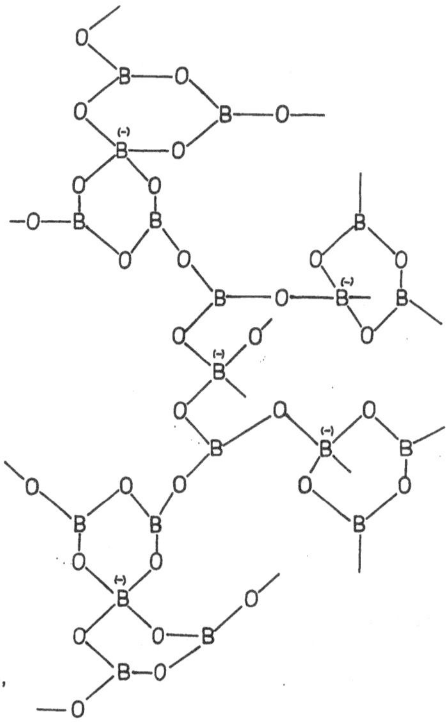

p $B_{19}O_{31}^{5-}$
19 : ∞_3 (2 x 5 : 4 Δ + T, 2 x 3 : 2 Δ + T, 2 Δ, 1 T),
network

2.4.1.2 LiBO$_2$-II Type

The tetragonal ($I\bar{4}2d$ = D_{2d}^{12}) γ-$LiBO_2$-II has the notation 3 : ∞_3(3T), network with tetrahedral boron (Fig. 3b)[432, 433].

2.4.1.3 LiBO$_2$-III Type

The monoclinic β-$LiBO_2$-III has the notation 2 : ∞($\triangle$ + T); chains containing trigonal and tetrahedral boron[433, 434].

2.4.1.4 $NaBO_2$ Type

To the trigonal ($R\bar{3}c = D_{3d}^6$) $NaBO_2$ type with $B_3O_6^{3-}$ (boroxine) rings belong α-$Na_3[BO_2]_3$ [435,436]; α-$K_3[BO_2]_3$ [437]; α-$Rb_3[BO_2]_3$ [437]; and α-$Cs_3[BO_2]_3$ [437].

2.4.1.5 $AgBO_2$ Type

The $AgBO_2$ type is represented only by orthorhombic ($Pbcn = D_{2h}^{14}$) $AgBO_2$ with the notation $2{:}\infty(\triangle + T)$ and containing a BO_2^- isopolyanion where boron is coordinated in equal parts tetrahedrally and trigonal-planar by oxygen [438].

2.4.1.6 $TlBO_2$ Type

The tetragonal ($P4_1 = C_4^2$) $TlBO_2$ with the notation $2{:}\infty(\triangle + T)$ consists of ofie boron-oxygen triangle and one boron-oxygen tetrahedron linked together to form an infinite $(B_2O_4)_n^{2n-}$ ion [213].

2.4.2 Metaborates With M^{2+} Cations, $M^{2+}[BO_2]_2$

2.4.2.1 $Mg[BO_2]_2$ Type

The polymer $(Mg[BO_2]_2)_n$ type has been investigated only by Debye-Scherrer photographs [439].

2.4.2.2 $Ca[BO_2]_2$-I Type

To the orthorhombic ($Pnca = D_{2h}^{14}$) $Ca[BO_2]_2$-I type with the short-hand notation $1{:}\infty(\triangle)$, chains, belong $Ca[BO_2]_2$-I [440]; $Sr[BO_2]_2$-I [441,442]; and $Eu[BO_2]_2$ = α-$Eu[B_2O_4]$ [443].

2.4.2.3 $Ca[BO_2]_2$-II = Calciborite Type

This type is represented only by the orthorhombic ($Pccn = D_{2h}^{10}$) $Ca[BO_2]_2$-II = mineral *calciborite* with the notation $2{:}\infty(\triangle + T)$, chains [410,443,444].

2.4.2.4 $Ca[BO_2]_2$-III Type

To the orthorhombic ($Pna2_1 = C_{2v}^9$) $Ca[BO_2]_2$-III type with the anion $B_6O_{12}^{6-}$ and with the notation $6{:}\infty_3[3T + (2\triangle + T)]$, network, belong $Ca[BO_2]_2$-III [445]; $Sr[BO_2]_2$-III [441]; and $Eu[BO_2]_2$-III [446].

2.4.2.5 $Ca[BO_2]_2$-IV Type

To the cubic ($Pa3 = T_h^6$) $Ca[BO_2]_2$-IV type with $B_3O_6^{3-}$ rings and the notation $3{:}\infty_3(3T)$, network, belong $Ca[BO_2]_2$-IV [447]; $Sr[BO_2]_2$-IV [441]; and $Eu[BO_2]_2$-IV [446].

2.4.2.6 $Ba[BO_2]_2$-I Type

This type is represented by the trigonal ($R\bar{3}c = D_{3d}^6$) high-temperature phase β-$Ba[BO_2]_2$-I with nearly planar $B_3O_6^{3-}$ rings and the notation $3{:}\infty_3(3\triangle)$, network [442,448].

2.4.2.7 *Ba[BO$_2$]$_2$-II Type*

The trigonal ($R3c = C_{3v}^6$) low-temperature phase α-Ba[BO$_2$]$_2$-II is also likely to contain $B_3O_6^{3-}$ rings [449].

2.4.2.8 *Cu[BO$_2$]$_2$ Type*

To the tetragonal ($I\bar{4}2d = D_{2d}^{12}$) Cu[BO$_2$]$_2$ type with $B_3O_6^{3-}$ rings and the notation $3{:}\infty_3(3T)$, network, belong Cu[B$_2$O$_4$] = Cu[BO$_2$]$_2$ [450,451]; and Pd[B$_2$O$_4$] = Pd[BO$_2$]$_2$ [452].

2.4.3 Metaborates With M^{3+} Cations, M[BO$_2$]$_3$

2.4.3.1 *La[BO$_2$]$_3$ Type*

To the monoclinic ($I2/a = C_{2h}^6$) La[BO$_2$]$_3$ type with $[B_6O_{12}]^{6-}$ ions combined to chains and the notation $6{:}\infty(4\triangle + 2T)$, chains (Fig. 3d), belong La[BO$_2$]$_3$ [453,454,455]; Ce[BO$_2$]$_3$ [456]; Pr[BO$_2$]$_3$ [456]; Nd[BO$_2$]$_3$ [321,456,457]; Sm[BO$_2$]$_3$ [456,458]; Eu[BO$_2$]$_3$ [456]; Gd[BO$_2$]$_3$ [456,458,459]; and Tb[BO$_2$]$_3$-I [456].

2.4.3.2 *Tb[BO$_2$]$_3$-II Type*

This orthorhombic ($Pbnm = D_{2h}^{16}$ or $Pbn2_1 = C_{2v}^9$) type is represented only by 3 B$_2$O$_3$ · Tb$_2$O$_3$-II = Tb[BO$_2$]$_3$-II [459].

2.4.3.3 *Bi[BO$_2$]$_3$ Type*

A new monoclinic ($C2 = C_2^3$) type has been found in Bi[B$_3$O$_6$] = 3 B$_2$O$_3$ · Bi$_2$O$_3$ = Bi[BO$_2$]$_3$ with $B_3O_6^{3-}$ rings and the notation $3{:}\infty_2(2\triangle + T)$, sheets [460].

2.4.4 Metaborates With $M^{6+}O_2$ Cations

2.4.4.1 *UO$_2$[BO$_2$]$_2$ Type*

UO$_2$[BO$_2$]$_2$ crystallizes in a monoclinic space group; the presence of twins prevented an assignment [461].

2.4.5 Metaborates With M^{2+} and M^{3+}

2.4.5.1 *CoSm[BO$_2$]$_5$ Type*

To the monoclinic ($P2_1/c = C_{2h}^5$) CoSm[B$_5$O$_{10}$] type with $B_5O_{10}^{5-}$ layers and the notation $5{:}\infty_2(2\triangle + 3T)$, layers (Fig. 3c), belong
LaMg[BO$_2$]$_5$ [462]; CeMg[BO$_2$]$_5$ [462] PrMg[BO$_2$]$_5$ [462];
NdMg[BO$_2$]$_5$ [462]; SmMg[BO$_2$]$_5$ [462]; EuMg[BO$_2$]$_5$ [462]
GdMg[BO$_2$]$_5$ [462]; TbMg[BO$_2$]$_5$ [462]; DyMg[BO$_2$]$_5$ [462];
HoMg[BO$_2$]$_5$ [462]; ErMg[BO$_2$]$_5$ [462]; YCo[BO$_2$]$_5$ [463,464];
LaCo[BO$_2$]$_5$ [465]; NdCo[BO$_2$]$_5$ [466]; SmCo[BO$_2$]$_5$ [467];
HoCo[BO$_2$]$_5$ [284,468].

2.4.6 Metaborates $M_4^{2+}[O(BO_2)_6]$ or $M_4^{2+}[S(BO_2)_6]$ or $M_4^{2+}[Se(BO_2)_6]$

2.4.6.1 *$Zn_4[O(BO_2)_6]$ Type*

The cubic ($I\bar{4}3m = T_d^3$) $Zn_4[O(BO_2)_6]$ type contains BO_4 tetrahedra in $[B_6O_{12}]^{6-}$ rings that build a network; the notation is $6{:}\infty_3(6T)$, network. To this type belong $Zn_4[O(BO_2)_6]$ [469,470,471]; $Hg_4[O(BO_2)_6]$ [472]; $Zn_4[S(BO_2)_6]$ [473,474,475]; Co_4-$[S(BO_2)_6]$ [473,474,475]; and $Zn_4[Se(BO_2)_6]$ [475].

2.4.7 Metaborate BO_2^- Besides PO_4^{3-}

In the trigonal ($P\bar{3} = C_{3i}^1$) $Sr_{9.4}Na_{0.3}(PO_4)_6(BO_2)$ linear BO_2 groups with two-coordinate boron have been found [476].

2.4.8 Metaborate BO_2^- Besides WO_4^{2-} or MoO_4^{2-}

Compounds of the types $Re[WBO_6]$, $(RE)_2[WB_2O_9]$, or $(RE)_4[WB_2O_{12}]$ had been discussed in Chapts. 2.2.1.11 and 2.2.1.15, respectively.

2.5 Diborates, $[B_2O_5]^{4-}$

2.5.1 $Na_4[B_2O_5]$ Type

The monoclinic ($C2/c = C_{2h}^6$) compound $Na_4[B_2O_5]$ has two nearly planar condensed BO_3^{3-} groups with an angle B–O–B of 120.6° [477]; see Fig. 3e; notation 2:2△, isolated.

2.5.2 Suanite Type

The monoclinic ($P2_1/a = C_{2h}^5$) $Mg_2[B_2O_5]$-II = mineral *suanite* has two planar BO_3^{3-} groups linked via O atoms with an angle B—O—B of 138.0° and 22.9° for Mg—O [478–480].

2.5.3 $Sr_2[B_2O_5]$ Type

To the monoclinic ($P2_1/a = C_{2h}^5$) $Sr_2[B_2O_5]$-type with two planar BO_3^{3-} groups linked via O atoms with an B–O–B angle of 138.9° belong the compound $Ca_2[B_2O_5]$ [224]; $Sr_2[B_2O_5]$ [481]; α-$Mn_2[B_2O_5]$ [482]; and $Eu_2[B_2O_5]$ [443].

2.5.4 $Co_2[B_2O_5]$ Type

To the triclinic $Co_2[B_2O_5]$ type with two planar BO_3^{3-} groups linked via O atoms with an B–O–B angle of 134.5° and 16° for Mg–O belong the compounds $Mg_2[B_2O_5]$-I [480,482]; β-$Mn_2[B_2O_5]$-I [482]; $Fe_2[B_2O_5]$ [482]; $Co_2[B_2O_5]$ [464,483]; and $Cd_2[B_2O_5]$ [484,485].

2.5.5 $MgCa[B_2O_5]$ Type

To the monoclinic ($P2_1 = C_{2h}^5$) $MgCa[B_2O_5]$ type with an angle B—O—B of 122 to 127.7° belong the compounds $MgCa[B_2O_5]$ [486]; $CaMn[B_2O_5]$ [487]; and (Mg,Fe)Ca-$[B_2O_5]$ = mineral *α-kurchatovite* [488].

2.5.6 β-Kurchatovite Type

The orthorhombic ($Pc2_1b = C_{2v}^5$) β-kurchatovite type is represented by the compound $Mg_5Ca_6Mn[B_2O_5]_6$ = mineral *β-kurchatovite* [489].

2.5.7 Other Not Fully Characterized Types

Other not fully structurally characterized compounds are the orthorhombic or monoclinic $Li_4[B_2O_5]$ [433]; monoclinic ($P2/m = C_{2h}^1$ or $P2 = C_2^1$ or $Pm = C_S^1$) $Ba_2[B_2O_5]$ [490]; and monoclinic ($C2/m = C_{2h}^3$) $Th[B_2O_5]$ [491].

2.5.8 The Ion $[B_2O_5]^{4-}$ Besides O^{2-} Ions

While the orthorhombic ($Pna2_1 = C_{2v}^9$) $TlNb[BO_3]_2 = Tl(NbO)[B_2O_5]$ contains trigonal boron [303], the hexagonal ($P\bar{6}2m = D_{3h}^3$) $NdAl_2[O_{0.5}(B_2O_5)_2] = NdAl_{1.67+0.67x}[B_4O_{10}]O_x$ ($0.5 \leqq x \leqq 0.6$) contains tetrahedral boron and consists of infinite $[B_2O_5]_n^{4n-}$ chains; the ideal composition $NdAl_3[B_4O_{10}]O_2$ corresponds with the mineral *muscovite*, $KAl_2[Si_3AlO_{10}](OH)_2$ [492–494].

2.5.9 The Ion $[B_2O_5]^{4-}$ Besides Halogenide Ions

For $Al_2[(B_2O_5)F_2]$, powder diffraction data have been given [495].

2.6 Triborates

2.6.1 The Ion $[B_3O_5]^-$

2.6.1.1 $Li[B_3O_5]$ Type

This orthorhombic ($Pna2_1 = C_{2v}^9$) type is represented only by $Li[B_3O_5]$, containing a six-membered $[B_3O_3O_{4/2}]$ ring (see Fig. 3f) with the short-hand notation $3\infty:2\triangle + T$, chains [496,497].

2.6.2 $Na[B_3O_5]$-I Type

To the monoclinic ($P2_1/c = C_{2h}^5$) $Na[B_3O_5]$-I type with $[B_9O_{15}]^{3-}$ rings (see Fig. 3o) and the notation $9:\infty_3(5:4\triangle + T, 4:2\triangle + 2T)$, network, belongs only α-$Na[B_3O_5]$-I $= Na_3[B_9O_{15}]$ [498].

2.6.3 $Na[B_3O_5]$-II Type

To the monoclinic ($P2_1/c = C_{2h}^5$) $Na[B_3O_5]$-II type with $[B_9O_{15}]^{3-}$ rings, but the notation $9:\infty_2(3:2\triangle + T, 5:4\triangle + T, 1T)$, belongs β-$Na[B_3O_5]$ = β-$Na_3[B_9O_{15}]$ [499].

2.6.4 $Cs[B_3O_5]$ Type

The orthorhombic ($P2_12_12_1 = D_2^4$) $Cs[B_3O_5]$ type contains $[B_3O_5]^-$ rings with the notation $3\infty:2\triangle + T$, chains; the structure type of $Cs[B_3O_5]$ [500] may be also in $Tl[B_3O_5]$ [501].

2.6.5 Other Not Fully Characterized Types

Triclinic γ-$Na[B_3O_5]$-III [498] and triclinic (but pseudohexagonal) $K[B_3O_5]$ [503] with $[B_9O_{15}]^{3-}$ rings are not yet fully characterized.

2.6.6 The Ion $[B_3O_7]^{5-}$

The ion $[B_3O_7]^{5-}$ with three four-coordinate boron atoms occurs in the orthothombic (Cmma = D_{2h}^{21}) $CaAl[B_3O_7]$ = mineral *johachidolite* [420, 421].

2.7 Tetraborates

2.7.1 The Ion $[B_4O_7]^{2-}$

2.7.1.1 $Li_2[B_4O_7]$ Type

The tetragonal ($I4_1cd$ = C_{4v}^{12}) $Li_2[B_4O_7]$ type contains $B_4O_7^{2-}$ groups building two three-dimensional connected networks, but with the notation 4:2△ + 2T, isolated; it is only represented by $Li_2[B_4O_7]$ [504, 505].

2.7.1.2 $Na_2[B_4O_7]$ Type

The triclinic $Na_2[B_4O_7]$ consists of $[B_8O_{14}]^{4-}$ ions with one non-bridging O atom; the notation is 8:∞_3(5:3△ + 2T, 3:2△ + T), network (Fig. 3h) [502].

2.7.1.3 $K_2[B_4O_7]$ Type

Triclinic $K_2[B_4O_7]$ also contains $[B_8O_{14}]^{4-}$ ions, but the notation is 8:∞_3(4-1:2△ + 2T, 3:△ + 2T, +1△), network [506].

2.7.1.4 $Cd[B_4O_7]$ Type

To the orthorhombic (Pbca = D_{2h}^{15}) $Cd[B_4O_7]$ type with $B_4O_7^{2-}$ ions that build up a framework (see Fig. 3g) with the notation 4∞_3:2△ + 2T, network, belong $Mg[B_4O_7]$ [507]; $Mn[B_4O_7]$ [508]; $Fe[B_4O_7]$ [509]; $Zn[B_4O_7]$ [510]; $Cd[B_4O_7]$ [511]; and $Hg[B_4O_7]$ [290].

2.7.1.5 $Ca[B_4O_7]$ Type

The monoclinic ($P2_1/n$ = C_{2h}^5) $Ca[B_4O_7]$ type contains $B_8O_{14}^{4-}$ ions, consisting of BO_2^-, $B_3O_5^-$, and $B_4O_7^{2-}$ ions with the notation 8:∞_3(3:2△ + T, 1T, 4:2△ + 2T), network (see Fig. 3i); it has been found only in $Ca[B_4O_7]$ [512, 513].

2.7.1.6 $Sr[B_4O_7]$ Type

To the orthorhombic ($P2_1nm$ = C_{2v}^7) $Sr[B_4O_7]$ type consisting of a three-dimensional network of BO_4 tetrahedra with one of the O atoms coordinated to three B atoms and the notation 4:∞_3(4T), network belong $Sr[B_4O_7]$ [514, 515] and $Pb[B_4O_7]$ [515].

2.7.1.7 $Ba[B_4O_7]$ Type

The monoclinic ($P2_1/c$ = C_{2h}^5) $Ba[B_4O_7]$ type contains (like the Ca compound) $B_8O_{14}^{4-}$ ions, but consisting of a pentaborate and a triborate unit with the notation 8:∞_3(3:△ + 2T, 5:3△ + 2T), network; see Fig. 3k [516].

2.7.1.8 α-Eu[B₄O₇] Type

A similar structure as $Sr[B_4O_7]$ has the orthorhombic ($Pnm2_1 = C_{2v}^2$) α-$Eu[B_4O_7]$ [517].

2.7.1.9 Other Not Fully Characterized Tetraborate Types

Not yet fully characterized are the structures of tetragonal $Cu_2[B_4O_7]$ [518]; hexagonal β-$Eu[B_4O_7]$ [519]; orthorhombic ($Ammm = D_{2h}^{19}$ or $Amm2 = C_2^{14}$) α-$Tl_2[B_4O_7]$; and trigonal-rhomboedric ($P\bar{3} = C_{3i}^1$ or $P3 = C_3^1$) β-$Tl_2[B_4O_7]$ [520].

2.7.2 The Ion $[B_4O_9]^{6-}$

The ion $[B_4O_9]^{6-}$ seems to exist in the two modifications of $Li_6[B_4O_9]$ [433] as well as in the orthorhombic $Na_4[Cd(B_4O_9)]$ [521].

2.7.3 The Ion $[(B_4O_{10})O_{0.5}]^{9-}$

The ion $[B_4O_{10})O_{0.5}]^{9-}$ can be formulated in $NdAl_{1.67+0.67x}(B_4O_{10})O_x$ [492–494]; see diborates.

2.7.4 The Ion $[B_4O_{11}]^{10-}$

The ion $[B_4O_{11}]^{10-}$ seems to exist in the triclinic $Li_4[Zn_3(B_4O_{11})]$ [300].

2.8 Pentaborates

2.8.1 The Ion $[B_5O_8]^-$

2.8.1.1 K[B₅O₈]-I Type

The orthorhombic ($Pbca = D_{2h}^{15}$) high-temperature phase α-$K[B_5O_8]$-I contains double ring $B_5O_8^-$ groups (Fig. 31) combined into two interlocking networks; the notation is 5:4△ + T, isolated [522].

2.8.1.2 K[B₅O₈]-II Type

To the orthorhombic ($Pbca = D_{2h}^{15}$) phase β-$K[B_5O_8]$-II which contains the double ring $B_5O_8^-$ with the notation 5:4△ + T, isolated, but in another three-dimensional network which can be described as consisting of helix chains, belong β-$K[B_5O_8]$-II [523]; β-$Rb[B_5O_8]$-II [524]; and perhaps $Tl[B_5O_8]$ [501].

2.8.1.3 Other Not Fully Characterized Types

Not fully characterized pentaborate types are the monoclinic low-temperature phase γ-$K[B_5O_8]$-III [503]; and the monoclinic ($P2_1/a = C_{2h}^5$ or $P222_1 = D_2^2$) α-$Cs[B_5O_8]$ [525].

2.8.2 The Ion $[B_5O_{11}]^{7-}$

The orthorhombic ($Pnma = D_{2h}^{16}$) $Bi_3O[B_5O_{11}]$ type = $Bi_3[B_5O_{12}]$ contains $[B_5O_{11}]^{7-}$ ions with the notation 5:3△ + 2T, isolated [526].

2.8.3 The Ion $[B_{4.67}O_{11}]^{8-}$

The monoclinic ($Cc = C_s^4$) $Na_2Zn_2Mn[B_{4.67}O_{11}]$ type consists of BO_4 tetrahedra and BO_3 triangles in ratio 1:1 [527].

2.8.4 The Ion $[B_5O_9X]^{4-}$ (X = Cl, Br)

2.8.4.1 $Ca_2[B_5O_9Br]$ Type

To the orthorhombic ($Pnn2 = C_{2v}^{10}$) $Ca_2[B_5O_9Br]$ type with the $[B_5O_9]^{3-}$ ion and the notation $5{:}\infty_3(2\triangle + 3T)$, network, belong the following compounds:
$Ca_2[B_5O_9Cl]$ [528, 529]; $Ca_2[B_5O_9Br]$ [528]; $Sr_2[B_5O_9Br]$ [529]; $Ba_2[B_5O_9Cl]$ [530]; $Ba_2[B_5O_9Br]$ [530]; $Eu_2[B_5O_9Cl]$-I [401, 531]; $Eu_2[B_5O_9Br]$ [401, 531, 532]; $Ca_{2-x}Eu_x[B_5O_9Cl]$ [530]; $Sr_{2-x}Eu_x[B_5O_9Br]$ [401]; $Sr_{2-x}Eu_x[B_5O_9I]$ [530]; $Pb_2[B_5O_9Cl]$ [530]; $Pb_2[B_5O_9Br]$ [530].

2.8.4.2 $Sr_2[B_5O_9Cl]$ Type

To the tetragonal ($P4_22_12 = D_4^6$) $Sr_2[B_5O_9Cl]$ type with the $[B_5O_9]^{3-}$ ion and the notation $5{:}\infty_3(2\triangle + 3T)$, network, belong $Sr_2[B_5O_9Cl]$ [529]; $Ba_2[B_5O_9Cl]$ [529]; $Eu_2[B_5O_9Cl]$-II [532]; $Sr_{2-x}Eu_x[B_5O_9Cl]$ [401]; $Ba_{2-x}Eu_x[B_5O_9Cl]$ [529]; $Ca_{2-x-y}Ba_xEu_y$-$[B_5O_9Cl]$ [529]; and $Sr_{2-x-y}Ba_xEu_y[B_5O_9Cl]$ [529].

2.8.4.3 Hilgardite Type

For $Ca_2[B_5O_9Cl] \cdot H_2O$ = mineral *hilgardite* [154÷159], see Chapt. 1.6.

2.9 Hexaborates

2.9.1 The Ion $[B_6O_{10}]^{2-}$

The ion $[B_6O_{10}]^{2-}$ seems to be present in $3\,B_2O_3 \cdot CaO = Ca[B_6O_{10}]$ [533].

2.9.2 The Ion $[B_6O_{11}]^{4-}$

The ion $[B_6O_{11}]^{4-}$ has been found in monoclinic ($P2_1/b = C_{2h}^5$) $3\,B_2O_3 \cdot 2\,CaO = Ca_2[B_6O_{11}]$ with the notation $6{:}\infty_3(5{:}2\triangle + 3T, 1T)$, A = B, network; a pentaborate unit and a single BO_4 tetrahedron are connected to build up a three-dimensional network [534].

2.9.3 The Ion $[B_6O_{17}]^{16-}$

It may be that the ion $[B_6O_{17}]^{16-}$ exists in monoclinic $Li_8Zn_4[B_6O_{17}]$ [300].

2.10 Heptaborates, Boracites

The boracites are three-dimensional, highly symmetrical, rigid boron-oxygen frame networks with tetrahedral boron, in the holes of which M^{2+} cations and X (or Y) anions are situated; the divalent cation is surrounded by four O^{2-} and two X^- ions constituting an irregular octahedron.

2.10.1 The Ion $[B_7O_{12}X]^{4-}$

2.10.1.1 The $Li_4[B_7O_{12}Cl]$-I Type

To the γ-$Li_4[B_7O_{12}Cl]$-I type, which is very similar to the cubic ($F\bar{4}3c = T_d^5$) high-boracite type, belong the high-temperature phase γ-$Li_4[B_7O_{12}Cl]$-I [535,536] and $Li_4[B_7O_{12}I]$ [537].

2.10.1.2 The $Li_4[B_7O_{12}Cl]$-II Type

This cubic ($P\bar{4}3n = T_d^4$) type is represented only by β-$Li_4[B_7O_{12}Cl]$-II [535].

2.10.1.3 The $Li_4[B_7O_{12}Cl]$-III Type

To this trigonal ($R3 = C_3^4$) type belong α-$Li_4[B_7O_{12}Cl]$-III [535,536]; $Li_4[B_7O_{12}Br]$ [535,537]; and $Li_4[B_7O_{12}Cl_{0.7}Br_{0.3}]$ [535].

2.10.2 The Ion $[B_7O_{12+x/2}X]^{(4+x)-}$

2.10.2.1 The $Li_5[B_7O_{12.5}Cl]$ Type

To the cubic ($F23 = T^2$) $Li_5[B_7O_{12.5}Cl]$ type belong the compounds $Li_5[B_7O_{12.5}Cl]$ [538]; $Li_5[B_7O_{12.5}Br]$ [538]; and $Li_{4+x}[B_7O_{12+x/2}Cl]$ [539,540].

2.10.3 The Ions $[B_7O_{12+x}Y_{1-y}]^{6-}$

2.10.3.1 The Ion $[B_7O_{12.5}\square_{0.5}Y_{1.0}]^{6-}$

This high-boracite type ($F\bar{4}3 = T_d^5$) is represented by $Mg_3[B_7O_{12.5}S]$ [474] and $Mn_3^{2+}[B_7O_{12.5}S]$ [474].

2.10.3.2 The Ion $[B_7O_{12.65}\square_{0.35}Y_{0.85}\square_{0.15}]^{6-}$

The following compounds belong to this high-boracite ($F\bar{4}3 = T_d^5$) type: $Mg_3[B_7O_{12.65}S_{0.85}]$ [473,475]; $Mn_3[B_7O_{12.65}S_{0.85}]$ [473,475]; $Fe_3[B_7O_{12.65}S_{0.85}]$ [473,475]; $Cd_3[B_7O_{12.65}S_{0.85}]$ [473,475]; $Mn_3[B_7O_{12.65}Se_{0.85}]$ [475]; $Fe_3[B_7O_{12.65}Se_{0.85}]$ [475]; $Cd_3[B_7O_{12.65}Se_{0.85}]$ [475,541]; $Mn_3[B_7O_{12.65}Te_{0.85}]$ [475].

2.10.4 The Ions $[B_7O_{13}X]^{6-}$

2.10.4.1 The High-Boracite = $Mg_3[B_7O_{13}Cl]$-I Type

The cubic ($F\bar{4}3 = T_d^5$) high-boracite type consists of a B—O network containing the B atoms and 12 of the 13 O atoms, while the M^{2+} and X^- ions are located in the holes; the last O atom is tetrahedrally coordinated by four B atoms. To this type belong: the high temperature phase α-$Mg_3[B_7O_{13}Cl]$-I = mineral α-boracite [474,542,543]; $Cr_3[B_7O_{13}F]$ [544]; $Cr_3[B_7O_{13}Cl]$-I [543,545]; $Cr_3[B_7O_{13}Br]$-I [543]; $Cr_3[B_7O_{13}I]$ [543]; $Mn_3[B_7O_{13}Cl]$-I [543]; $Mn_3[B_7O_{13}Br]$-I [543]; $Mn_3[B_7O_{12}I]$-I [543]; $(Mg_{1-x}Mn_x)_3[B_7O_{13}Br]$ [546]; $Fe_3[B_7O_{13}Cl]$-I [543,547]; $Fe_3[B_7O_{13}Br]$-I [543,547]; $Fe_3[B_7O_{13}I]$-I [543,547]; $(Mg_{1-x}Fe_x)_3[B_7O_{13}Br]$ [546]; $Co_3[B_7O_{13}Cl]$-I [543]; $Co_3[B_7O_{13}Br]$-I [543]; $Co_3[B_7O_{13}I]$-I [543,548];

$Ni_3[B_7O_{13}Cl]$-I [543]; $Ni_3[B_7O_{13}Br]$-I [543] $Ni_3[B_7O_{13}I]$-I [549, 550]; $Ni_3[B_7O_{13}Cl_{0.5}Br_{0.5}]$-I [543]; $Ni_3[B_7O_{13}Br_{0.9}I_{0.1}]$-I [543]; $Ni_3[B_7O_{13}I_{1-x}(OH)_x]$ [551]; $(Co,Ni)_3[B_7O_{13}(NO_3)]$-I [544]; $Cu_3[B_7O_{13}Cl]$-I [543]; $Cu_3[B_7O_{13}Br]$-I [543, 548]; $Cu_3[B_7O_{13}(NO_3)]$-I [544]; $Zn_3[B_7O_{13}Cl]$-I [543, 547]; $Zn_3[B_7O_{13}Br]$-I [543]; $Zn_3[B_7O_{13}I]$-I [543]; $Cd_3[B_7O_{13}Cl]$-I [543]; $Cd_3[B_7O_{13}Br]$-I [543]; $Cd_3[B_7O_{13}I]$-I [543].

2.10.4.2 The Low-Boracite = $Mg_3[B_7O_{13}Cl]$-II Type

The orthorohombic ($Pca2_1 = C^5_{2v}$) low-boracite type is similar to $Mg_3[B_7O_{13}Cl]$-I, but the Mg^{2+} ions are shifted 24 pm along the z-axis and the Cl^- ions 21 pm along the cubic (111) direction. To this type belong:
the low-temperature phase β-$Mg_3[B_7O_{13}Cl]$-II [543, 552)
$Cr_3[B_7O_{13}Cl]$-II [543];
$Cr_3[B_7O_{13}Br]$-II [543]; $Mn_3[B_7O_{13}Cl]$-II = mineral *chambersite* [474, 545, 553]; $Mn_3[B_7O_{13}Br]$-II [474]; $Mn_3[B_7O_{13}I]$-II [474, 543]; $Fe_3[B_7O_{13}Cl]$-II [543, 547]; $Fe_3[B_7O_{13}Br]$-II [543, 547]; $Fe_3[B_7O_{13}I]$-II [543, 554]; $(Mn_{1-x}Fe_x)_3[B_7O_{13}Br]$-II [546]; $Co_3[B_7O_{13}Cl]$-II [543]; $Co_3[B_7O_{13}Br]$-II [543]; $Co_3[B_7O_{13}I]$-II [543]; $Co_3[B_7O_{13}(NO_3)]$-II [544]; $Ni_3[B_7O_{13}Cl]$-II [543]; $Ni_3[B_7O_{13}Br]$-II [543, 555]; $Ni_3[B_7O_{13}I]$-II [543]; $Ni_3[B_7O_{13}Cl_{0.5}Br_{0.5}]$-II [543, 555]; $Ni_3[B_7O_{13}Br_{0.9}I_{0.1}]$-II [543]; $Ni_3[B_7O_{13}(NO_3)]$-II [544]; $(Co,Ni)_3[B_7O_{13}(NO_3)]$-II [544]; $Cu_3[B_7O_{13}Cl]$-II [543]; $Cu_3[B_7O_{13}Br]$-II [543]; $Cu_3[B_7O_{13}(NO_3)]$-II [544]; $Zn_3[B_7O_{13}Cl]$-II [543, 547]; $Zn_3[B_7O_{13}Br]$-II [543]; $Zn_3[B_7O_{13}I]$-II [543]; $Zn_3[B_7O_{13}(NO_3)]$-II [544]; $Cd_3[B_7O_{13}Cl]$-II [543]; $Cd_3[B_7O_{13}Br]$-II [543]; $Cd_3[B_7O_{13}I]$-II [543]; $Cd_3[B_7O_{13}(NO_3)]$-II [544].

2.10.4.3 The Ericaite = $(Mg,Fe)_3[B_7O_{13}Cl]$-III Type

To the trigonal ($R3c = C^6_{3v}$) ericaite type belong $Mg_3[B_7O_{13}F]$ [544]; $Mn_3[B_7O_{13}F]$ [544]; $Mn_3[B_7O_{13}F_{0.44}(OH)_{0.56}]$ [544]; $Mn_3[B_7O_{13}(OH)]$ [474]; $Fe_3[B_7O_{13}F]$ [544]; $Fe_{2.4}Mg_{0.6}$-$[B_7O_{13}Cl]$ = mineral *ericaite* [553]; $Fe_{2.676}Mg_{0.243}Mn_{0.081}[B_7O_{13}Cl]$ = mineral *congolite* [556]; $Fe_3[B_7O_{13}Cl]$-III [547, 557]; $Fe_3[B_7O_{13}Br]$-III [557]; $Fe_3[B_7O_{13}I]$-V [547, 554]; $Co_3[B_7O_{13}F]$ [544]; $Zn_3[B_7O_{13}F]$ [544]; and $Zn_3[B_7O_{13}Cl]$-III [547, 557].

2.10.4.4 $Fe_3[B_7O_{13}I]$-III and $Fe_3[B_7O_{13}I]$-IV

$Fe_3[B_7O_{13}I]$-III as well as $Fe_3[B_7O_{13}I]$-IV crystallize in the monoclinic $Pc = C^2_s$ space group [554].

2.11 Higher Borates

2.11.1 The Ion $[B_8O_{13}]^{2-}$

2.11.1.1 The $Na_2[B_8O_{13}]$ Type

This monoclinic ($P2_1/a = C^5_{2h}$) type with the notation $8{:}\infty_2(5{:}4\triangle + T, 3{:}2\triangle + T)$, layers (see Fig. 3m), is represented by α-4 $B_2O_3 \cdot Na_2O$ = α-$Na_2[B_8O_{13}]$ [558].

2.11.1.2 The $Ag_2[B_8O_{13}]$ Type

To this monoclinic ($P2_1 = C_{2h}^5$) type with the notation $8{:}\infty_3(5{:}4\triangle + T, 3{:}2\triangle + T)$, network, belong β-$Ag_2[B_8O_{13}]$ and the mixed $Ag_{0.6}Na_{0.4}[B_8O_{13}]$ [559].

2.11.1.3 The $Ba[B_8O_{13}]$-II Type

This orthorhombic ($P222_1 = D_2^2$) type with pseudotetragonal symmetry and the notation $8{:}\infty_2(5{:}4\triangle + T, 3{:}2\triangle + T)$, chains that build up three-dimensional networks, is represented by $4\ B_2O_3 \cdot BaO$-II = $Ba[B_8O_{13}]$-II [560].

2.11.1.4 Other Types

Not yet fully characterized are pseudocubic $Cs_2[B_8O_{13}]$ [561] and tetragonal $Ba[B_8O_{13}]$-I [562].

2.11.2 The Ion $[B_8O_{14}]^{4-}$

For compounds with the anion $[B_8O_{14}]^{4-}$, see $[B_4O_7]^{2-}$, above.

2.11.3 The Ion $[B_8O_{15}]^{6-}$

The ion $[B_8O_{15}]^{6-}$ occurs in monoclinic ($P2_1/c = C_{2h}^5$ or $Pc = C_s^2$) $8\ B_2O_3 \cdot 3\ K_2O \cdot Al_2O_3 = K_3[AlB_8O_{15}]$; the aluminoborate polyanion is built up from the double-ring pentaborate group, the single-ring triborate group and a single AlO_4 tetrahedron, connected to an infinite two-dimensional double layer; the notation is therefore $8{:}\infty_2(5{:}4\triangle + T, 3{:}2\triangle + T, Al{:}1T)$, layers, modified; it is similar to β-$Na_3[B_9O_{15}]$ crystals, except for replacement of B by Al at the tetrahedral sites [563].

2.11.4 The Ion $[B_8O_{17}]^{8-}$

The ion $[B_8O_{17}]^{8-}$ may exist in $Na_4Ga_4[B_8O_{17}]$, for which only the powder diagram has been given [564].

2.11.5 The Ion $[B_9O_{14}]^-$

The pseudotetragonal, orthorhombic refined ($P4_122 = D_4^3$ or $P222_1 = D_2^2$) α-$Cs[B_9O_{14}]$ contains planar boroxine groups and one $[B_3O_5]^-$ group; the notation is therefore $9{:}\infty_3(2\times 3{:}3\triangle, 3{:}2\triangle + T)$, network (Fig. 3n); the monoclinic β-$Cs[B_9O_{14}]$ has not yet been fully characterized [561, 565].

2.11.6 The Ion $[B_9O_{15}]^{3-}$

For compounds with the anion $[B_9O_{15}]^{3-}$, see $[B_3O_5]^-$ above.

2.11.7 The Ion $[B_{10}O_{21}]^{12-}$

The triclinic $5\ B_2O_3 \cdot 6\ PbO = Pb_6[B_{10}O_{21}]$ contains the ion $[B_{10}O_{21}]^{12-}$. It is built up from two $[B_4O_7]^{2-}$ groups which are connected by two planar BO_3^{3-} triangles; the notation is $10{:}(2\times 4{:}2\triangle + 2T, 2\triangle)$, isolated [566].

2.11.8 The Ion $[B_{10}O_{30}]^{30-}$

The triclinic compound $B_2O_3 \cdot 3\,CuO = Cu_3[B_2O_6]$ is better formulated as $Cu_{15}[B_{10}O_{30}] = Cu_{15}[O_2(B_2O_5)_2(BO_3)_6]$; it forms a layer structure consisting of almost planar $[B_2O_5]^{4-}$ groups, planar isolated BO_3^{3-} groups, and isolated O^{2-} ions; the notation is $10{:}\,\infty_2(6 \times \triangle, 2 \times 2\triangle, 2\,O)$, layers, modified; it should have also been cited in the Chapts. 2.2.1 and/or 2.5 [567].

2.11.9 The Ion $[B_{12}O_{29}]^{22-}$

The ion $[B_{12}O_{29}]^{22-}$ may exist in triclinic $Ba_5[Al_4B_{12}O_{29}]$, studied only by powder diffraction [381].

2.11.10 The Ion $[B_{19}O_{31}]^{5-}$

The monoclinic ($C2/c = C_{2h}^6$) compound $3.8\,B_2O_3 \cdot K_2O = 19\,B_2O_3 \cdot 5\,K_2O = K_5[B_{19}O_{31}]$ contains a three-dimensional network of two connected pentaborate, two triborate groups, two BO_3 triangles and one BO_4 tetrahedron; the notation is, therefore, $19{:}\,\infty_3(2 \times 5{:}4\triangle + T,\ 2 \times 3{:}2\triangle + T,\ 2\triangle,\ 1T)$, network; see Figure 3p [568].

3 References

1. Heller, G.: Top. Curr. Chem. *15* (2), 206/80 (1970)
2. Christ, C. L.: Am. Mineralogist *45*, 334/40 (1960)
3. Tennyson, Ch.: Fortschr. Mineral. *41*, 64/91 (1963)
4. Edwards, J. O., Ross, V. F.: J. Inorg. Nucl. Chem. *15*, 329/37 (1960)
5. Clark, J. R., Christ, C. L.: Am. Mineralogist *56*, 1934/54 (1971)
6. Christ, C. L., Clark, J. R.: Phys. Chem. Min. *2*, 59/87 (1977)
7. Zachariasen, W. H.: Acta Crystallogr. *7*, 305/10 (1954)
8. Peters, C. R., Milberg, M. E.: ibid. *17*, 229/34 (1964)
9. Zachariasen, W. H.: ibid. *16*, 385/9 (1963)
10. Zachariasen, W. H.: ibid. *16*, 380/4 (1963)
11. Höhne, E.: Z. Anorg. Allg. Chem. *342*, 188/94 (1966); Z. Chemie [Leipzig] *4*, 431/2 (1964)
12. Zviedre, I. I., Ozols, Ya. K., Ievins, A. F.: Latvijas PSR Zinatnu Akad. Vestis, Kim. Ser. *1974*, 3/4
13. Zachariasen, W. H.: Acta Crystallogr. *17*, 749/51 (1964); Am. Cryst. Assoc. Meeting, Abstr., Austin, 1966, 59
14. Nakamura, S., Hayashi, H.: J. Ceram. Soc. Japan, Yogyo Kyo Kaishi *83*, 38/45 (1975)
15. Block, S., Perloff, A.: Acta Crystallogr. *16*, 1233/6 (1963)
16. Zviedre, I. I., Ievins, A. F.: Latvijas PSR Zinatnu Akad. Vestis, Kim. Ser. *1974*, 401/5; Zviedre, I. I., Ozols, Ya. K., Ievins, A. F.: ibid. *1973*, 387/9 C.A. *79*, No. 130124 (1973)
17. Zeigan, D., Kutschabsky, L.: Monatsber. Deutsch. Akad. Wiss. Berlin *7*, 867/9 (1965) (C. *1967*-15-285)
18. Zeigan, D.: Acta Crystallogr. *21*, A 74 (1966)
19. Zeigan, D.: Z. Chemie [Leipzig] *7*, 241/2 (1967)
20. Simonov, M. A., Kazanskaya, E. V., Egorov-Tismenko, Yu. K., Zhelezin, E. P., Belov, N. V.: Soviet Phys.-Dokl. *21*, 471 (1976/77); Dokl. Akad. Nauk SSSR *230*, 91/4 (1976)
21. Ozols, Ya. K., Ievins, A. F.: Latvijas PSR Zinatnu Akad. Vestis, Kim. Ser. *1964*, 137/43
22. Simonov, M. A., Yamnova, N. A., Kazanskaya, E. V., Egorov-Tismenko, Yu. K., Belov, N. V.: Soviet Phys.-Dokl. *21*, 314/7 (1976/77); Dokl. Akad. Nauk SSSR *228*, 1337/40 (1976) C.A. *85*, No. 102787 (1976)
23. Sedlacek, P., Dornberger-Schiff, K.: Acta Crystallogr. *B 27*, 1532/41 (1971)

24. Wang, N.: Neues Jahrb. Mineral., Monatsh. *1971*, 315/25
25. Kutschabsky, L.: Z. Chem. [Leipzig] *5*, 110/1 (1965)
26. Kravchenko, N. B.: Zh. Strukt. Khim. *6*, 872/7 (1965) C.A. *64*, No. 10500 (1966)
27. Kravchenko, N. B.: ibid. *6*, 724/8 (1965)
28. Gode G. K., Ivchenko, N. P., Kurkutova, E. N.: Russ. J. Inorg. Chem. *20*, 1736 (1975/76); Zh. Neorgan. Khim. *20*, 3136/7 (1975)
29. Kutschabsky, L.: Acta Crystallogr. *B 25*, 1811/6 (1969)
30. Ross, V. F., Edwards, J. O.: Am. Mineralogist *44*, 875/7 (1959)
31. Effenberger, H.: Acta Crystallogr. *B 38*, 82/5 (1982)
32. Collin, R. L.: ibid *4*, 204/9 (1951)
33. Iorysh, Z. I., Rumanova, I. M., Belov, N. V.: Dokl. Akad. Nauk SSSR *228*, 1076/9 (1976) C.A. *85*, No. 102724 (1976)
33a Giese, R. F., Penna, G.: Am. Mineralogist *68*, 255/61 (1983)
34. Moore, P. B., Ghose, S.: ibid. *56*, 1527/38 (1971)
35. Prewitt, Ch. T., Buerger, M. J.: ibid. *46*, 1077/85 (1961)
36. Embrey, P. G.: Mineral. Mag. *32*, 666/8 (1960)
37. Wuhan College of Geology (K'o Hsueh T'ung Pao *22*, 361/3 and 364/6 (1977) C.A. *87* No. 144358 (1977) and C.A. *87*, No. 125760 (1977)
38. Heller, G., Horbat, F.: Z. Naturforsch. *32b*, 989/91 (1977)
39. Alcock, N. W., Hagger, R. M., Harrison, W. D., Wallbridge, M. G. H.: Acta Crystallogr. *B 38*, 676/7 (1982)
40. Clark, M. J. R., Lynton, H.: Can. J. Chem. *48*, 405/9 (1970)
41. Menchetti, S., Sabelli, C.: Acta Crystallogr. *B 38*, 1282/5 (1982)
42. Zachariasen, W. H., Plettinger, H. A., Marezio, M.: ibid. *16*, 1144/6 (1963)
43. Nefedov, Ye. I.: Dokl. Akad. Nauk SSSR *174*, 189/92 (1967)
44. Schlatti, M.: Naturwissenschaften *54*, 578 (1967); Tschermaks Mineral. Petrogr. Mitt. [3] *12*, 463/9 (1968)
45. Dal Negro, A., Tadini, C.: ibid. *21*, 94/100 (1974) C.A. *82*, No. 24721 (1975)
46. Moore, P. B., Araki, T.: Nature Phys. Sci. *239*, 25/6 (1972); Am. Mineralogist *59*, 985/1004 (1974)
47. Murdoch, J.: Am. Mineralogist *47*, 718/22 (1962)
48. Pertsev, N. N., Malinko, S. V., Vakhrushev, V. A., Fitsev, B. P., Sokolova, E. V., Nikitina, I. B.: Zap. Vses. Mineral. Obshchestva [Leningrad] *109*, 569/73 (1980) C.A. *94*, No. 68732 (1981)
49. Franz, G., Ackermand, D., Koch, E.: Am. Mineralogist *66*, 872/7 (1981)
50. Golovastikov, N. I., Belova, J. N., Belov, N. V.: Zap. Vses. Mineral. Obshchestva [Leningrad] , 405/15 (1955) C.A. *53*, No. 993i (1959)
51. Capponi, J.-J., Chenavas, J., Joubert, J. C.: Bull. Soc. France Mineral. Crist. *95*, 413/8 (1972) C.A. *77*, No. 169817 (1972)
52. Kazanskaya, E. V., Egorov-Tismenko, Yu. K., Simonov, M. A., Belov, N. V.: Dokl. Akad. Nauk SSSR *240*, 1100/3 (1978) C.A. *89*, No. 121257 (1978)
53. Paton, F., Mc Donald, S. G. G.: Acta Crystallogr. *10*, 653/6 (1957)
54. Krogh-Moe, J.: ibid. *23*, 500/1 (1967)
55. Kazanskaya, E. V., Chemodina, T. N., Egorov-Tismenko, Yu. K., Simonov, M. A., Belov, N. V.: Soviet Phys.-Cryst. *22*, 35/6 (1977); Kristallografiya *22*, 66/8 (1977)
56. Fujiwara, T., Takada, M., Masutomi, K., Isobe, T., Okada, H., Nakai, I., Nagashima, K.: Chigaku Kenkyo *33*, 11/20 (1982) C.A. *98*, No. 19605 (1983)
57. Simonov, M. A., Egorov-Tismenko, Yu. K., Belov, N. V.: Soviet Phys.-Cryst. *21*, 332/3 (1976); Kristallografiya *21*, 592/4 (1976)
58. Kudoh, Y., Takéuchi, Y.: Cryst. Struct. Commun. *2*, 595/8 (1973)
59. Takéuchi, Y., Kudoh, Y.: Am. Mineralogist *60*, 273/9 (1975)
60. Grigor'ev, A. P., Brovkin, A. A., Nekrasov, I. Ya.: Dokl. Akad. Nauk SSSR *166*, 937/40 (1966)
61. Epprecht, W. T., Schaller, W. T., Vlisidis, A. C.: Schweiz. Mineral. Petrogr. Mitt. *39*, 85/104 (1959)
62. Dal Negro, A., Martin-Pozas, J. M., Ungaretti, L.: Am. Mineralogist *60*, 879/83 (1975)
63. Corazza, E.: Acta Crystallogr. *B 32*, 1329/33 (1976)
64. Corazza, E.: ibid. *B 30*, 2194/9 (1974)
65. Christ, C. L., Clark, J. R.: Z. Kristallogr. *114*, 341/2 (1960)

66. Clark, J. R., Appleman, D. E., Christ, C. L.: J. Inorg. Nucl. Chem. *26*, 73/95 (1964)
67. Clark, J. R., Christ, C. L.: Acta Crystallogr. *10*, 776/7 (1957); Z. Kristallogr. *112*, 213/33 (1959)
68. Rumanova, I. M., Genkina, E. A.: Latvijas PSR Zinatnu Akad. Vestis, Kim. Ser. *1981*, 643/53
69. Kurkutova, E. N., Rumanova, I. M., Belov, N. V.: Soviet Phys.-Dokl. *10*, 808/10 (1965/66); Dokl. Akad. Nauk SSSR *164*, 90/3 (1965)
70. Kurkutova, E. N.: Acta Crystallogr. *21*, A 61 (1966)
71. Ozols, Ya. K., Tetere, I. V., Ievins, A. F.: Izv. Akad. Nauk Latv. SSR, Ser. Kim. *1973*, 3/7 C.A. *78*, No. 165580 (1973)
72. Egorov, V. A., Batsanova, L. R.: Russ. J. Inorg. Chem. *12*, 144 (1967); Zh. Neorgan. Khim. *12*, 279/80 (1967)
73. Simonov, M. A., Egorov-Tismenko, Yu. K., Kazanskaya, E. V., Belokoneva, E. L., Belov, N. V.: Soviet Phys.-Dokl. *23*, 159/61 (1978/79); Dokl. Akad. Nauk SSSR *239*, 326/9 (1978)
74. Touboul, M., Bois, C., Mangin, D., Amoussou, D.: Acta Crystallogr. *C 39*, 685/9 (1983)
75. Christ, C. L., Clark, J. R., Evans, H. T.: ibid. *11*, 761/70 (1958)
76. Bondars, A.: Latvijas PSR Zinatnu Akad. Vestis, Kim. Ser. *1981*, 580/92
76A. Gode, H., Veveris, A.: ibid. *1984*, 11/4, C.A. *100*, No. 166930 (1984)
77. Hainsworth, F. N., Petch, H. E.: Can. J. Phys. *44*, 3083/107 (1966)
78. Rumanova, I. M., Ashirov, A.: Soviet Phys.-Cryst. *8*, 665/80 (1963/64); Kristallografiya *8*, 824/45 (1963)
79. Sabelli, C., Stoppioni, A.: Can. Mineralogist *16*, 75/80 (1980)
80. Clark, J. R., Christ, C. L., Appleman, D. E.: Acta Crystallogr. *15*, 207/13 (1962)
81. Konnert, J. A., Clark, J. R., Christ, C. L.: Z. Kristallogr. *132*, 241/54 (1970)
82. Erd, R. C., Eberlein, G. D., Christ, C. L.: Can Mineralogist *10*, 108/12 (1969)
83. Zviedre, I. I., Ozols, Ya. K., Ievins, A. F.: Latvijas PSR Zinatnu Akad. Vestis, Kim. Ser. *1974*, 387/94, C.A. *81*, No. 178301 (1974)
84. Ozols, Ya. K.: ibid. *1977*, 356/7
85. Clark, J. R., Christ, C. L.: Acta Crystallogr. *B 33*, 3272/3 (1977)
86. Zviedre, I. I., Ievins, A. F.: Latvijas PSR Zinatnu Akad. Vestis, Kim. Ser. *1974*, 395/400 C.A. *81*, No. 178302 (1974)
87. Yamnova, N. A., Egorov-Tismenko, Yu. K., Simonov, M. A., Belov, N. V.: Soviet Phys.-Dokl. *19*, 326/7 (1974); Dokl. Akad. Nauk SSSR *216*, 1281/4 (1974); Yamnova, N. A., Simonov, M. A., Belov, N. V.: Soviet Phys.-Cryst. *22*, 356/7 (1977/78); Kristallografiya *22*, 624/6 (1977)
88. Corazza, E., Menchetti, S., Sabelli, C.: Acta Crystallogr. *B 31*, 1993/7 (1975)
89. Corazza, E., Menchetti, S., Sabelli, C., Stoppioni, A.: Neues Jahrb. Mineral., Abhandl. *131*, 208/33 (1977)
90. Shashkin, D. P., Simonov, M. A., Belov, N. V.: Soviet Phys.-Cryst. *16*, 186/9 (1971); Kristallografiya *16*, 231/5 (1971)
91. Simonov, M. A., Egorov-Tismenko, Yu. K., Belov, N. V.: Soviet Phys.-Dokl. *22*, 277/9 (1977/78); Dokl. Akad. Nauk SSSR *234*, 822/5 (1977) C.A. *87*, No. 61097 (1977)
92. Abdulla'ev, G. K., Mamedov, Kh. S.: Russ. J. Strukt. Chem. *7*, 831 (1966); Zh. Struct. Khim. *7*, 896 (1966); Inorgan. Materialy USSR *5*, 28/32 (1969); Izv. Akad. Nauk SSSR, Neorgan. Materialy *5*, 56/63 (1969)
93. Brown, G. E., Clark, J. R.: Am. Mineralogist *63*, 814/23 (1978)
94. Giacovazzo, C., Menchetti, S., Scordari, F.: ibid. *58*, 523/30 (1973)
95. Giese, R. F.: Can. Mineralogist *9*, 573 (1967/68)
96. Levy, H. A., Lisensky, G. C.: Acta Crystallogr. *B 34*, 3502/10 (1978)
97. Marezio, M., Plettinger, H. A., Zachariasen, W. H.: ibid. *16*, 975/80 (1963)
98. Janda, R., Heller, G., Pickardt, J.: Z. Kristallogr. *154*, 1/9 (1981)
99. Silins, E. Ya., Bel'skii, V. K., Ozolins, G. V.: Latvijas PSR Zinatnu Akad. Vestis, Kim. Ser. *1982*, 375/6 C.A. *97*, No. 102045 (1982)
100. Batsanov, A. S., Nava, E. U., Struchkov, Yu. T., Akimov, V. M.: Cryst. Struct. Commun. *11*, 1629/30 (1982); Nava, E. U., Batsanov, A. S., Struchkov, Yu. T., Akimov, V. M., Molodkin, A. K., Skvortsov, V. G., Rodionov, N. S.: Russ. J. Inorg. Chem. *28*, 962/5 (1983/84); Zh. Neorgan. Khim. *28*, 1706/10 (1983)
101 Wan, C., Ghose, S.: Am. Mineralogist *62*, 1135/43 (1977)
102. Erd, R. C., McAllister, J. F., Eberlein, G. D.: ibid. *64*, 369/75 (1979)
103. Sauka, Ya. Ya.: Zh. Strukt. Khim. *1*, 453/7 (1960) C.A. *56*, No. 5476 (1962)

104. Sokolova, E. V., Egorov-Tismenko, Yu. K., Simonov, M. A., Belov, N. V.: Mineral. Zh. [Kiev] *2* (6), 58/61 (1981) C.A. *94*, No. 112836 (1981)
105. Sokolova, E. V., Yamnova, N. A., Simonov, M. A., Belov, N. V.: Soviet Phys.-Cryst. *24*, 669/73 (1979/80); Kristallografiya *24*, 1169/76 (1979) C.A. *92*, No. 68043 (1980)
106. Solans, X., Font-Altaba, M., Solans, J., Domenech, M. V.: Acta Crystallogr. *B 38*, 2438/41 (1982)
107. Gode, G. K., Zuika, I. V., Adijane, G.: Latvijas PSR Zinatnu Akad. Vestis, Kim. Ser. *1971*, 538/40 C.A. *76*, No. 39484 (1972)
108. Berzina, I. R.: Tezisy Dokl., VI. Konf. Molodykh. Nauk Rabotn. Inst. Neorgan. Khim., Moskva, 1977, p. 18 C.A. *90*, No. 46855 (1979)
109. Ivchenko, N. P., Kurkutova, E. N.: Soviet Phys.-Cryst. *20*, 326/8 (1975); Kristallografiya *20*, 533/7 (1975), C.A. *83*, No. 106635 (1975)
110. Gode, G. K., Majore, I. V.: Latvijas PSR Zinatnu Akad. Vestis, Kim. Ser. *1976*, 344/6 C.A. *85*, No. 86527 (1976)
111. Berzina, I. R., Ozols, Ya. K., Ievins, A. F.: Soviet Phys.-Cryst. *20*, 255/6 (1975); Kristallografiya *20*, 419/22 (1975)
112. Moore, P. B., Araki, T.: Am. Mineralogist *59*, 60/5 (1974)
113. Yamnova, N. A., Simonov, M. A., Belov, N. V.: Soviet Phys.-Dokl. *20*, 244/6 (1975); Dokl. Akad. Nauk SSSR *221*, 1326/9 (1975) C.A. *83*, No. 69496 (1975)
114. Giese, R. F.: Science *154*, 1453/4 (1966)
115. Chialdi, G., Corazza, E., Sabelli, C.: Atti Accad. Nazl. Lincei Rend., Classe Sci., Fis. Mat. Nat. [8] *42*, 236/51 (1967)
116. Cooper, W. F., Larsen, F. K., Coppens, P., Giese, R. F.: Am. Mineralogist *58*, 21/31 (1973)
117. Corazza, E., Menchetti, S., Sabelli, C., Stoppioni, A.: Neues Jahrb. Mineral., Abhandl. *131*, 208/33 (1977)
118. Menchetti, S., Sabelli, C.: Acta Crystallogr. *B 34*, 1080/4 (1978)
119. Woller, K. H., Heller, G.: Z. Kristallogr. *156*, 151/7 (1981)
120. Touboul, M., Bois, C., Amoussou, D.: J. Solid State Chem. *48*, 412/9 (1983)
120A. Touboul, M., Ingrain, D., Amoussou, D.: J. Less-Common Met. *96*, 213/21 (1984) C.A. *100*, No. 149969 (1984)
121. Yamnova, N. A., Simonov, M. A., Kazanskaya, E. V., Belov, N. V.: Soviet Phys.-Dokl. *20*, 799/801 (1975/76); Dokl. Akad. Nauk SSSR *225*, 823/6 (1975)
122. Menchetti, S., Sabelli, C.: Acta Crystallogr. *B 34*, 45/9 (1978)
123. Merlino, S., Sartori, F.: ibid. *B 28*, 3559/67 (1972)
124. Zachariasen, W. H., Plettinger, H. A.: ibid. *16*, 376/9 (1963)
125. Merlino, S., Sartori, F.: Contrib. Mineral. Petrol. [Berlin] *27*, 159/65 (1970)
126. Ashmore, J. P., Petch, H. G.: Can. J. Phys. *48*, 1091/7 (1970)
126A. Behm, H.: Acta Crystallogr. *C 40*, 217/20 (1984) C.A. *100* No. 148852 (1984)
127. Cook, W. R., Jaffe, H.: Acta Crystallogr. *10*, 705/7 (1957)
128. Solans, X., Font-Altaba, M., Aguilo, M., Solans, J., Domenech, M. V.: J. Appl. Crystallogr. 637/40 (1983)
129. Domenech, M. V., Solans, J.: Trab. Geol. *11*, 55/60 (1981); Domenech, M. V., Solans, J., Solans, X.: Acta Crystallogr. *B 37*, 643/5 (1981)
130. Merlino, S.: Atti Accad. Nazl. Lincei Rend., Classe Sci., Fis. Mat. Nat. [8] *47*, 85/99 (1969) C.A. *72*, No. 71772 (1970)
131. Vineyard, B. D., Godt, H. C.: Inorg. Chem. *3*, 1144/7 (1964)
132. Frohnecke, J., Hartl, H., Heller, G.: Z. Naturforsch. *32b*, 268/74 (1977)
133. Amoussou, D., Wandji, R., Touboul, M.: Compt. Rend. *C 290*, 391/2 (1980) C.A. *93* No. 123947 (1980)
134. Woller, K. H., Heller, G.: Z. Kristallogr. *156*, 159/66 (1981)
135. Merlino, S., Sartori, F.: Acta Crystallogr. *B 25*, 2264/70 (1969)
136. Hurlbut, C. S., Aristarain, L. F.: Am. Mineralogist *52*, 1048/59 (1967)
137. Canillo, E., Dal Negro, A., Ungaretti, L.: ibid. *58*, 110/5 (1973)
138. Corazza, E., Menchetti, S., Sabelli, C.: Acta Crystallogr. *B 31*, 2405/10 (1975)
139. Marezio, M.: ibid. *B 25*, 1787/95 (1969)
140. Konnert, J. A., Clark, J. R., Christ, C. L.: Am. Mineralogist *57*, 381/96 (1972)
141. Clark, J. R., Christ, C. L.: ibid. *56*, 1934/54 (1971)

142. Rumanova, I.M., Gandymov, O., Belov, N. V.: Soviet Phys.-Cryst. *16*, 236/40 (1971/72); Kristallografiya *16*, 286/91 (1971)
143. Rumanova, I. M., Gandymov, O.: Soviet Phys.-Cryst. *16*, 75/81 (1971); Kristallografiya *16*, 99/106 (1971)
144. Ghose, S., Wan, C., Clark, J. R.: Am. Mineralogist *63*, 160/71 (1978)
145. Kurbanov, Kh. M., Rumanova, I. M., Belov, N. V.: Dokl. Akad. Nauk SSSR *152*, 1100/3 (1963) C.A. *60* No. 4898 (1964)
146. Rumanova, I. M., Kurbanov, Kh. M., Belov, N. V.: Soviet Phys.-Cryst. *10*, 513/22 (1965/66); Kristallografiya *10*, 601/13 (1965) C.A. *64* No. 1439 (1966)
147. Menchetti, S., Sabelli, C., Trosti-Ferrari, R.: Acta Crystallogr. *B 38*, 3072/5 (1982)
148. Bondareva, O. S., Egorov-Tismenko, Yu. K., Simonov, M. A., Belov, N. V.: Soviet Phys.-Dokl. *23*, 806/8 (1978/79); Dokl. Akad. Nauk SSSR *243*, 641/4 (1978) C.A. *90* No. 95886 (1979)
149. Shevyrev, A. A., Murdyan, L. A., Simonov, V. A., Egorov-Tismenko, Yu. K., Simonov, M. A., Belov, N. V.: Dokl. Akad. Nauk SSSR *257*, 111/4 (1981) C.A. *94*, No. 181092 (1981)
150. Sokolova, E. V., Yamnova, N. A., Belov, N. V.: Soviet Phys.-Cryst. *25*, 411/5 (1980/81); Kristallografiya *25*, 716/21 (1980), *93* No. 141285 (1980)
151. Menchetti, S., Sabelli, C.: Acta Crystallogr. *B 33*, 3730/3 (1977)
152. Burzlaff, H.: Neues Jahrb. Mineral., Monatsh. *1967*, 157/9, *67* No. 120755 (1967)
153. Menchetti, S., Sabelli, C., Trosti-Ferrari, R.: Acta Crystallogr. *B 38*, 2987/91 (1982)
154. Ghose, S., Wan, C.: Am. Mineralogist *64*, 187/95 (1979)
155. Rumanova, I. M., Iorysh, Z. I., Belov, N. V.: Dokl. Akad. Nauk SSSR *236*, 91/4 (1977)
156. Wan, C., Ghose, S.: Am. Mineralogist *68*, 604/13 (1983); Ghose, S.: ibid. *67*, 1265/72 (1982); Ghose, S., Wan, C.: Nature *270*, 594/5 (1977)
157. Braitsch, O.: Beitr. Mineral. Petrogr. *6*, 233/47 (1959)
158. Davies, W. O., Machin, M. P.: Am. Mineralogist *53*, 2084/7 (1968)
159. v. Hodenberg, R., Kühn, R.: Kali Steinsalz *7*, 165/70 (1977)
160. Corazza, E., Sabelli, C.: Atti Accad. Nazl. Lincei Rend., Classe Sci., Fis. Mat. Nat. [8] *41*, 527/52 (1966)
161. Corazza, E., Menchetti, S., Sabelli, C.: Am. Mineralogist *59*, 1005/15 (1974)
162. Merlino, S., Sartori, F.: Science [2] *171*, 377/9 (1971)
163. Hanic, F., Lindqvist, O., Nyborg, J., Zedler, A.: Coll. Czech. Chem. Commun. *36*, 3678/701 (1971)
164. Dal Negro, A., Ungaretti, L., Sabelli, C.: Am. Mineralogist *56*, 1553/66 (1971)
165. Dal Negro, A., Ungaretti, L., Della Giusta, A.: Cryst. Struct. Commun. *5*, 427/31 (1976)
166. Genkina, E. A., Rumanova, I. M., Belov, N. V.: Soviet Phys.-Cryst. *21*, 111 (1976/77); Kristallografiya *21*, 209/10 (1976)
167. Dal Negro, A., Ungaretti, L., Basso, R.: Cryst. Struct. Commun. *5*, 433/6 (1976)
168. Walenta, K.: Tschermaks Mineral. Petrogr. Mitt. *26*, 69/77 (1979)
169. Dal Negro, A., Sabelli, C., Ungaretti, L.: Atti Accad. Nazl. Lincei Rend., Classe Sci. Fis. Mat. Nat. [8] *47*, 353/64 (1969) C.A. *72*, No. 126011 (1970)
170. Aristarain, L. F., Hurlbut, C. S.: Am. Mineralogist *52*, 1776/84 (1967)
171. Dal Negro, A., Ungaretti, L., Sabelli, C.: Naturwissenschaften *60*, 350A (1973)
172. Silins, E. Ya., Ozols, Ya. K., Ievins, A. F.: Latvijas PSR Zinatnu Akad. Vestis, Kim. Ser. *1971*, 375/6, C.A. *75* No. 102311 (1971)
173. Silins, E. Ya., Ozols, Ya. K., Ievins, A. F.: ibid. *1974*, 115/6, C.A. *81* No. 7181 (1974)
174. Silins, E. Ya., Ievins, A. F.: ibid. *1975*, 747/8; C.A. *84* No. 114498 (1976) Soviet Phys.-Cryst. *22*, 288/91 (1977/78); Kristallografiya *22*, 505/9 (1977) C.A. *87*, No. 32420 (1977)
175. Silins, E. Ya., Ozolins, G. V.: Latvijas PSR Zinatnu Akad. Vestis, Kim. Ser. *1979*, 643/5 C.A. *92*, No. 119996 (1980)
176. Silins, E. Ya., Dzene, A. A., Ievins, A. F.: ibid. *1975*, 751/2 C.A. *84*, No. 126056 (1976)
177. Silins, E. Ya., Ievins, A. F.: ibid. *1976*, 375/82 C.A. *85*, No. 200800 (1976)
178. Silins, E. Ya., Ievins, A. F.: Boraty Boratnye Sist. *1978*, 72/5 C.A. *91*, No. 66627 (1979)
179. Silins, E. Ya., Ozols, Ya. K., Ievins, A. F.: Soviet Phys.-Cryst. *18*, 317/9 (1973/74); Kristallografiya *18*, 503/7 (1973) C.A. *79*, No. 58769 (1973)
180. Heller, G., Schellhaas, J.: Z. Kristallogr. *164*, 237/46 (1983) C.A. *100*, No. 165807 (1984)
181. Behm, H.: Acta Crystallogr. *C 39*, 1156/9 (1983)

182. Katseva, G. N.: Izv. Vysshikh. Uchebn. Zavedenii Khim. i Khim. Tekhnol *14*, 1622/4 (1971); C.A. *76*, No. 64189 (1972)
183. Hurlbut, C. S., Erd, R. C.: Am. Mineralogist *59*, 647/51 (1974)
184. Ghose, S., Wan, C.: ibid. *62*, 979/89 (1977)
185. Clark, J. R.: ibid. *49*, 1549/68 (1964)
186. Erd, R. C., McAllister, J. F., Vlisidis, A. C.: ibid. *46*, (560/1 (1961)
187. Brovkin, A. A., Zayakina, N. V., Brovkina, V. S.: Soviet Phys.-Cryst. *20* 563/6 (1975/76); Kristallografiya *20*, 911/6 (1975)
188. Zayakina, N. V., Brovkin, A. A.: Soviet Phys.-Cryst. *23*, 659/61 (1978/79); Kristallografiya *23*, 1167/70 (1978) C.A. *90*, No. 113328 (1979)
189. Kondrat'eva, V. V., Ostrovskaya, I. V., Yarzhemskii, Ya. Ya.: Zap. Vses. Mineral. Obshch. *95*, 45/50 (1966) C. A. *64*, No. 17260e (1966)
190. Konnert, J. A., Clark, J. R., Christ, C. L.: Am. Mineralogist *55*, 1911/31 (1970)
191. Razmanova, Z. P., Rumanova, I. M., Belov, N. V.: Soviet Phys.-Dokl. *24*, 234/7 (1979); Dokl. Akad. Nauk SSSR *245*, 1112/5 (1979) C.A. *91*, No. 81742 (1979)
192. Dal Negro, A., Kumbasar, I., Ungaretti, L.: Am. Mineralogist *58*, 1034/43 (1973)
193. Simonov, M. A., Chichagov, A. V., Egorov-Tismenko, Yu. K., Belov, N. V.: Soviet Phys.-Dokl. *20*, 161/2 (1975/76); Dokl. Akad. Nauk SSSR *221*, 87/90 (1975)
194. Yamnova, N. A., Simonov, M. A., Belov, N. V.: Zh. Strukt. Khim. *17*, 489/95 (1976)
195. Rumanova, I. M., Razmanova, Z. P., Belov, N. V.: Soviet Phys.-Dokl. *16*, 518/21 (1971/72); Dokl. Akad. Nauk SSSR *199*, 592/5 (1971)
196. Gode, G. K., Majore, I. V.: Latvijas PSR Zinatnu Akad. Vestis, Kim. Ser. *1981* (3), 369/70
197. Saiko, I. G., Kononova, G. N., Petrov, V. I., Tavrovskaya, A. Ya.: Russ. J. Inorg. Chem. *26*, 1732/5 (1981/82); Zh. Neorgan. Khim. *26*, 3231/5 (1981); Russ. J. Inorg. Chem. *27*, 190/2 (1982); Zh. Neorgan. Khim. *27*, 335/8 (1982)
198. Menchetti, S., Sabelli, C.: Acta Crystallogr. *B 35*, 2488/93 (1979)
199. Behm, H.: ibid. *C 39*, 20/2 (1983)
200. Heller, G., Pickardt, J.: Z. Naturforsch. *40b*, 462/6 (1985)
201. Behm, H.: Acta Crystallogr. *C 41*, 64215 (1985)
202. de C. T. Carrondo, M. A. A. F., Skapski, A. C.: ibid. *B 34*, 3551/4 (1978)
202A. Griffith, W. P., Skapski, A. C., West, A. P.: Chem. Ind. [London] *1984* (5), 185/6, C.A. *100*, No. 148921 (1984)
203. Pawel, A., Heller, G., Pickardt, J.: Z. Kristallogr. *157*, 251/7 (1981)
204. van Geldern, D. W.: Rec. Trav. Chim. *75*, 117/26 (1956)
205. Baptista, A., Baptista, N. R.: Anais Acad. Brasil. Cienc. *40*, 157/65 (1968)
206. van Geldern, D. W.: Rec. Trav. Chim. *77*, 739/45 (1958)
207. Bretschneider, G.: DBP 1011858 (1954); ASTM 8-208
208. Bretschneider, G.: DBP 940348 (1953); ASTM 8-220
209. DeWolff, P. M.: Acta Crystallogr. *10*, 590/5 (1957)
210. Benhassaine, A.: Compt. Rend. *C 274*, 1442/5 and 1516/9 (1972)
211. Nakamura, S.: Japan Kokai Tokkyo Koho 75-28497 (1973/75)
212. Krć, J.: Anal. Chem. *23*, 806 (1951)
213. Touboul, M., Amoussou, D.: J. Less-Common Metals *56*, 39/46 (1977); Rev. Chim. Miner. *15*, 223/32 (1978) C.A. *90*, No. 64842 (1979)
214. Kondrat'eva, V. V.: Soviet Phys.-Cryst. *9*, 616/7 (1964/65); Kristallografiya *9*, 735/6 (1964) C.A. *62*, No. 1144 (1964)
215. Ferraris, G., Francini-Angela, M., Orlandi, P.: Can. Mineralogist *16*, 69/73 (1978)
216. Braitsch, O.: Beitr. Mineral. Petrog. *8*, 60/66 (1961)
217. Povarennykh, A. S.: Idei E. S. Federova Sovrem. Kristallogr. Mineral. *1972*, 118/36
218. Erd, R. C., McAllister, J. F., Vlisidis, A. C.: Am. Mineralogist *55*, 349/57 (1970)
219. Shichagov, A. V., Simonov, M. A., Belov, N. V.: Dokl. Akad. Nauk SSSR *218*, 576/9 (1974)
220. Yakubovich, O. V., Egorov-Tismenko, Yu. K., Simonov, M. A., Belov, N. V.: ibid. *239*, 1103/6 (1978); Soviet Phys.-Dokl. *23*, 225/6 (1978)
221. Shipovalov, Yu. V., Avrova, N. P.: Issled. Obl. Khim. Fiz. Metodov Anal. Mineral. Syr'ya *1971*, 176/9, C.A. *77* No. 144984 (1972)
222. Avrova, N. P., Bocharov, V. M., Khalturina, I. I., Yunosova, Z. R.: Geol. Razved. Mestorozhd. Tverd. Polez. iskop. Kaz. *1968*, 169/73 after Am. Mineralogist *56*, 1122 (1971)

223. Hsieh, H.-T., Chien, T.-Ch., Liu, L.-P.: Scientia Sinica *13*, 813/21 (1964) after Am. Mineralogist *50*, 262/3 (1965)
224. Schäfer, U. L.: Neues Jahrb. Min., Monatsh. *1968*, 75/80
225. Strunz, H.: Mineralogische Tabellen, 5. Aufl., Akadem. Verlagsgesellschaft, Frankfurt/Main 1970
226. Kramer, H., Allen, R. D.: Am. Mineralogist *41*, 689/700 (1956)
227. Meixner, H.: Beitr. Mineral. Petrogr. *3*, 445/55 (1953); Meixner, H., Mönke, H.: Kali Steinsalz *3*, 228 (1961)
228. Bogomolov, M. A., Nikitina, I. B., Pertsev, N. N.: Dokl. Akad. Nauk SSSR *184*, 1398/401 (1969); Dokl. Acad. Sci., Earth Sci. Sect. U.S.S.R. *184*, 127/31 (1969)
229. Kondrat'eva, V. V., Ostrovskaya, I. V., Yarzhemskii, Ya. Ya.: Zap. Vses. Min. Obshch. [Leningrad] *95*, 45/50 (1966) from. Am. Mineralogist *51*, 1550 (1966)
230. Yarzhemskii, Ya. Ya.: Mineral. Sbornik Lvovskoe Geol. Obshchestvo No. *6*, 169/74 (1952) from Am. Mineralogist *40*, 941 (1955)
231. Malinko, S. V., Fitsev, B. P., Kuznetsova, N. N., Cherkasova, L. E.: Zap. Vses. Min. Obshch. [Leningrad] *109*, 469/76 (1980) from Min. Abstracts *32*, 327 (1981)
232. Mokievskii, V. A.: Zap. Vses. Min. Obshch. [Leningrad] *82*, 317 (1953) from Am. Mineralogist *40* 551/2 (1953)
233. Gorbov, A. F.: Tr. Vses. Nauchn. Issled. Proektn. Inst. Galurgii No. *64*, 54/62 (1973) from C.A. *83*, No. 118676 (1975)
234. Barrer, M., Freund, E. F.: J. Chem. Soc. Dalton Trans. *1974*, 2054/60
235. Taxer, K. J., Bürger, M. J.: Z. Kristallogr. *125*, 423/36 (1967)
236. Richter, L.: Fortschr. Mineral. *55*, Beiheft No. 1, 113/5 (1977)
237. Bocharov, V. M., Khalturina, I. I., Avrova, N. P., Shinovalov, Yu. V.: Tr. Mineralog. Muzeya, Akad. Nauk SSSR *19* 121/5 (1969) from Am. Mineralogist *55*, 1069 (1970)
238. Ostrovskaya, I. V.: Tr. Mineralog. Muzeya, Akad. Nauk SSSR *19*, 205 (1969) from C.A. *73*, No. 57855 (1970)
239. Raup, O. B., Gude, A. J., Dwornik, E. J., Luttitta, F., Rose, H. J.: Am. Mineralogist *53*, 1081/95 (1968)
240. Tourné, C. M., Tourné, G. F.: Compt. Rend. C *266*, 1363/5 (1968); Bull. Soc. Chin. France *1969*, 1124/36; Tourné, C. M., Tourné, G. F., Malik, S. A., Weakley, T. J. R.: J. Inorg. Nucl. Chem. *32*, 3875/90 (1970)
240A. Weakley, T. J. R.: Acta Crystallogr. *C 40*, 16/8 (1984)
241. Leonyuk, N. I., Maltseva, L. I.: 6th Eur. Crystallogr. Meet., Abstracts, Barcelona 1980, p. 117
242. Leonyuk, N. I.: Vestn. Mosk. Univ., Ser. 4: Geol. *1983* (3), 28/37; C.A. *99*, No. 74069 (1983); Leonyuk, N. I., Leonyuk, L. I.: Izdatel. Mosk. Gos. Univ., Moscow 1983, 1/215; C.A. *99*, No. 185471 (1983)
243. Osaka, A., Takahashi, K.: J. Ceram. Soc. Japan, Yogyo Kyo Kaishi *91*, 374/7 (1983); C.A. *99*, No. 127131 (1983); Takahashi, K., Osaka, A.: ibid. *91*, 516/20 (1983); C.A. *100*, No. 13409 (1984)
244. Gode, G. K.: Latvijas PSR Zinatnu Akad. Vestis, Kim. Ser. *1981*, 517/24; C.A. *95*, No. 214226 (1981)
245. García-Blanco, S.: Rev. Port. Quim. *23*, 231/4 (1981)
246. Vegas, A., Cano, F. H., García-Blanco, S.: Acta Crystallogr. *B 33*, 3607/9 (1977)
247. Abdulla'ev, G. K., Mamedov, Kh. S.: Soviet Phys.-Cryst. *22*, 154/6 (1977/78); Kristallografiya *22*, 271/4 (1977)
248. Bither, T. A., Young, H. S.: J. Solid State Chem. *6*, 502/8 (1973) US. Pat. 37 55536 (1971/73) C.A. *79*, No. 127891 (1973)
249. Murthy, G. S., Venudhar, Y. C., Murthy, K. S., Rao, K. V. K.: J. Mater. Sci. *15*, 248/9 (1980)
250. Levin, E. M., Roth, R. S., Martin, J. B.: Am. Mineralogist *46*, 1030/55 (1961)
251. Pottier, M. J.: Compt. Rend. C *282*, 157/9 (1976); Acad. Roy. Belg. Mem. Classe Sci. Coll. 8° [2] *43*, No. 1 (1978) 1/75; C.A. *90*, No. 63777 (1979); Bull. Soc. Chim. Belg. *83*, 235/8 (1974); C.A. *84*, No. 129794 (1976)
252. Ross, M., Evans, H. T., Appleman, D. E.: Am. Mineralogist *49*, 1603/21 (1964)
253. Pottier, M. J., Schroeder, U., Tarte, P.: Compt. Rend. C *284*, 407/9 (1977)
254. Biedl, A.: Am. Mineralogist *51*, 521/2 (1966)
255. Lewis, J. F.: ibid. *52*, 42/54 (1967)

256. Levin, E. M., Roth, R. S., Martin, J. B.: ibid. *46*, 1055 (1961); note added in press
257. Meyer, H. J., Skokan, A.: Naturwissenschaften *58*, 566 (1971)
258. Abrahams, S. C., Bernstein, J. L., Keve, E. T.: J. Appl. Crystallogr. *4*, 284/90 (1971)
259. Schmid, H.: Acta Crystallogr. *17*, 1980/1 (1964)
260. Bither, T. A., Frederick, C. G., Gier, T. E., Weiher, J. F., Young, H. S.: Solid State Commun. *8*, 109/12 (1970) C.A. *72*, No. 60719 (1970)
261. Tombs, N. C., Croft, W. J., Mattraw, H. C.: Inorg. Chem. *2*, 872/3 (1963)
262. Diehl, R.: Solid State Commun. *17*, 743/5 (1965)
263. Bernal, I., Struck, C. W., White, J. G.: Acta Crystallogr. *16*, 849/50 (1963)
264. Muller, O., O'Neill, J. F.: Rare Earths Mod. Sci. Technol., Proc. 13th Rare Earths Res. Conf., Wheeling, Va., 1977 (Publ. 1978), pp. 173/9; C.A. *92*, No. 13933 (1980)
265. Walter, K.-H.: Ber. Kernforschungszentrum Karlsruhe, KFK *280*, 1965
266. Morgan, P. E. D., Carrol, P. J., Lange, F. F.: Mater. Res. Bull. *12*, 251/60 (1977)
267. Newnham, R. E., Redman, M. J., Santero, R. P.: J. Am. Ceram. Soc. *46*, 253/6 (1963)
268. Henry, J.-Y.: Mater. Res. Bull. *11*, 577/84 (1976)
269. Bradley, W. F., Graf, D. L., Roth, R. S.: Acta Crystallogr. *20*, 283/7 (1966)
270. Kindermann, B.: Dissertation Münster, 1974
271. Abdullaev, G. K., Dzhafarov, G. G., Mamedov, Kh. S.: Azerb. Khim. Zh. *1976* (3), 117/20, C.A. *86*, No. 131445 (1977)
272. Böhlhoff, R., Bambauer, H. U., Hoffmann, W.: Z. Kristallogr. *133*, 386/95 (1971); Naturwissenschaften *57*, 129 (1970)
273. Swanson, H. E., Morris, M. C., Stickfield, R., Evans, E. H.: Natl. Bur. Stands. (US), Monograph 25, Sect. 1 1962
274. Weidelt, J., Bambauer, H. U.: Naturwissenschaften *55*, 342 (1968)
275. Dzhurinskii, B. F., Gokhman, L. Z., Osiko, A. V., Zorina, L. N., Soshchin, N. P.: Izv. Akad. Nauk SSSR, Neorgan. Mater. *18*, 1739/42 (1982)
276. Meyer, H. J.: Naturwissenschaften *56*, 458/9 (1969)
277. Weigel, F., Scherer, V.: Radiochim. Acta *7*, 50/5 (1967)
278. Meyer, H. J.: Naturwissenschaften *59*, 215 (1972)
279. Palkina, K. K., Kuznetsov, V. G., Butman, L. A., Dzhurinskii, B. F.: Soviet J. Koord. Chem. *2*, 215/7 (1976); Koord. Khim. *2*, 286/9 (1976), C.A. *84*, No. 129141 (1976)
280. Sadanaga, R.: X-Rays *5*, 2/7 (1948), C.A. *1950*, No. 5765h
281. Berger, S. V.: Acta Chem. Scand. *3*, 660/75 (1949)
282. Sadanaga, R., Nishimura, T., Watanabe, T.: Mineral. J. [Tokyo] *4*, 380/8 (1965), C.A. *67*, No. 26835 (1967)
283. Bondareva, O. S., Simonov, M. A., Belov, N. V.: Soviet Phys.-Cryst. *23*, 272/3 (1978/79); Kristallografiya *23*, 491/3 (1978), C.A. *89*, No. 51794 (1978)
284. Aliev, O. A., Zul'fugarly, D. I., Guseinova, G. A.: Dokl. Akad. Nauk Azerb. SSR *34*, (5), 47/52 (1978), C.A. *90* No. 95682 (1979)
285. Pardo, J., Martinez-Ripoll, M., García-Blanco, S.: Anales Fiz. *67*, 399/400 (1971); Acta Crystallogr. *B 30*, 37/40 (1974)
286. Aliev, O. A., Zul'fugarly, D. I.: Russ. J. Inorg. Chem. *26*, 1795/6 (1981/82); Zh. Neorgan. Khim. 3347/9 (1981)
287. Maijling, J., Figusch, V., Hanic, F., Wiglasz, V., Corba, J.: Mater. Res. Bull. *9*, 1379/82 (1974), C.A. *81*, No. 160659 (1974)
288. Vegas, A., Cano, F. H., García-Blanco, S.: Acta Crystallogr. *B 31*, 1416/9 (1975)
289. Richter, L., Müller, F.: Z. Anorg. Allg. Chem. *467*, 123/5 (1980), C.A. *93*, No. 177751 (1980)
290. Chrétien, A., Priou, R.: Compt. Rend. C *271*, 1310/2 (1970), C.A. *74*, No. 37848 (1971)
291. Machida, K., Adachi, G., Hata, H., Shiokawa, J.: Bull. Chem. Soc. Japan *54*, 1052/5 (1981)
292. García-Blanco, S., Fayos, J.: Z. Kristallogr. *127*, 145/59 (1968)
293. Baur, W. H., Tillmanns, E.: ibid. *131*, 213/21 (1970)
294. Stewner, F.: Acta Crystallogr. *B* 27, 904/10 (1971)
295. König, H., Hoppe, R.: Z. Anorg. Allg. Chem. *434*, 225/32 (1977)
296. Jansen, M., Scheld, W.: ibid. *477*, 85/9 (1981)
297. Jansen, M., Brachtel, G.: ibid. *489*, 42/6 (1982)

298. Marchand, R., Piffard, Y., Tournoux, M.: Compt. Rend. *C 276*, 177/9 (1973), C.A. *78*, 103076 (1973); Piffard, Y., Marchand, R., Tournoux, M.: Rev. Chim. Minerale *12*, 210/7 (1975), C.A. *83*, No. 200522 (1975)
299. Bondareva, O. S., Simonov, M. A., Egorov-Tismenko, Yu. K., Velov, B. V.: Soviet Phys.-Cryst. *23*, 269/72 (1978/79); Kristallografiya *23*, 487/90 (1978), C.A. *89* No. 68937 (1978)
300. Bondareva, O. S., Ivashenko, A. N., Mel'nikov, O. K., Malinovskii, Yu. A., Belov, N. V.: Soviet Phys.-Cryst. *27*, 102/6 (1982); Kristallografiya *27*, 170/7 (1982)
301. Sokolova, E. N., Simonov, M. A., Belov, N. V.: Soviet Phys.-Cryst. *25*, 733/4 (1980/81); Kristallografiya *25*, 1285/6 (1980), C.A. *94*, No. 75088 (1981)
302. Kazanskaya, E. V., Sandomirskii, P. A., Simonov, M. A., Belov, N. V.: Soviet Phys.-Dokl. *23*, 108/10 (1978); Dokl. Akad. Nauk SSSR *238*, 1340/3 (1978), C.A. *89*, No. 121141 (1978)
303. Gasperin, M.: Acta Crystallogr. *B 30*, 1181/3 (1974)
304. Sokolova, E. N., Boronikhin, V. A., Simonov, M. A., Belov, N. V.: Soviet Phys.-Dokl. *24*, 417/20 (1979); Dokl. Akad. Nauk SSSR *246*, 1126/9 (1979), C.A. *91*, No. 100261 (1979)
305. Belov, N. V.: Mineral. Sb. [Lvov] *33*, 6/8 and 12/13 (1979) from C.A. *94*, No. 165905 (1981)
306. Lehmann, H.-A., Schadow, H.: Z. Anorg. Allg. Chem. *348*, 42/9 (1966)
307. Mascetti, J., Vlasse, M., Fouassier, C.: J. Solid State Chem. *39*, 288/93 (1981), C.A. *95* No. 195573 (1981)
308. Mascetti, J., Fouassier, C., Hagenmuller, P.: ibid. *50*, 204/12 (1983)
309. Verstegen, J. M. P. J.: J. Electrochem. Soc. *121*, 1631/3 (1974)
310. Vicat, J., Aléonard, S.: Bull. Soc. France Mineral. Crist. *91*, 293/5 (1968), C.A. *69*, No. 102589 (1968)
311. Vicat, J., Aléonard, S.: Compt. Rend. *C 266*, 1046/9 (1968), C.A. *69*, No. 31093 (1968); Mater Res. Bull. *3*, 611/20 (1968), C.A. *69*, No. 81438 (1968)
312. Bayer, G.: Z. Kristallogr. *133*, 85/90 (1971)
313. Pottier, M.-J.: Spectrochim. Acta *A 33*, 625/30 (1977), C.A. *87*, No. 191405 (1977)
314. Aléonard, S., Vicat, J.: Bull. Soc. France Mineral. Crist. *89*, 271/2 (1966), C.A. *65*, No. 14551b (1966)
315. Konovalenko, S. I., Voloshin, A. V., Pakhomovskii, Ya. A., Anan'ev, S. A., Perlina, G. A., Rogachev, D. L., Kusnetsov, V. Ya.: Dokl. Akad. Nauk SSSR *272*, 1449/53 (1983)
316. Schultze, D.: Z. Anorg. Allg. Chem. *386*, 227/8 (1971)
317. Baucher, A., Gasparin, M.: Mater. Res. Bull. *10*, 469/75 (1975), C.A. *83*, No. 69521 (1975); Baucher, A., Gasparin, M., Cervelle, B.: Acta Crystallogr. *B 32*, 2211/5 (1976)
318. Rza-Zade, P. F., Ismailova, F. A., Ganf, K. L., Safarov, M. I., Ali-Zade, M. Z.: Inorgan. Materials USSR *15*, 223/5 (1979); Izv. Akad. Nauk SSSR, Neorgan. Materialy *15*, 283/5 (1979), C.A. *90*, No. 196637 (1979)
319. Guseinova, Sh. A., Rza-Zade, P. F., Mamedova, E. D.: Azerb. Khim. Zh. *1977* (2), 86/9; C.A. *87* No. 161006 (1977)
320. Abdulla'ev, G. K., Mamedov, Kh. S., Amiraslanov, I. R., Magerramov, A. I.: Zh. Strukt. Khim. *18*, 410/3 (1977), C.A. *87*, No. 32346 (1977)
321. Abdulla'ev, G. K., Mamedov, Kh. S., Dzhafarov, G. G.: Azerb. Khim. Zh. *1977* (2), 115/9; C.A. *87*, No. 125673 (1977)
322. Abdulla'ev, G. K., Mamedov, Kh. S., Rza-Zade, P. F., Guseinova, Sh. A., Dzhafarov, G. G.: Azerb. Khim. Zh. *1978* (2), 125/9; C.A. *89*, No. 207567 (1979); Russ. J. Inorg. Chem. *22*, 1765/6 (1977/78); Zh. Neorgan. Khim. *22*, 3239/42 (1977)
323. Abdulla'ev, G. K., Mamedov, Kh. S.: Soviet Phys.-Cryst. *22*, 220/2 (1977/78); Kristallografiya *22*, 389/92 (1977)
324. Abdulla'ev, G. K., Mamedov, Kh. S.: Soviet Phys.-Cryst. *27*, 229/30 (1982/83); Kristallografiya *27*, 381/3 (1982), C.A. *96*, No. 208790 (1982)
325. Abdulla'ev, G. K., Mamedov, Kh. S.: J. Struct. Chem. USSR *13*, 881/3 (1972); Zh. Strukt. Khim. *13*, 943/6 (1972), C.A. *78*, No. 34904 (1973)
326. Bambauer, H. U., Khodaverdi, A., Kindermann, B., Steuhl, H. H.: Z. Kristallogr. *146*, 53/60 (1977)
327. Bambauer, H. U., Kindermann, B.: ibid. *147*, 63/74 (1978)
328. Dzhurinskii, B. F., Tananaev, I. V., Aliev, O. A.: Inorganic Materials USSR 5, 1682/5 (1969), Izv. Akad. Nauk. SSSR, Neorgan. Materialy *5* 1978/81 (1969)

329. Abdulla'ev, G. K., Aliev, C. A.: Soviet Phys.-Cryst. *15*, 693/4 (1970/71); Kristallografiya *15*, 812/3 (1970)
330. Abdulla'ev, G. K., Mamedov, Kh. S., Amirov, S. T.: Soviet Phys.-Cryst. *18*, 675/6 (1973/74); Kristallografiya *18*, 1075/7 (1973)
331. Palkina, K. K., Kuznetsov, V. G., Dzhurinskii, B. P., Moruga, L. G.: Russ. J. Inorg. Chem. *17*, 341/3 (1972); Zh. Neorgan. Khim. *17*, 652/6 (1972)
332. Palkina, K. K., Kuznetsov, V. G., Moruga, L. G.: J. Struct. Chem. USSR *13*, 317/8 (1972); Zh. Strukt. Khim. *13*, 341/2 (1972)
333. Palkina, K. K., Kuznetsov, V. G., Dzhurinskii, B. P., Moruga, L. G.: Zh. Strukt. Khim. *14*, 1053/7 (1973)
334. Abdulla'ev, G. K., Mamedov, Kh. S.: Soviet Phys.-Cryst. *27*, 478/80 (1982/83); Kristallografiya *27*, 795/7 (1982), C.A. *97*, No. 102151 (1982)
335. Abdulla'ev, G. K., Mamedov, Kh. S.: Zh. Strukt. Khim. *15*, 157/9 (1974)
336. Abdulla'ev, G. K., Mamedov, Kh. S.: Zh. Strukt. Khim. *17*, 188/91 (1976), C.A. *84*, No. 158380 (1976)
337. Il'in, V. K., Dzhurinskii, B. F., Novotortsev, V. M., Tananaev, I. V.: Izv. Akad. Nauk SSSR, Neorgan. Materialy *16*, 1304/5 (1980), C.A. *93*, No. 142123 (1980)
338. Leonyuk, N. I., Pashkova, A. V., Timchenko, T. I.: Soviet Phys. Dokl. *24*, 233/4 (1979); Dokl. Akad. Nauk SSSR *245*, 1109/12 (1979), C.A. *91*, No. 66405 (1979)
339. Mills, A. D.: Inorg. Chem. *1*, 960/1 (1962)
340. Al'shinskaya, L. I., Leonyuk, N. I., Timchenko, T. I.: Deposited Doc. *1977* [VINITI-4078-77] 18/22 from C.A. *90*, No. 171637 (1979)
341. Ballman, A. A.: Am. Mineralogist *47*, 1380/3 (1962)
342. Belokoneva, E. L., Azizov, A. V., Leonyuk, N. I., Simonov, M. A., Belov, N. V.: Zh. Strukt. Khim. *22*, 196/9 (1981), C.A. *95*, No. 89347 (1981)
343. Leonyuk, N. I., Pashkova, A. V., Belov, N. V.: Krist. Tech. *14*, 47/50 (1979), C.A. *91*, No. 30657 (1979)
344. Burkov, V. I., Kizel, V. A., Leonyuk, N. I., Sitnikov, N. M.: Soviet Phys.-Cryst. *27*, 121/2 (1982); Kristallografiya *27*, 196/7 (1982), C.A. *96*, No. 19617 (1982)
345. Kong, H. Y. P., Dwight, K.: Mater. Res. Bull. *9*, 1661/5 (1974), C.A. *82*, No. 37543 (1975)
346. Tate, I., Oishi, S.: Chem. Lett. *1981*, 1301/4
347. Kuroda, R., Mason, S. F., Rosini, R.: J. Chem. Soc., Faraday Trans. II, *77*, 2125/40 (1981)
348. Blasse, G., Bril, A.: J. Inorg. Nucl. Chem. *29*, 266/7 (1967)
349. Al'shinskaya, L. I., Leonyuk, N. I., Timchenko, T. I.: Cryst. Res. Technol. *14*, 897/903 (1979); C.A. *92*, No. 50164 (1980)
350. Al'shinskaya, L. I., Leonyuk, N. I.: Deposited Doc. *1978* [VINITI-482-78] 184/90, C.A. *91*, No. 115612 (1979)
351. Joubert, J.-C., White, W. B., Roy, R.: J. Appl. Crystallogr. *1*, 318/9 (1968), C.A. *71*, No. 34037 (1969)
352. Takahashi, T., Yamada, O., Ametani, K.: Mater. Res. Bull. *10*, 153/6 (1975)
353. Mal'tseva, L. I., Leonyuk, N. I., Timchenko, T. I.: Krist. Techn. *15*, 35/42 (1980), C.A. *93*, No. 17059 (1980)
354. Belokoneva, E. L., Al'shinskaya, L. I., Simonov, M. A., Leonyuk, N. I., Timchenko, T. I., Belov, N. V.: Zh. Strukt. Khim. *19*, 382/4 (1978), C.A. *89*, No. 121171 (1978)
355. Belokoneva, E. L., Al'shinskaya, L. I., Simonov, M. A., Leonyuk, N. I., Timchenko, T. I., Belov, N. V.: ibid. *20*, 542/4 (1979), C.A. *91*, No. 166730 (1979)
356. Belokoneva, E. L., Simonov, M. A., Pashkova, A. V., Timchenko, T. I., Belov, N. V.: Dokl. Akad. Nauk SSSR *255*, 854/8 (1980), C.A. *94*, 112860 (1981)
357. Belokoneva, E. L., Pashkova, A. V., Timchenko, T. I., Belov, N. V.: ibid. *261*, 361/5 (1981), C.A. *96*, No. 44231 (1982); Belokoneva, E. L., Timchenko, T. I.: Sov. Phys.-Cryst. *28*, 658/61 (1983/84); Kristallografiya *28*, 1118/23 (1983)
358. Bartsch, H.-H., Jarchow, O.: Z. Kristallogr. *159*, 13/4 (1982)
359. Kindermann B.: Z. Kristallogr. *146*, 67/72 (1977)
360. Capponi, J. J., Chenavas, J., Joubert, J. C.: J. Solid State Chem. *7*, 49/54 (1973)
361. Blum, P., Bozon, H.: Compt. Rend. *239*, 811/2 (1954), C.A. *49*, No. 5164c (1955)
362. Bertaut, E. F.: Acta Crystallogr. *3*, 473/4 (1950); Bertaut, E. F., Bochirol, L., Blum, P.: Compt. Rend. *230*, 764/5 (1950)

363. Mikhov, V. T., Apostolov, A., Taraleshkova, V. S., Andreevska, V. G.: Dokl. Bolg. Akad. Nauk *26*, 859/62 (1973)
364. Blasse, G.: J. Inorg. Nucl. Chem. *31*, 1519/21 (1969)
365. Venkatakrishnan, V., Bürger, M. J.: Z. Kristallogr. *135*, 321/38 (1972)
366. Blasse, G.: J. Inorg. Nucl. Chem. *32*, 700 (1970)
367. Takéuchi, Y.: Mineral. J. [Sapporo] *2*, 245/68 (1958), C.A. *54*, No. 12906 (1960); Takéuchi, Y., Watanabe, T., Ito, T.: Acta Crystallogr. *3*, 98/107 (1950)
368. Da Silva, J. C., Clark, J. R., Christ, C. L.: Bull. Geol. Soc. Am. *66*, 1540/1 (1956)
369. Konev, A. A., Lebedeva, V. S., Kashaev, A. A., Ushchapovskaya, Z. F.: Zap. Vses. Mineralog. Obshch. *99*, 225/31 (1970) from Am. Mineralogist *56*, 360 (1971) and Bull. Soc. France Mineral. Crist. *94*, 570 (1971)
370. Swinnea, J. S., Steinfink, H.: Am. Mineralogist *68*, 827/32 (1983)
371. Lopez Ruiz, J., Salvador, P. S.: ibid. *56*, 2149/51 (1971)
372. Brovkin, A. A., Aleksandrov, S. M., Nekrasov, I. Ya.: Rentgenogr. Mineral. Syr'ya, Akad. Nauk SSSR *1963*, (3), 16/34 from C.A. *60*, No. 8724h (1964)
373. Schwab, A. M., Bertaut, E. F.: Bull. Soc. France Mineral. Crist. *93*, 255/7 (1970)
374. Götz, W., Herrmann, V.: Naturwissenschaften *53*, 475 (1966)
375. Yamnova, N. A., Simonov, M. A., Belov, N. V.: Soviet Phys.-Cryst. *20*, 89/90 (1975); Kristallografiya *20*, 156/9 (1975); Dokl. Akad. Nauk SSSR *238*, 1094/7 (1978), C.A. *83*, No. 19517 (1975)
376. Konnert, J. A., Appleman, D. E., Clark, J. R., Finger, L. W., Kato, T., Miúra, Y.: Am. Mineralogist *61*, 116/22 (1976)
377. Takéuchi, Y., Haga, N., Kato, T., Miúra, Y.: Can. Mineralogist *16*, 475/85 (1978), C.A. *89*, No. 189349 (1978)
378. Bovin, J.-O., O'Keeffe, M.: Am. Mineralogist *65*, 1130/3 (1980); Acta Crystallogr. *A 37*, 28/35 (1981)
379. Stenger, C. G. F., Verschoor, G. C., Ijdo, D. J. W.: Mater. Res. Bull. *8*, 1285/95 (1973)
380. Schäfer, U.-L., Kuzel, H.-J.: Neues Jahrb. Mineral., Monatsh. *1967*, 131/6, C.A. *67*, No. 77037 (1967)
311. Hübner, K. H.: Neues Jahrb. Mineral., Abhandl. *112*, 150/6 (1970)
382. Hyman, A., Perloff, A.: Acta Crystallogr. *B 28*, 2007/11 (1972)
383. Choisnet, J., Groult, D., Raveau, B., Gasperin, M.: Acta Crystallogr. *B 33*, 1841/5 (1977)
384. Abrahams, S. C., Zyontz, L. E., Bernstein, J. L., Remeika, J. P., Cooper, A. S.: J. Chem. Phys. *75*, 5456/60 (1981)
385. Palkina, K. K., Saifuddinov, V. Z., Kusnetsov, V. G., Dzhurinskii, B. F., Lysanova, G. V., Reznik, E. M.: Russ. J. Inorg. Chem. *24*, 663/6 (1979/80); Zh. Neorgan. Khim. *24*, 1193/8 (1979), C.A. *91*, No. 80170 (1979)
386. Lysanova, G. V., Dzhurinskii, B. F., Komova, M. G., Tananaev, I. V.: Russ. J. Inorg. Chem. *28*, 1344/9 (1983/84); Zh. Neorgan. Khim. *28*, 2369/76 (1983)
387. Sokolova, E. N., Azisov, A. V., Simonov, M. A., Leonyuk, N. I., Belov, N. V.: Soviet Phys.-Dokl. *23*, 814/6 (1978); Dokl. Akad. Nauk SSSR *243*, 655/8 (1978), C.A. *90*, No. 79450 (1979)
388. Ihara, M., Imai, K., Fukunaga, J., Yoshida, N.: J. Ceram. Soc. Japan, Yogyo Kyo Kaishi *88*, 77/84 (1980); C.A. *92*, No. 139149 (1980)
389. Ysker, J. St.: Dissertation Münster, 1972
390. Moore, P. B., Araki, T.: Am. Mineralogist *61*, 88/94 (1976)
391. Bovin, J.-O., O'Keeffe, M., O'Keefe, M. A.: Acta Crystallogr. *A 37*, 28/35 (1981)
392. Dzhurinskii, B. F., Reznik, E. M., Tananaev, I. V.: Russ. J. Inorg. Chem. *25*, 1639/4 (1980/81); Zh. Neorgan. Khim. *25*, 2981/7 (1980)
393. Reznik, E. M., Dzhurinskii, B. F., Tananaev, I. V.: Russ. J. Inorg. Chem. *27*, 118/22 (1982); Zh. Neorgan. Khim. *27*, 212/8 (1982)
394. Baidina, I. A., Bakakin, V. V., Podberezskaya, N. V., Alekseev, V. I., Batsanova, L. R., Pavlyuchenko, V. S.: Zh. Strukt. Khim. *19*, 125/9 (1978)
395. Solov'eva, L. P., Bakakin, V. V.: Soviet Phys.-Cryst. *15*, 802/5 (1970/71); Kristallografiya *15*, 922/5 (1970)
396. Baidina, I. A., Bakakin, V. V., Batsanova, L. R., Palchik, N. A., Podberezskaya, N. V., Solov'eva, L. P.: Zh. Strukt. Khim. *16*, 1050/3 (1975), C.A. *84*, No. 158359 (1975)
397. Okazaki, H., Nakajima, S., Mizuno, H.: J. Chem. Soc. Japan, Nippon Kagaku Zasshi *86*, 1015/8 (1965), C.A. *65*, No. 1507c (1966)

398. Brovkin, A. A., Nishikova, L. V.: Soviet Phys.-Cryst. *20*, 452/5 (1975/76); Kristallografiya *20*, 740/5 (1975)
399. Brovkin, A. A., Nishikova, L. V.: Soviet Phys.-Cryst. *20*, 252/4 (1975/76); Kristallografiya *20*, 415/8 (1975)
400. Majling, J., Figusch, V., Čorba, J., Hanić, F.: J. Appl. Cryst. *7*, 402 (1974); Žak, Z., Hanić, F.: Acta Crystallogr. *B 32*, 1784/7 (1976)
401. Machida, K., Adachi, G., Shiokawa, J.: Chem. Lett. *1982*, 41/4; Machida, K., Adachi, G., Moriwaki, Y., Shiokawa, J.: Bull. Chem. Soc. Japan *54*, 1048/51 (1981)
402. Behm, H.: Acta Crystallogr. *C 39*, 1317/9 (1983)
403. Foord, F. E., Erd, R. C., Hunt, G. R.: Can. Mineralogist *19*, 303/10 (1982)
404. Rodellas, C., García-Blanco, S., Vegas, A.: Z. Kristallogr. *165*, 255/60 (1983)
405. Yakubovich, O. V., Simonov, M. A., Belov, N. V.: Soviet Phys.-Cryst. *20*, 87/8 (1975); Kristallografiya *20*, 152/5 (1975)
406. Aurivillius, B.: Chem. Scripta *22*, 168/70 (1983), C.A. *99*, No. 222757 (1983)
407. Lumpkin, G. R., Ribbe, P. H.: Am. Mineralogist *68*, 164/76 (1983)
408. Capponi, J. J., Chenavas, J., Joubert, J. C.: Mater. Res. Bull. *8*, 275/8 (1973)
409. Fang, J. H., Newnham, R. E.: Mineral. Mag. *35*, 269 (1965) 196/9; C.A. *63*, No. 1284b (1965)
410. Egorov-Tismenko, Yu. K., Simonov, M. A., Belov, N. V.: Dokl. Akad. Nauk SSSR *251*, 1122/3 (1980), C.A. *93*, No. 86056 (1980)
411. Nagai, T., Ihara, M.: J. Chem. Soc. Japan, Yogyo Kyo Kaishi *80*, 432/7 (1972)
412. van de Spijker, W. H. M. M., Konijnendijk, W. L.: Inorg. Nucl. Chem. Lett. *14*, 389/92 (1978)
413. Ihara, M.: J. Ceram. Soc. Japan, Yogyo Kyo Kaishi *79*, 152/5 (1971)
414. Zaslovskii, A. I., Zvinchuk, R. A.: Dokl. Akad. Nauk SSSR *90*, 781/3 (1953)
415. Mrose, M. E., Rose, W. J.: Am. Mineral. Soc., Am. Geol. Soc., Abstracts Annual Meet. *1961*, 111A
416. Diehl, R., Brandt, G.: Acta Crystallogr. *B 31*, 1662/6 (1975)
417. Levin, E. M., Robbins, C. R., Waring, J. L.: J. Am. Ceram. Soc. *44*, 87/91 (1961)
418. Ansell, G. B., Leonowicz, M. E., Modrick, M. A., Wanklyn, B. M., Wondre, F. R.: Acta Crystallogr. *B 38*, 892/3 (1982)
419. Capponi, J. J., Marezio, M.: ibid. *B 31*, 2440/3 (1975)
420. Moore, P. B., Araki, T.: Nature Phys. Sci. *240*, No. 99, 63/5 (1972)
421. Aristarain, L. F., Erd, R. C.: Am. Mineralogist *62*, 327/9 (1977)
422. Chackraburtty, D. M.: Acta Crystallogr. *10*, 199/200 (1957); Indian J. Phys. *31*, 235/41 (1957)
423. Clark, M. J. R., Lynton, H.: Can. J. Chem. *47*, 2943/6 (1969)
424. Chenavas, J., Grey, I. E., Guitel, J. C., Joubert, J. C., Marezio, M., Remeika, J. P., Cooper, A. S.: Acta Crystallogr. *B 37*, 1343/6 (1981)
425. Bauer, H.: Z. Anorg. Allg. Chem. *337*, 183/90 (1965)
426. Blasse, G., Bril, A., de Vries, J. I.: J. Inorg. Nucl. Chem. *31*, 568/70 (1969)
427. Tarte, P., De Wispelaere-Schröder, U.: Compt. Rend. (2) *295*, 351/4 (1982)
428. Liebertz, J., Stähr, S.: Z. Kristallogr. *160*, 135/7 (1982)
429. Prewitt, C. T., Shannon, R. D.: Acta Crystallogr. *B 24*, 869/74 (1968)
430. Berger, S. V.: Acta Chem. Scand. *7*, 611/22 (1953)
431. Will, G., Kirfel, A., Josten, B.: J. Less-Common Met. *82*, 255/67 (1981); Kirfel, A., Will, G., Stewart, R. F.: Acta Crystallogr. *B 39*, 175/85 (1983)
432. Marezio, M., Remeika, J. P.: J. Chem. Phys. *44*, 749/51 (1966)
433. Maraine-Giroux, C., Bouaziz, R., Perez, G.: Rev. Chim. Minerale *9*, 779/87 (1972); Bouaziz, R., Maraine, C.: Compt. Rend. *C 274*, 390/3 (1972)
434. Chang, C. H., Margrave, J. L.: Mater. Res. Bull. *2*, 929/33 (1967)
435. Marezio, M., Plettinger, H. A., Zachariasen, W. H.: Acta Crystallogr. *16*, 594/5 (1963)
436. Borisova, N. N., Vedishcheva, N. M., Pivovarov, M. M.: Russ. J. Inorg. Chem. *23*, 388/9 (1978); Zh. Neorgan. Khim. *23*, 703/6 (1978)
437. Schneider, W., Carpenter, G. B.: Acta Crystallogr. *B 26*, 1189/91 (1970)
438. Jansen, M., Brachtel, G.: Naturwissenschaften *67*, 606 (1980), C.A. *94*, No. 57243 (1981); Brachtel, G., Jansen, M.: Z. Anorg. Allg. Chem. *478*, 13/9 (1981), C.A. *95*, No. 124461 (1981)
439. Lehmann, H.-A., Papenfuß, H.-J.: Z. Anorg. Allg. Chem. *298*, 130/3 (1959)
440. Marezio, M., Plettinger, H. A., Zachariasen, W. H.: Acta Crystallogr. *16*, 390/2 (1963)
441. Dernier, P. D.: ibid. *B 25*, 1001/3 (1969)

442. Block, S., Perloff, A., Weir, C. E.: ibid. *17*, 314/5 (1964); Mighell, A. D., Perloff, A., Block, S.: Acta Crystallogr. *20*, 819/23 (1966)
443. Machida, K., Adachi, G., Shiokawa, J.: Acta Crystallogr. *B 35*, 149/51 (1979)
444. Zachariasen, W.: ibid. *23*, 390/2 (1967)
445. Marezio, M., Remeika, J. P., Dernier, P. D.: ibid. *B 25*, 955/64 (1969)
446. Adachi, G., Machida, K., Shiokawa, J.: J. Less-Common Metals *93*, 389/98 (1983)
447. Marezio, M., Remeika, J. P., Dernier, P. D.: Acta Crystallogr. *B 25*, 965/70 (1969)
448. Block, S., Perloff, A., Weir, C. E.: ibid. *17*, 314/5 (1964)
449. Liebertz, J., Stähr, S.: Z. Kristallogr. *165*, 91/3 (1983), C.A. *100*, No. 183710 (1984)
450. Martinez-Ripoll, M., Martinez-Carrera, S., Garcia-Blanco, S.: Acta Crystallogr. *B 27*, 677/81 (1971)
451. Abdulla'ev, G. K., Mamedov, Kh. S.: Zh. Strukt. Khim. *22*, 184/7 (1981), C.A. *95*, No. 195622 (1981)
452. Depmeier, W., Schmid, H., Hänssler, F.: Naturwissenschaften *67*, 456 (1980); Depmeier, W., Schmid, H.: Acta Crystallogr. *B 38*, 605/6 (1982)
453. Bambauer, H. U., Weidelt, J., Ysker, J.-St.: Naturwissenschaften *55*, 81 (1968); Z. Kristallogr. *130*, 207/13 (1969)
454. Ysker, J.-St., Hoffmann, W.: Naturwissenschaften *57*, 129/30 (1970); Ysker, J.-St., Böhlhoff, R., Bambauer, H. U., Hoffmann, W.: Z. Kristallogr. *132*, 457 (1970)
455. Abdulla'ev, G. K., Mamedov, Kh. S., Dzhafarov, G. G.: Soviet Phys.-Cryst. *26*, 473/4 (1981/82); Kristallografiya *26*, 837/40 (1981), C.A. *95*, No. 142360 (1981)
456. Weidelt, J.: Z. Anorg. Allg. Chem. *374*, 26/34 (1970)
457. Pakhomov, V. I., Sil'nitskaya, G. B., Medvedev, A. V., Dzhurinskii, B. F.: Inorg. Materials USSR *8*, 1036/9 (1972), C.A. *77*, No. 119411 (1972); Izv. Akad. Nauk SSSR, Neorgan. Mater. *8*, 1259/63 (1972)
458. Abdulla'ev, G. K., Mamedov, Kh. S., Dzhafarov, G. G.: Soviet Phys.-Cryst. *20*, 161/3 (1975); Kristallografiya *20*, 265/9 (1975)
459. Pakhomov, V. I., Sil'nitskaya, G. B., Dzhurinskii, B. F.: Inorg. Materials USSR *7*, 476/7 (1971); Izv. Akad. Nauk SSSR, Neorgan. Mater. *7*, 539/41 (1971), C.A. *74*, No. 147494 (1971)
460. Liebertz, J.: Z. Kristallogr. *158*, 319 (1982), C.A. *97*, No. 31728 (1982); Fröhlich, R., Bohaty, L., Liebertz, J.: Acta Crystallogr. *C 40*, 343/4 (1984), C.A. *100*, No. 165812 (1984)
461. Holcombe, C. E., Johnson, D. H.: J. Cryst. Growth *49*, 207/10 (1980)
462. Saubat, B., Vlasse, M., Fouassier, C.: J. Solid State Chem. *34*, 271/7 (1980), C.A. *93*, No. 214556 (1980)
463. Abdulla'ev, G. K., Mamedov, Kh. S., Dzhafarov, G. G., Aliev, O. A.: Russ. J. Inorg. Chem. *25*, 198/200 (1980/81); Zh. Neorgan. Khim. *25*, 364/7 (1980), C.A. *92*, No. 156206 (1980)
464. Aliev, O. A., Zul'fugarly, D. I.: Dokl. Akad. Nauk Azerb. SSR [Baku] *35*, (8), 59/63) (1979), C.A. *92*, No. 136130 (1980)
465. Abdulla'ev, G. K., Mamedov, Kh. S., Dzhafarov, G. G.: Zh. Strukt. Khim. *16*, 71/6 (1975)
466. Abdulla'ev, G. K.: ibid. *17*, 1128/31 (1976)
467. Abdulla'ev, G. K., Mamedov, Kh. S., Dzhafarov, G. G.: Soviet Phys.-Cryst. *19*, 457/9 (1974/75); Kristallografiya *19*, 737/40 (1974)
468. Abdulla'ev, G. K., Mamedov, Kh. S., Amiraslanov, I. R., Dzhafarov, G. G., Aliev, O. A., Usubaliev, B. T.: Russ. J. Inorg. Chem. *23*, 1286/8 (1978/79); Zh. Neorgan. Khim. *23*, 2332/5 (1978)
469. Smith, P., García-Blanco, S., Rivoir, L.: Z. Kristallogr. *119*, 375/83 (1964); An. Fisica Quim. *57*, 263 (1961); Smith-Verdier, P., García-Blanco, S.: Z. Kristallogr. *151*, 175/7 (1980), C.A. *93*, 35385 (1980)
470. Bondareva, O. S., Egorov-Tismenko, Yu. K., Simonov, M. A., Belov, N. V.: Dokl. Akad. Nauk SSSR *241*, 815/7 (1978)
471. Bohaty, L., Haussühl, S., Liebertz, J., Stähr, S.: Z. Kristallogr. *161*, 157/8 (1982)
472. Chang, C. H., Margrave, J. L.: Inorg. Chim. Acta *1*, 378/80 (1967)
473. Fouassier, C., Levasseur, A., Joubert, J. C., Muller, J., Hagenmuller, P.: Z. Anorg. Allg. Chem. *375*, 202/8 (1970)
474. Joubert, J. C., Muller, J., Pernet, M., Ferrand, B.: Bull. Soc. France Minéral. Crist. *95*, 68/74 (1972)
475. Levasseur, A., Rouby, B., Fouassier, C.: Compt. Rend. *C 277*, 421/3 (1973)

476. Calvo, C., Faggiani, R., Krishnamachari, N.: Acta Crystallogr. *B 31*, 188/92 (1971)
477. König, H., Hoppe, R., Jansen, M.: Z. Anorg. Allg. Chem. *449*, 91/101 (1979)
478. Takéuchi, Y.: Acta Crystallogr. *5*, 574/81 (1952)
479. Mrose, M. E., Fleischer, M.: Am. Mineralogist *48*, 915/24 (1963)
480. Il'in, Yu. N., Kravchenko, V. V., Petrov, K. I.: Russ. J. Inorg. Chem. *28*, 909/11 (1983/84); Zh. Neorgan. Khim. *28*, 1609/12 (1983)
481. Bartl, H., Schuckmann, W.: Neues Jahrb. Mineral., Monatsh. *1966*, 253/8
482. Block, S., Burley, G., Perloff, A., Mason, R. D.: J. Res. Nat. Bur. Stand. *62*, 95/100 (1959)
483. Berger, S. V.: Acta Chem. Scand. *4*, 1054/65 (1950)
484. Hand, W. D., Krogh-Moe, J.: J. Amer. Ceram. Soc. *45*, 197 (1962)
485. Sokolova, E. V., Simonov, M. A., Belov, N. V.: Soviet Phys.-Dokl. *24*, 524/6 (1979); Dokl. Akad. Nauk SSSR *247*, 603/6 (1979), C.A. *91*, No. 166725 (1979)
486. Yakubovich, O. V., Yamnova, N. A., Shchedrin, B. M., Simonov, M. A., Belov, N. V.: Dokl. Akad. Nauk SSSR *228*, 842/5 (1976), C.A. *85*, No. 134868 (1975)
487. Yakubovich, O. V., Simonov, M. A., Belov, N. V.: ibid. *238*, 98/100 (1978), C.A. *88*, No. 144595 (1978)
488. Simonov, M. A., Egorov-Tismenko, Yu. K., Yamnova, N. A., Belokoneva, E. L., Belov, N. V.: ibid. *251*, 1125/8 (1980), C.A. *93*, No. 86057 (1980)
489. Yakubovich, O. V., Simonov, M. A., Belokoneva, E. L., Egorov-Tismenko, Yu. K., Belov, N. V.: ibid. *230*, 837/40 (1976), C.A. *85*, No. 200865 (1976)
490. Hübner, K. H.: Neues Jahrb. Mineral., Monatsh. *1969*, 335/42
491. Baskin, Y., Harada, Y., Handwerk, J. H.: J. Amer. Ceram. Soc. *44*, 456/9 (1961)
492. Pushcharovskii, D. Yu., Karpov, O. G., Leonyuk, N. I., Belov, N. V.: Dokl. Akad. Nauk SSSR *241*, 91/4 (1978)
493. Timchenko, T. I., Leonyuk, N. I., Pashkova, A. V., Zhuravleva, O. L.: Soviet Phys.-Dokl. *24*, 336/7 (1979); Dokl. Akad. Nauk SSSR *246*, 613/5 (1979), C.A. *91*, No. 100009 (1979)
494. Pashkova, A. V., Sorokina, O. V., Leonyuk, N. I., Timchenko, T. I., Belov, N. V.: Dokl. Akad. Nauk SSSR *258*, 103/6 (1981)
495. Teske, K., Lehmann, H.-A.: Z. Chem. [Leipzig] *6*, 230 (1966)
496. Ihara, M., Yuge, M., Krogh-Moe, J.: J. Ceram. Assoc. Japan, Yogyo Kyo Kaishi *88*, 179/84 (1980); C.A. *92*, No. 224667 (1980)
497. König, H., Hoppe, R.: Z. Anorg. Allg. Chem. *434*, 71/9 (1978)
498. Krogh-Moe, J.: Acta Crystallogr. *B 30*, 747/52 (1974)
499. Krogh-Moe, J.: ibid. *B 28*, 1571/6 (1972)
500. Krogh-Moe, J.: ibid *B 30*, 1178/80 (1974)
501. Touboul, M.: Compt. Rend. *C 277*, 1025/7 (1973)
502. Krogh-Moe, J.: Acta Crystallogr. *B 30*, 578/82 (1974)
503. Krogh-Moe, J.: ibid. *14*, 68 (1961)
504. Krogh-Moe, J.: ibid. *B 24*, 179/81 (1968)
505. Natarjan, M., Faggiani, R., Brown, I. D.: Cryst. Struct. Commun. *8*, 367/70 (1979)
506. Krogh-Moe, J.: Acta Crystallogr. *B 28*, 3089/93 (1972)
507. Bartl, H., Schuckmann, W.: Neues Jahrb. Mineral., Monatsh. *1966*, 142/8
508. Abrahams, S. C., Bernstein, J. L., Gibart, P., Robbins, M., Sherwood, R. C.: J. Chem. Phys. *60*, 1899/905 (1974)
509. Rumanova, I. R., Genkina, E. A.: Latvijas PSR Zinatnu Akad. Vestis, Kim. Ser. *1981*, 571/9, C.A. *95*, No. 229841 (1981)
510. Martinez-Ripoll, M., Martinez-Carrera, S., García-Blanco, S.: Acta Crystallogr. *B 27*, 672/7 (1971); Ann. Fis. Quim. *66*, 5/6 (1970)
511. Ihara, M., Krogh-Moe, J.: Acta Crystallogr. *20*, 132/4 (1966)
512. Kindermann, B.: Z. Kristallogr. *146*, 61/6 (1977)
513. Zayakina, N. V., Brovkin, A. A.: Soviet Phys.-Cryst. *22*, 156/9 (1977/78); Kristallografiya *22*, 275/80 (1977)
514. Krogh-Moe, J.: Acta Chem. Scand. *18*, 2055/60 (1964)
515. Perloff, A., Block, S.: Acta Crystallogr. *20*, 274/9 (1966)
516. Block, S., Perloff, A.: ibid. *19*, 297/300 (1965)
517. Machida, K., Adachi, G., Shiokawa, J.: ibid. *B 36*, 2008/11 (1980), C.A. *93*, No. 177696 (1980)

518. Rza-Zade, P. F., Abdulla'ev, G. K., Eyubova, N. A., Samedov, F. R.: Inorg. Materials USSR *7*, 1872/4 (1971); Izv. Akad. Nauk SSSR, Neorgan. Mater. *7*, 2098/100 (1971), C.A. *76*, No. 51082 (1972)
519. Churilova, N. N., Serebrennikov, V. V.: Trudy Tomsk Univ. No. *204*, 277/81 (1971); C.A. *77*, No. 108869 (1972)
520. Touboul, M., Amoussou, D.: Compt. Rend. *C 285*, 145/7 (1977)
521. Abdulla'ev, G. K., Rza-Zade, P. F., Mamedov, Kh. S.: Izv. Akad. Nauk SSSR, Neorgan. Mater. *17*, 456/8 (1981), C.A. *94*, No. 198262 (1981)
522. Krogh-Moe, J.: Acta Crystallogr. *B 28*, 168/72 (1972)
523. Krogh-Moe, J.: ibid. *18*, 1088/9 (1965)
524. Krogh-Moe, J.: Arkiv for Kemi *14*, 439/49 (1979)
525. Krogh-Moe, J.: ibid. *14*, 451/9 (1959)
526. Vegas, A., Cano, F. H., García-Blanco, S.: J. Solid State Chem. *17*, 151/5 (1976)
527. Bondareva, O. S., Malinovskii, Yu. A., Belov, N. Y.: Soviet Phys.-Chryst. *25*, 541/4 (1980/81); Kristallografiya *25*, 944/9 (1980), C.A. *94*, No. 93972 (1981); Bondareva, O. S., Ivashchenko, A. N., Mel'nikov, O. K., Malinovskii, Yu. A., Belov, N. V.: Soviet Phys.-Cryst. *26*, 283/6 (1981/82); Kristallografiya *26*, 499/504 (1981), C.A. No. 50318 (1981)
528. Lloyd, D. J., Levasseur, A., Fouassier, C.: J. Solid State Chem. *6*, 179/86 (1973); Compt. Rend. *C 274*, 1684/7 (1972)
529. Peters, T. E., Baglio, J.: J. Inorg. Nucl. Chem. *32*, 1089/95 (1970)
530. Levasseur, A., Fouassier, C.: Compt. Rend. *C 272*, 80/2 (1971); Fouassier, C., Levasseur, A., Hagenmuller, P.: J. Solid State Chem. *3*, 206/8 (1971)
531. Machida, K., Adachi, G., Yasuoka, N., Kasai, N., Shiokawa, J.: Inorg. Chem. *19*, 3807/11 (1980), C.A. *94*, No. 10160 (1981)
532. Machida, K., Ishino, T., Adachi, G., Shiokawa, J.: Mater. Res. Bull. *14*, 1529/34 (1979), C.A. *92*, No. 120911 (1980)
533. Nakamura, S.: J. Japan. Ceram. Soc., Yogyo Kyokaishi *86*, 437/43 (1978), C.A. *90*, No. 132099 (1979)
534. Zayakina, N. V., Brovkin, A. A.: Soviet Phys.-Cryst. *21*, 277/9 (1976); Kristallografiya *21*, 502/6 (1976)
535. Jeitschko, W., Bither, T. A., Bierstedt, P. E.: Acta Crystallogr. *B 33*, 2767/75 (1977)
536. Levasseur, A., Lloyd, D. J., Fouassier, C., Hagenmuller, P.: J. Solid State Chem. *8*, 318/24 (1973), C.A. *80*, No. 41816 (1974)
537. Jeitschko, W., Bither, T. A.: Z. Naturforsch. *27b*, 1423 (1972)
538. Vlasse, M., Levasseur, A., Hagenmuller, P.: Solid State Ionics *2*, 33/7 (1981)
539. Villeneuve, G., Echegut, P., Reau, J. M., Levasseur, A., Brethous, J. C.: J. Solid State Chem. *30*, 275/81 (1979), C.A. *92*, No. 68180 (1980)
540. Berger, G., Vignaud, G., Levasseur, A.: J. Phys. Chem. Solids *41*, 1223/9 (1980)
541. Gould, R. O., Nelmes, R. J., Gould, S. E. B.: J. Phys. *C 14*, 5259/67 (1981)
542. Sueno, S., Clark, J. R., Papike, J. J., Konnert, J. A.: Am. Mineralogist *58*, 691/7 (1973)
543. Schmid, H.: J. Phys. Chem. Solids *26*, 973/88 (1965)
544. Bither, T. A., Young, H. S.: J. Solid State Chem. *10*, 302/11 (1974)
545. Nelmes, R. J., Thornley, F. R.: J. Phys. *C 7*, 3855/74 (1974)
546. Heyde, F., Beyrich, H.: Naturwissenschaften *52*, 6170/3 (1965)
547. Schmid, H., Trooster, J. M.: Solid State Commun. *5*, 31/5 (1967)
548. Nelmes, R. J., Hay, W. J.: J. Phys. *C 14*, 5247/57 (1981)
549. Nelmes, R. J., Thornley, F. R.: ibid. *C 9*, 665/80; 681/92 (1976); Thornley, F. R., Nelmes, R. J.: Acta Crystallogr. *A 30*, 748/57 (1974), C.A. *81*, No. 178277 (1974)
550. Will, G., Morche, H.: J. Phys. *C 10*, 1389/94 (1977)
551. Joubert, J. C., Muller, J., Fouassier, C., Levasseur, A.: Krist. Techn. *6*, 65/8 (1971), C.A. *79*, No. 92263 (1973)
552. Dowty, E., Clark, J. R.: Z. Kristallogr. *138*, 64/99 (1973); Solid State Commun. *10*, 543/8 (1972); Am. Mineralogist *58*, 1098/9 (1973)
553. Belov, V. F., Pyl'nev, V. G., Zheludov, I. S., Korovushin, V. V., Korneev, E. V., Yarmukhamedov, Yu. N.: Soviet Phys.-Cryst. *20*, 96/7 (1975); Kristallografiya *20*, 167/8 (1975)
554. Kobayashi, J., Sato, Y., Schmid, H.: Phys. Status Solidi (a) *10*, 259/70 (1972)
555. Abrahams, S. C., Bernstein, J. L., Svensson, C.: J. Chem. Phys. *75*, 1912/8 (1981)

556. Wendling, G. E., v. Hodenberg, R., Kühn, R.: Kali Steinsalz *6*, 1/3 (1972)
557. Schmidt, H.: Phys. Status Solidi *37*, 209/23 (1970)
558. Hyman, A., Perloff, A., Mauer, F., Block, S.: Acta Crystallogr. *22*, 815/21 (1967)
559. Krogh-Moe, J.: ibid. *18*, 77/81 (1965)
560. Krogh-Moe, J., Ihara, M.: ibid. *B 25*, 2153/4 (1969)
561. Krogh-Moe, J.: Arkiv for Kemi *12*, 247/9 (1958)
562. Robbins, C. R., Levin, E. M.: J. Res. Natl. Bur. Std. *73A*, 79/91 (1969)
563. Tanaka, Y., Fukunaga, J., Setoguchi, M., Higashi, T., Ihara, M.: J. Ceram. Soc. Japan, Yogyo Kyo Kaishi *90*, 458/63 (1982), C.A. *98*, No. 21013 (1983)
564. Rza-Zade, P. F., Kulieva, S. A., Ganf, K. L., Samedov, F. R.: Izv. Akad. Nauk SSSR, Neorgan. Mater. *8*, 1866/7 (1972), C.A. *78*, No. 34529 (1973)
565. Krogh-Moe, J., Ihara, M.: Acta Crystallogr. *23*, 427/30 (1967)
566. Krogh-Moe, J., Wold-Hansen, P. S.: ibid. *B 29*, 2242/6 (1973)
567. Behm, H.: ibid. *B38*, 2781/4 (1982), C.A. *98*, No. 25828 (1983)
568. Krogh-Moe, J.: Acta Crystallogr. *B 30*, 1827/32 (1974)

Molecular and Electronic Structure of Penta- and Hexacoordinate Silicon Compounds

Stanislav N. Tandura[1], Mikhail G. Voronkov[2], and Nikolai V. Alekseev[1]

1 NMR Spectroscopy Center, Ministry of Chemical Industry, Shosse Entuziastov 36, Moscow, 111123/USSR
2 Institute of Organic Chemistry, Siberian Division of the USSR Academy of Sciences, Irkutsk, 664033/USSR

Table of Contents

1 Introduction

Oxygen compounds of silicon constitute three fourths of the earth's crust. In practically all rocks and minerals as well as constructing materials based on silicon, the element is present in a tetra-covalent state. Due to its electron-acceptor ability, however, a silicon atom can be five-, six-, and, possibly, even seven-coordinate.

In the early 19th century, Gay-Lussac discovered $F_4Si \cdot 2\,NH_3$ [1]. Three years later Davy obtained the same compound [2], and Davy and Berzelius synthesized other compounds of six-coordinate silicon: fluorosilicic acid, $H_2[SiF_6]$, and its salts, the hexafluorosilicates [3]. In the second half of the last century adducts of silicon tetrahalides with tertiary amines were synthesized [4]. In 1903 Dilthey investigated silicon diketonates containing six-coordinate silicon [5,6]. In the first half of our century the possibility of the presence of $Si(OH)_6^{2-}$ ions in silica was widely discussed [7]. Interest in five- and six-coordinate silicon compounds increased in the second half of the present century. As a result, new classes of silicon compounds with an expanded coordination sphere were discovered and were studied by various physical and chemical methods. Specific biological activity was observed for some of the compounds and promising ways of their practical application were planned [8-12].

Donor-acceptor interactions play an important role in many physical, chemical, and biological processes. In spite of numerous studies in this field, the nature and mechanism of the formation of inter- and intra-molecular coordination are not quite clear yet. This is especially true for the chemistry of silicon where the problems of the electron-acceptor ability of the silicon atom and of its role in penta- and hexa-coordinate states are still open to question. In the usual compounds of four-coordinate silicon, its atom is in the sp^3 hybridization state which causes tetrahedral arrangement of the valente bonds. Nevertheless, the similarity between silicon and carbon is, to a great extent, formal both in chemical and physical aspects. A great number of carbon compounds do not have stable silicon analogs and multiple bonds involving silicon have long been sought in vain. During the recent past, numerous investigations have been concerned with tetravalent intermediates with an unusual degree of coordination in which a silicon atom participated in the formation of Si=Si (disilene), Si=C (silaethylene or silene) [13-19], Si=O (silanone) [20-24] and Si=S (silathione) [25,26] double bonds, triple bonds [27-30], or entered into an aromatic system (silabenzene) [31-37]. Only in 1981 has it been possible to isolate and characterize Si=C [32] and Si=Si double bonds [39]. A number of other structural reports containing double bonds to silicon have been published as well [40-51]. Silicon—metal double bonds or silylene—metal complex [52] and triple bonds containing silicon have yet to be found. On the other hand, many organosilicon compounds have no carbon analogs (e.g., polyorganylsesquioxanes [53], polyorganylsilazanes, and so on).

Structural parameters of the majority of molecules containing Si—O and Si—N bonds differ greatly from those of the carbon analogs. For silicon compounds enhanced values of the SiOSi angle (130–180°) and the tendency to a planar configuration of nitrogen bonds in the NSi_3 grouping are typical [54-65]. Silicon bonds with electro-negative atoms (F, O, N, and Cl) are much shorter than the sum of covalent radii, even when corrected for differences in electronegativity [55,66].

Numerous investigations have shown that in silicon compounds the tendency to expand the coordination sphere of the silicon atom is more typical than to its narrowing. In contrast to germanium, tin, and lead [67-69], compounds of silicon of lower valency are metastable under ordinary conditions [76-72]. Shortlived divalent silicon compounds of the types X_2Si (silylenes) [73-76] and Si=Y where Y = O and S (silicon monoxide and monosulfide) [77] are formed only at high temperatures and are thermodynamically unstable. Compounds of mono- and trivalent silicon, namely, diatomic molecules of the XSi type, tricoordinate silyl anions SiR_3^- [78-80] and radicals (for example $[(Me_3Si)_2CH]_3Si\cdot$ [81]), are, as expected, highly reactive. Neutral and ionic SiH, SiH_2, and SiH_3 particles have frequently been postulated as intermediates in various silane reactions. Direct evidence for their brief existence came from spectroscopic measurements, e.g., electronic spectra of flash-photolysis, from optical emission of excited particles in discharge plasma, low-temperature ESR and matrix IR spectra, electron-impact mass spectra and ion cyclotron resonance [76].

Since the silicon atom is less electronegative than carbon, it should carry a positive charge more readily[1]. Trivalent silyl cations are energetically similar to the corresponding trivalent carbocations [85-86], but are substantially less stable than their carbon analogs [87-89]. However, the existence of silicenium ions R_3Si^+, the silicon analogs of stable carbenium ions, seems to be reliably well established in the gas phase [90-92]. An ion cyclotron resonance study of fluoromethylsilanes has shown the following order of decreasing stability for fluoromethylsilicenium ions: $Me_3Si^+ > Me_2FSi^+ > MeF_2Si^+ > F_3Si^+$ [93]. Until recently, there has been no convinving evidence for their existence in condensed media. Reports about the preparation of silicenium ion in an inert gas matrix at low temperatures [94,95] have been shown to be erroneous [96]. However, in accordance with previous abinitio calculations [97,98], the choice of an alkylthio substituent (i-PrS) on silicon with high polarizability, low electronegativity, and 3p lone-pair donating ability of sulfur has permitted the silicenium ion to be observed in solution [99]. Abinitio calculations predict the increasing stabilization of the positive charge on the R_3Si^+ under successive vinylization of silicenium ions as well [100,101].

In the R_3MX (M = SiC) molecules triorganylsilyl groups donate electrons easier than their more electronegative carbon analogs. It was assumed that silicon atom could retain the positive charge in four-coordinate ions. Thus, for instance, five-coordinate adducts, $X_4Si \cdot D$, seemed to be ionic compounds of silicon with a smaller coordination number i.e., $[X_3Si \leftarrow D]^+X^-$. A conductivity study gave reason to assume that adducts of trimethylsilane iodide and bromide with hexamethylphosphorus-triamide (D = $(Me_2N)_3PO$; X = Br, I), unlike a pure covalent complex with trimethylfluorosilane [102], have an ionic structure [103-106]. A kinetic study of the racemization of optically active triorganohalosilanes indicated a strongly negative activation entropy and second order of racemization in the nucleophile. These results have been interpreted [107] in terms of an extension of coordination at silicon without existence of ionic adducts. However, 1:1 adducts of bromo- or iodotrimethylsilane and pyridine do not occur as higher coordination complexes in the solid state. From X-ray diffraction data the bromine and iodine ions are displaced from the coordination sphere of the silicon atom, possibly due to a higher polarizability which leads to a four-coordinate ionic structure $[Me_3Si(py)]^+X^-$. The Si—Br and Si—I distances

1 The little known species SiH_5^+ may be regarded as a weaker SiH_3^+—H_2 complex [82-84]

(435.9 and 455.9 pm, respectively [108]) are longer than the sum of the van der Waals radii of the atoms concerned (405 and 425 pm). From ^{29}Si NMR data on N-(dimethylbromosilylmethyl)lactames, a reversible Si—Br dissociation increasing with strengthening the Si←O interaction at low temperature has been evidenced from lower ^{29}Si shielding with decreasing the temperature [109,110].

The melting-point diagram of the Me_3SiCl-pyridine system gives no information about the formation of an adduct [111]. Even with HMPT, which is a stronger Lewis base than pyridine, chlorotrimethylsilane does not form a solid adduct at room temperature. Reactions of triorganochlorosilanes with various nitrogen donors have been interpreted in terms of formation of both an ionic species and coordination compounds [112-120]. However, silicon tetrachloride and strong Lewis bases tend to produce only adducts [119-121]. There are many examples of adduct formation between silicon-transition-metal compounds and Lewis bases involving charge separation [105,122]. Based on X-ray diffraction data, for example, the adduct $Me_3SiCo(CO)_4 \times PMe_2(SiMe_3)$ consists of a separated $[Me_2SiP(SiMe_3)_2]^+$ cation and a $Co(CO)_4^-$ anion [123].

It has been shown by various chemical methods that, in contrast to carbon, silicon in many substitution reactions does not form carbenium ion analogs (electrophilic substitution) and its nucleophilic reactions proceed via the formation of transition states in which silicon is penta-coordinated. The importance of the pentavalent state in which the silicon atom experiences octet expansion may readily be seen by considering a broad scope of substitution reactions at silicon, permutational isomerism in penta-coordinate intermediates upon solvolysis of optically active derivates, racemization of the configuration of an initially chiral silicon reactant, sigmatropic reactions and thermal rearrangement involving migration of silyl groups [124-133].

There are known many adducts of SiF_4 with nitrogen-containing bases [134,146], the majority of these being 1:2 adducts. Little is known about complexes of SiF_4 with oxygen-containing bases, presumably owing to a much weaker interaction [120,147]. The heat of dissociation of the $SiF_4 \cdot 2\,(CH_3)_2O$ complex at substantially below room temperature was reported to be 9.0 kcal · mole^{-1} [147] as compared to 40.5 kcal · mole^{-1} for the $SiF_4 \cdot 2\,N(CH_3)_3$ complex [148].

There is good reason to assign to many stable organosilicon compounds a structure in which the silicon coordination polyhedron represents either an octahedron or a trigonal bipyramid, for instance, to SiF_6^{2-} and to adducts of silicon tetrafluoride with tertiary amines which are stable on heating to 300 °C [149]. Nevertheless, although the octahedral structure of SiF_6^{2-} was established long ago [150-153], direct evidence for the silicon ability to exist in a penta-coordinate state was first obtained fairly recently when a dimethylsilylamine pentamer was studied by X-ray crystallography [154-156]. Earlier, systematic and comprehensive studies of a vast class of penta-coordinate silicon compounds, i.e., silatranes [11,157-159], with various physical and chemical methods had begun. Investigations of the structure of silicon compounds with higher coordination numbers were hampered by the instability of these species. Recently, matrix isolation techniques have been employed to reinvestigate the 1:1 molar adducts between SiF_4 and nitrogen- or oxygen-containing bases such as $SiF_4 \cdot NH_3$ [145,146] and $SiF_4 \cdot (CH_3)_2O$ [160]. Up to now silatranes and phthalocyanines are the largest known groups of trigonal-bipyramidal and octahedral silicon compounds. Recently a stable penta-coordinate silicon compound of square (or tetragonal) pyramidal structure has been found [161-163]. The possibility of the existence of seven- and eight-coordinate silicon complexes was suggested [164-166]. For a review of silicon bonding for different coordination numbers, see [11, 134-143, 157-159, 167-179].

For the last two decades crystal and electronic structures of penta- and hexa-coordinate compounds of silicon have been studied rather thoroughly and made a great contribution to both the chemistry of silicon and that of coordination compounds (with secondary bonds, i.e., with interactions intermediate in strength between covalent and van der Waals forces). The present review is devoted to the results obtained by X-ray and electron diffraction techniques and will attempt to demonstrate the practical application of NMR spectroscopy to different aspects of the coordination chemistry of silicon. There are no specific claims as to literature coverage, but references are quoted covering the period up to and including 1983, with an occasional 1984 entry.

2 Formation of Penta- and Hexa-coordinate Silicon Compounds

From the standpoint of the atomic structure the differences between carbon and silicon are as follows (Table 1):

1. The covalent radius of the silicon atom is one and one half that of the carbon atom thus rendering the silicon atom more sterically accessible.
2. The positive charge of a nucleus in the silicon atom is shielded with an additional shell of 8 electrons and the electronegativity of silicon is lower than that of carbon.
3. In contrast to a carbon atom, which does not have vacant d-orbitals, the outer valente shell of silicon has vacant 3d atomic orbitals (AO).
4. The polarizability of silicon is greatly facilitated by a larger (as compared to the carbon atom) size of valent AO and, consequently, by a higher diffuseness of its electron cloud.
5. Valence electrons of silicon are at a much greater distance from the nucleus than in the carbon atom due to which the ionization potential of the silicon atom, in spite of a greater nuclear charge, is about 3.1 eV less than that of the carbon atom.

The silicon atom is characterized by a higher electron affinity than carbon (by 0.27 eV) and by a greater ability to coordination, i.e. by a pronounced electrophilicity[1]. Germanium, another homolog of silicon in Group IVA, differs only slightly from silicon in its chemical and physical properties [210, 211]. All other things being equal, there is an increasing tendency on proceeding down Group IVA for the stabilization of a trigonal-bipyramidal intermediate or transition state relative to the tetrahedral

---

1 The electron-acceptor ability of the Group IVA elements increases with a rise in their atomic number and, consequently, in the atomic radius. From the standpoint of structural chemistry this is, for instance, clearly illustrated by the structure of $(CH_3)_3MCN$ crystals (M = Si, Ge, or Sn) in which a minimum intermolecular non-covalent contact between M and N atoms is 366 [207], 357 [208] and 249 pm [209], respectively. When M = Si this distance corresponds to the sum of van der Waals radii of both atoms, when M = Ge it is by 15 pm less than this sum. For M = Sn the environment of the tin atom is trigonal bipyramidal.

Table 1. Some Fundamental Properties of C and Si Atoms

	C	Si
Atomic number	6	14
Atomic weight[a] [180]	12.011	28.0855
Radii of the principal maxima of outer orbitals, pm [181]	62.0 (2s) 59.6 (2p)	90.4 (3s) 106.8 (3p)
Atomic radius (Bragg-Slater), pm [182]	70	110
Covalent radius, pm [183]	77.2	116.9
Ionic radius, pm [184, 185]	260 (−4)	41 (+4)
Ionization energies, eV [186]		
E_1	11.26	8.15
E_2	24.38	16.34
E_3	47.89	34.49
Electron affinity, eV [187]	1.12	1.39
Electronegativity[b]		
1932 Pauling [188–190]	2.5	1.8
1934 Mulliken [191–193]	2.63	2.44
1946 Gordy [194]	2.52	1.82
1958 Allred-Rochov [195]	2.60	1.90
1961 Allred [196]	2.55	1.90
1965 Voronkov-Kovalev [197]	2.46	1.89
1966 Sanderson [198]	3.79	2.62
1977 Mande-Deshmukh [199]	2.73	1.87
1981 Blustin-Raynes [200]	2.30	1.88
1983 Ohwada [201]	2.5	1.8
1983 Hargittai [202]	2.6	1.9
1983 Sanderson [183]	2.746	2.138
Hardness Parameter, eV [203, 204]	5.00	3.38
Van der Waals radius, pm		
Pauling [189]	170	217
Bondi [205]	170	210
Glidewell[c] [206]	125	155

[a] Scaled to the relative atomic mass ^{12}C; [b] in sp^3 valence states; [c] Intramolecular non-bonded atomic radius

reactant. It is believed that for a silicon atom to exhibit complex formation ability it must have at least one electronegative substituent. Nevertheless, it has been reported recently that a $[(CH_3)_3\overline{Si^{-}CH_2CH_2C}H]$ ion exists in the gaseous phase in which the silicon atom forms five Si—C bonds [212].

2.1 Structural Features

The bond angles of silicon in the molecules of its compounds are more readily distorted than the corresponding bond angles of carbon in isostructural organic analogs [213–217]. Thus, a tetrahedral molecule of silane SiH_4 (T_d) can be transformed into a planar form (D_{4h}) with a much lower energy requirement than a molecule of methane CH_4 [218–220]. A high lability of X—Si—X and X—Si—Y angles is observed not only when X is a π-donor σ-acceptor atom (F, O, N)[2] but even when X = C [222-225]. In some cases the energy needed for rearrangement of silicon bond angles can be less than the energy of an additional coordinative interaction. This must prompt the formation of structures containing a silicon atom with an expanded coordination sphere.

Semiempirical MNDO calculations indicate that the difference in energy between the tetrahedral and planar forms of bis(ethylenedioxy)silane [219] and bis(o-phenylenedioxy)silane is considerably smaller than that for the corresponding carbon

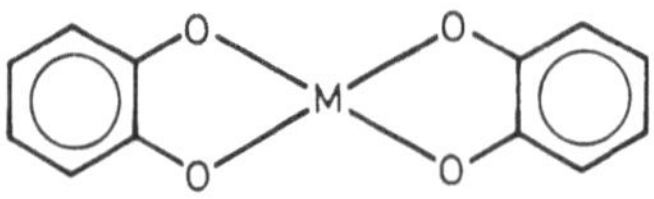

compounds. X-Ray analyses of the spiro compounds show that the M = C compounds have a slightly distorted tetrahedral arrangement about a spiro carbon atom but in the silicon analog the spiro atom has a planar environment [226]. Nevertheless, experimental evidence provided from a crystallographic symmetry argument (the space group determination) is insufficient to establish that the molecule is planar in the crystalline state [227-229]; in the crystal structure of bis(1.8-naphthalenedioxy)silane (six-membered rings) the angles at the spiro-silicon atom are tetrahedral [230]. A strong distortion arises from the incorporation of the Si atom into the five-membered rings of bis(tetramethylethylenedioxy)silane [231]. The angle of 88.3° between the planes through each ring shows that no significant planarization has occurred. However, the strong decrease in the endocyclic OSiO angles is probably the reason for the stability of the compounds of penta-coordinate silicon atom in the spiro[4.4]octane system.

Organosilicon spirocyclic compounds consisting of two five-membered rings linked through the silicon atom, readily form rather stable complexes with bases; the central silicon atom in such complexes is penta-coordinate. Thus, bis(ethylene-1,2-dihydroxy)silane forms adducts with alkali metal alkoxides [232, 233], which event is not

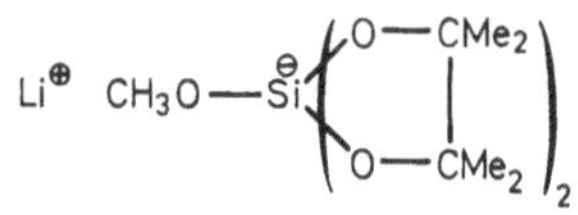

observed for the acyclic silicon analog $Si(OCH_3)_4$. The stabilization of trigonal-bipyramidal compounds of silicon relative to the corresponding four-coordinate tetrahedral species by bridging of an apical and an equatorial position with a five-membered ring has been well documented [234-236].

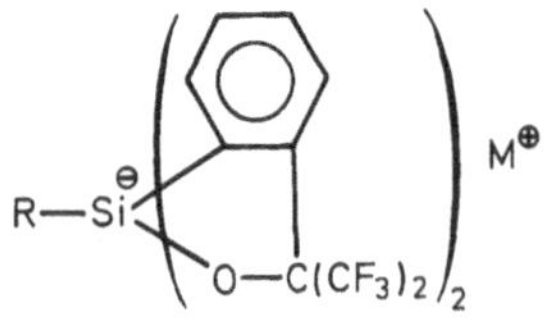

2 In contrast, the lability of carbon bond angles rises when carbon is bonded to π-acceptor σ-donor atoms (Li, Be, B, and so on) [221].

The bidentate fluoroalkoxy ligand also provides stabilization of the siliconate species by the five-membered rings and the enhanced electronegativity of the apical oxygen. The equatorial electropositive carbon ligand to the central atom enhances the difference in electronegativity between the central atom and the apical ligand and leads to a stable hypervalent bond. These ligands are exceptionally well suited to stabilize penta-coordinate silicon, but less capable for stabilizing hexa-coordinate structures [237–240].

The effect of the geometry of ligands on the coordination stability of silicon compounds is diverse and can be evaluated only qualitatively [241]. Complexes of multidentate ligands exhibit some remarkable properties, in particular an unusual stability as compared to corresponding monodentate ligands, i.e., a chelate effect [242–245]. The nature of the latter is the detailed balance between entropy and enthalpy contributions. The extra enthalpy stabilization is due to the fact that the cyclic ligand is already "prestrained", i.e., it does not need to be contracted into another conformation more suited for complex formation; this is in marked contrast to the behavior of open chain analogs. The entropy contribution arises from the smaller configurational entropy at the cyclic ligand. As to a separation of the effects, the available data are far from conclusive [244–246]. The chelate effect decreases with increasing the saturated ring size: five-membered structures are more stable than the corresponding six- or seven-membered rings [247–249].

IR [250] and NMR [251] spectroscopic data indicate that for $F_3Si(CH_2)_nOC(O)C_6H_5$ with n = 1 the intramolecular Si←O bond closing the five-membered ring exists in non-polar and polar media as well as in the crystalline and gaseous (to 150 °C) states. With n = 2, however, the six-membered cyclic form with penta-coordinate silicone is present in non-polar solvents in equilibrium with the acyclic form, whereas with n = 3 the molecules occur in only the acyclic form.

Lehn and co-workers described the increased stability of complexes by several orders of magnitude over monocyclic analogs observed upon the addition of another connecting bridge onto the macrocyclic ring to form macrobicyclic ligands or cryptands (cryptate or macrobicyclic effect) [252]. It is reasonable, therefore, to suggest that the stability of the complexes increases for the four bond formations in the following sequence:

As based on NMR [255,256] and IR [257,258] spectral data, no Si←N coordinative interaction is observed in solutions of acyclic derivatives of (2-dimethylaminoethoxy)-silane, $X_{4-n}Si(OCH_2CH_2NMe_2)_n$. In the crystal structures of monocyclic derivatives, however, such interaction does exist (Sect. 4.4.3). The stability of this bond is favored by transannular interaction of the opposite atoms, characteristic of eight-membered rings [259]. The NMR spectra [260–267], dipole moment measurements [268–270] and mass spectral data [271,272] of $R_2Si(OCH_2CH_2)_2X$ indicate a transannular Si←N interaction for nitrogen (X = NR), but no interaction for X = O.

Among the stable compounds of penta-coordinate silicon, the most studied are silatranes $X\overset{\downarrow}{Si}(OCH_2CH_2)_3N$ (tricyclic organosilicon esters of tris-(2-hydroxy-alkyl)amines, 5-aza-2,8,9-trioxy-1-silabicyclo[3.3.3]-undecanes [11,159].

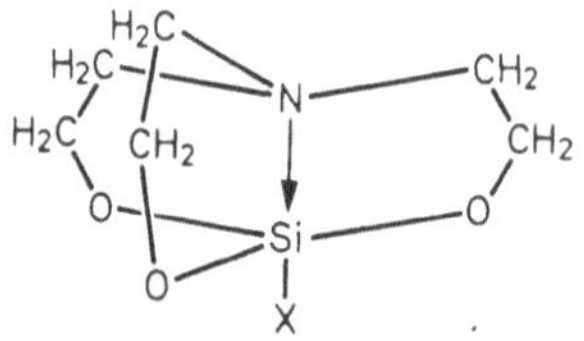

The basal nitrogen atom is linked through three $(CH_2)_2$ links to oxygen atoms at the triangular base. The tetradentate tripodal ligand occupies four coordination positions around the central atom, but for penta-coordinate species the steric requirements of the ligands are such that the trigonal bipyramidal structure results [253,254]. In these compounds there is an intramolecular donor-acceptor interaction between the nitrogen and silicon atoms.

Every fragment (OCH_2CH_2N) of the atrane skeleton can be considered as a deprotonated molecule of ethanolamine, $HOCH_2CH_2NH_2$. In the gaseous phase this molecule has a staggered conformation with respect to the C—C bond [273,274]. The arrangement of the OH and NH_2 groups corresponds to the gauche-form stabilized not only by intramolecular interaction between the most basic center (amino group) and the hydroxy group [275–278], but by the gauche effect as well [279–283]. This conformation is retained in the solution state [281,282]. In addition, a chelate effect occurs for ethanolamine complexes [283]. It should be expected that the macrocyclic and macrobicyclic effects of other hydroxyethyl substituted amines such as diethanolamine and triethanolamine are more important. In molecules of triethanolamine [284] (precursor of silatranes) and its cations $[(HOCH_2CH_2)_3NH]^+X^-$ (X = SH [284], $OCOCH_2OC_6H_4$-2-CH_3 [285] and $OCOCH_2SC_6H_4$-4-Cl [286]) a claw-shaped structure consisting of three chains is achieved. In each of the three chains the C—O and C—N bonds are in the gauche position with respect to one another, the nitrogen atom being inside the 2-hydroxyethyl groups:

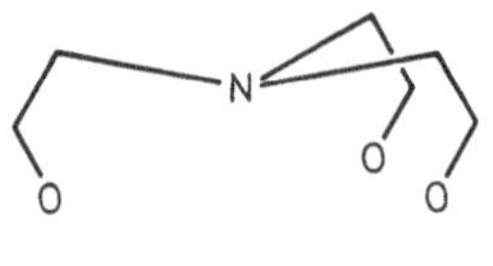

X-ray investigations have indicated a very similar structure of the free ligand as compared to their complexes with silicon (see Sect. 4.4.3) and germanium [287].

An unshared electron pair of the nitrogen atom in such a molecule is oriented inside the cage formed by three carbon-oxygen chains. A similar picture is observed in cryptands. In their molecules the unshared electron pairs of nitrogen atoms are also oriented inside the heterocyclic skeleton [289–290]. The outside orientation of the electron pair of the nitrogen atom in a cryptand [1.1.1.], $N(CH_2CH_2OCH_2CH_2)_3N$, is realized in a complex with BH_3 [291]. In an ammonium cation of cryptand[$H^+ \subset$ 1.1.1.]

the proton is located inside the cage[3] [291]. The formation of a cryptand[2.2.2.], $N(CH_2CH_2OCH_2CH_2OCH_2CH_2)N$, complex with two molecules of BH_3 in which two nitrogen atoms are inverted is accompanied with great space shifts of oxygen atoms as compared with a free ligand [293]. If in the triethanolamine molecule three oxygen atoms are bonded to the central atom not reacting with the nitrogen atom, the orientation of the unshared electron pair of the latter is retained. This is observed in molecules of $(Ph_2P)ClPtSi(OCH_2CH_2)_3N$ [294,295], $S{=}P(OCH_2CH_2)_3N$ [296], and in the gaseous phase in molecules of $CH_3Si(OCH_2CH_2)_3N$ [297]. In these molecules the configuration of the nitrogen bond is very close to planarity and the nitrogen atom is displaced only slightly inside the skeleton with respect to the neighboring carbon atoms. This structure of the nitrogen coordination polyhedron of a tripode ligand is, most likely, due to steric reasons. Consequently, the skeleton conformation of silatrane molecules plays an important role in the formation of the intramolecular Si←N bond.

Tetracyclic tetradentate ligands (porphyrins, phthalocyanines and the like) change considerably the direction of the valent bonds of the central silicon atom [298,299], transforming it into the hexa-coordinate form [300–304].

Thus, certain structural features, in particular incorporation of the central atom into a five-membered ring, chelate, macrocyclic and cryptate effects may stabilize the geometry of penta- and hexa-coordinate silicon species. Much of the insight as to which structural features might be expected to stabilize this species comes from studies of species involving higher valent states of other nonmetallic elements of the third row of the periodic table [305–307].

Indeed, the majority of the higher-covalent silicon compounds studied which have an increased coordination number are formed by bi-, tri-, and tetradentate ligands; similar effects are apparent with other Group IVA elements. It is likely that the structural factors are responsible for the existence of pentacoordinate pentavalence carbon species [308] expanding the valence shell to 10 (Sect. 3.2). For the same reason that there are about twenty structures of the organotin complexes with seven- and eight-coordination [309,310].

2.2 Medium Effect

The structure of coordination compounds can vary with a change of the state of aggregation [311]. The strength of donor-acceptor bonds, both intra- and intermolecular, increases when passing from the gaseous to the solution and then to the crystalline state, and with decreasing temperature and increasing pressure [312].

As follows from the thermodynamic data on the dissociation of SiF_4 adducts, the crystal lattice energy of the complex is much higher that the enthalpy of their formation in the gaseous phase [148]. Complexes of organyltrichlorosilanes with tertiary amines exist only at low temperature and dissociate into free components at room temperature [313]. The well studied complexes $X_4Si \cdot 2\,Py$ (X = F, Cl, Br) are dissociated completely in the gaseous phase [314]. The anomalies of concentration and temperature dependencies in the IR-spectrum of $SiCl_4$ are caused by intermolecular inter-

[3] Cryptand [1.1.1.] is thermodynamically very strong and a kinetically extremely slow base [292].

action in the solution state [315]. The enthalpy of dimerization 2 $SiCl_4 \rightleftarrows (SiCl_4)_2$ in the gaseous phase is 1.4 kcal · mol^{-1} which exceeds the enthalpy of van der Waals intermolecular interaction (0.5 kcal · mol^{-1}) [316]. Association of $SiCl_4$ molecules in the liquid phase is also evidenced from the cryoscopic studies [317, 318].

In crystals of dimethylsilylamine (at −120 °C) and 1-methylsilatrane, the silicon atom is penta-coordinated [154, 156]; whereas in the gas phase the structure of the silicon coordination polyhedron is close to tetrahedral [319]. Intermolecular non-bonded Si ... D interactions are observed only in crystals of disiloxane and isocyanatosilane [320, 321]. The interactions are strong enough to affect the mutual orientation of the molecules, but too weak to change their geometry.

The nature of the solvent as well as its aggregate state considerably affects intermolecular interactions in silicon coordination compounds. Thus, in the IR spectra of crystalline complexes of $OPPh_3$ and $OP(NMe_2)_3$ with silicon tetrachloride, the P=O bond stretching vibration frequency, ν(PO), is by 45 cm^{-1} lower than that of the free bases [322]; in the spectra of their diluted solutions (both in polar and non-polar solvents), however, the frequency is the same [323, 324]. In crystals of (aroyloxymethyl)trifluorosilanes and in their solutions (with low ε of the solvent) there is an intramolecular Si←O=C interaction. In pyridine solution, however, it strongly weakens and is completely absent in the gaseous phase (the frequency ν(CO) decreases by 60 cm^{-1}) [325–328]. Adduct formation of trialkoxysilanes $(RO)_3SiH$ (R = CH_3, C_2H_5, C_3H_7) with 18 solvents was evaluated from the very large (>40 cm^{-1}) change in Si—H IR stretching mode [329]. The integrated intensities of ν(SiH) bands indicate that the interaction of triethylsilane with electron-donating solvents is of the donor-acceptor type [330].

Measurements of the ^{15}N and ^{23}Si chemical shifts (see Sect. 4.5.5) and IR absorptions [331] of silatranes and 2.3-dioxa-6-aza-2-silacyclooctanes in various aggregate states have shown that the extent of Si←N coordination increases in passing from the gaseous to the crystalline state as well as with deacreasing the temperature of solution their polarity, polarizability and electrophilicity (see Sect. 4.5.5). The temperature dependence of ^{29}Si chemical shifts of stereochemically nonrigid intramolecular silicon complexes has been explained in terms of the existence of equilibrium between tetra- and penta-coordinate forms [263, 266, 332] or the substituent position exchange in the trigonal bipyramidal stereoisomers [261, 264].

In all silicate minerals formed under crustal conditions silicon is coordinated to four oxygen atoms. In high-pressure transformations, silicon commonly increases its coordination number. The longer the Si—O distances in tetrahedral silicates the higher the pressure transformations to phases with octahedral silicon. The average Si—O bond distance for the pressure transformation is 159 pm. This distance is achieved at room temperature at pressures in all measured silicates and may be a minimum for tetrahedryl Si—O bonds; 300 kbar is an upper pressure limit for the silicon tetrahedron and 80kbar is a lower pressure limit for octahedral silicon. Temperature has little effect on Si—O bond distances in either tetrahedra or octahedra [333].

Thus, in cases when weak interactions can affect the molecular geometry, the packing of the molecules in the crystal lattice and the solvent effect acquire an important role. In solutions a structure may exist which differs from that achieved in the gaseous phase under the action of intramolecular forces only.

3 Theoretical Concepts of an Increase in the Silicon Coordination Number

Since an interaction between two molecules generally involves charge transfer, it may be considered as a donor-acceptor interaction in the widest sense of the word [334-339]. The formation of coordination donor-acceptor bonds between acceptors with low-lying vacant AOs and donors with accessible electron pairs is typical of the majority of elemnts of the periodic system, e.g., the silicon atom can form more bonds than appears to be allowed by the octet rule [340-342]. An eminent problem in the theory of electronic structure involves the role of d orbitals in the ground states of second row elements.

3.1 Participation of d Orbitals

The clearly pronounced acceptor ability of the tetravalent silicon atom, due to which its valence shell may contain 10 or even 12 electrons, as well as similar structures of compounds of second-row elements containing a penta- and hexa-coordinate central atom and of the corresponding compounds of transition metals[1] is usually interpreted on the assumption that the bonding process involves not only s and p AOs but vacant 3d AOs localized within the valence shell [55, 128]. The silicon atom has five vacant 3d AOs the participation of which can, in principle, lead to penta- and hexa-coordinate states. The trigonal bipyramid and tetragonal pyramid are the most symmetric configurations of a SiX_5 molecule. The formation of five Si—X bonds in a trigonal bipyramid requires, according to the above concepts, the participation of $3d_z$ AO, and in a square pyramide of $3d_{x^2-y^2}$ AO (sp^3d-hydridization). In both cases two sets of non-equivalent bonds arise. Thus, s^xpd^{1-x} and $s^{1-x}p^2d^x$ hybridizations for axial equatorial positions in a trigonal bipyramid occur [343, 344].

In octahedral silicon complexes, two AO (d_{z^2}, $d_{x^2-y^2}$, e_z symmetry) participate simultaneously in the σ-bonding (sp^3d^2-hybridization). The d_{xy}, d_{yz} and d_{xz} orbitals with t_{2q} symmetry can be used for π-bonding with the appropriate orbitals of the substituents [345, 346]. It was put forward that an additional donor-acceptor interaction which increases the coordination number of the second row Main Group element involves continuum of the energetic states lying above the ionization potential [347].

Although the possibility of participation of vacant 3d AOs in chemical bonding in compounds of non-transition elements has been widely studied by quantum chemistry [348-351], it is now open to question whether or not they really contribute to the formation of chemical bonds or their role is reduced to an additional polarization effect. It was doubted that the role of 3d AOs of the Period III elements (Si, P, S) in increasing their coordination number or in $(p\text{-}d)_\pi$ interaction should be taken into account [352-354]. It was stated, for instance, that in some states of the atoms of the Period III elements 3d AOs were so diffuse that they could not participate

1 It is generally accepted that geometry and the bonding nature in complex compounds of transition metals are determined by hybridization of AOs of the central atom, including filled 3d AOs.

noticeably in the bonding. In order describe the physicochemical properties of compounds of the Group IV A elements, instead of the concept of $(p\text{-}d)_\pi$ conjugation, attention is more and more directed to the ideas of hypervalent bonding, which suggests that a good first order representation of the structures of SiF_5 species may be constructed without an appreciable contribution from the silicon 3d orbitals.

3.2 Hypervalent Bonding

In contrast to 3d orbital approaches, Rundle proposed that of electron-rich three-center bonding [355]. The concept of hypervalent bonding or valence-shell expansion, initially designed to explain the nature of bonds in polyhalide ions and rare gas fluorides [355], was further developed for other orbital-deficient compounds and is given in the most generalized form by Musher [356]. When higher row atoms "expand their valence by adding ligands, they add them colinearly along the axis of one of the (previously unshared) pairs of p electrons", thereby forming a three-center four-electron system. Hypervalent bonds in coordinate silicon compounds differ from similar bonds of sulfur, phosphorus, and chlorine since the lone electron pair is, in this case, supplied by the ligand.

The bonding in SiF_5^- is electron rich, involving two more electrons than are normally accommodated in the valence shell of the central silicon. The formation of the axial fragment F—Si—F in SiF_5^- involves silicon $3p_z$ AO. The bonding molecular orbital along the F—Si—F axis is formed by the phase overlap of the central atom p_z orbital and a p orbital from each fluorine. From four valence electrons that are available, two occupy the bonding three-centre MO and two an approximately nonbonding MO. Thus, one may expect that the three-center four-electron axial Si—F bond is of a lower order of magnitude than the covalent two-electron bond. The experimental data agree well with these assumptions: the axial SiF bonds in SiF_5^- are longer and more polar than the equatorial bonds.

Hypervalent bonds in compounds of hexacoordinated silicon are formed in a similar fashion. Thus, in the anions of the SiF_5^- and SiF_6^{2-} type there are two three-center four-electron bonds:

Hypervalent MOs are formed by mixing of bonding and nonbonding AOs which results in a shift of the electron density from the central silicon atom to the ligands. This event increases the positive charge on the silicon and negative charge on the fluorine atoms. In the formation of a hypervalent bond with the silicon atom both similar and different ligands can participate. The following conditions of the formation of hypervalent bonds have been stated [355, 356]:

1. These bonds can be formed when ligands are more electronegative than the central silicon atom.

2. The most electronegative substituents of those surrounding the silicon atom participate in a hypervalent bond or tend to occupy the axial positions in a trigonal-bipyramidal molecule.
3. The Si—X bond length in penta- and hexa-coordinate silicon compounds is longer then that in similar tetrahedral molecules.
4. The concept of hypervalent bonding is assumed to decrease the electron density on the central silicon atom upon complexation.

In spite of some critical remarks [357], the hypervalent model is successfully used for interpreting the results of physiscochemical investigations [358,359], for example, for silatranes [360,361]. For organosilicon complex anions of the SiX_5^- type both models predict (in a good agreement with experimental data) the existence of two types of bonds: ordinary covalent with a more pronounced s-character, and weakened with an excess electron density.

The much discussed alternative three-center four-electron and d orbital hydridization schemes are not mutually exclusive and do not exclude other models [357]. The same trends are found in the molecular orbital theory approach considering the relative energy separation of the 3s to 3p atomic orbitals on the central atom [362]. Most of the theoretical treatments suggest relatively small bonding contributions from d orbitals and outer s orbitals. On that basis hypervalent species might exist with central atoms from the first row atoms in which low-lying d orbitals are not available. The considered electron-rich bonds control the structure of coordinate compounds originating from central atom-ligand interactions. On the other hand, closed systems of atoms in the form of clusters or cages are associated with direct ligand-ligand interactions and different considerations will apply.

3.3 Hypermetallic Bonding

A wide variety of molecules appear to form more bonds than they have electron pairs (i.e. the electron-deficient compounds) and seemingly violate the Lewis definition of an electron pair bond. These molecules necessitate the involvement of two-electron multi-center bonding with atoms (always a metal) containing more low-energy orbitals. A large number of MX_n molecules comprised of first row elements [363–365] including carbon [365–374] fall into this category. In carbide clusters such as $Fe_5(CO)_{15}C$ [375], $Ru_6(CO)_{17}C$ [376], or $Fe_6(CO)_{16}C^{2-}$ [377] the interstitial hole of the clusters is occupied by a C atom with a coordination number 5 and 6. Crystal structures of $Li_8O[C_6H_3\text{-}2,6\text{-}(OMe)_2]_6$ [378,379] and $Cu_4(C_6H_4\text{-}2\text{-}CH_2NMe_2)_4$ [380] contain three- and four-center two-electron bonded C(aryl) atoms. In any case, the "expanded" bonding to carbon is found with the 3–21G basis, which has no d functions. Thus, the CLi_5 and CLi_6 molecules exhibit a hypermetalated bonding, but not hypervalency in the strict sense. The extra electrons (beyond the usual octet) are not associated with carbon but involved with metal-metal bonding [381,382]. Hypermetalation should be a remarkably general phenomenon that can involve not only first [383] but also second row elements. The models of electronic structure which have proved to be useful for molecules of first row elements are likely to be more readily adapted for use with molecules of second row elements. The Si sp^3 (and possible sp^2) orbitals are suited for multi-center bonding.

4 Structure of Silicon Compounds with Coordination Numbers 5 and 6

In the course of the formation of silicon complexes, the charge density patterns reorganize. Although the so-called partial charge of an atom in a molecule can not be defined unambiguously, comparisons of the results of quantum chemical calculations for reactants and reaction products as well as the consideration of the bond length valence angle variations and the NMR parameters of the five- and six-coordinate silicon compounds have proved to be of great qualitative value to chemists.

4.1 Stereochemistry

From a geometrical point of view it is impossible to place five points in equivalent positions on a spherical surface excepting a planar arrangement. There are two types of favored structure, the trigonal bipyramid (TBP) with D_{3h} symmetry and the square (or tetragonal) pyramid (SPY) with C_{4v} symmetry:

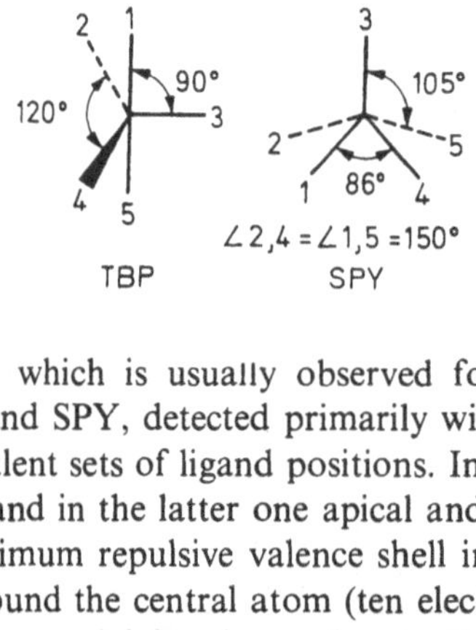

Both TBP, which is usually observed for acyclic penta-coordinated main group elements, and SPY, detected primarily with transition metals derivatives, have two non-equivalent sets of ligand positions. In the former there are three equatorial and two axial, and in the latter one apical and four basal substituents.

The minimum repulsive valence shell interactions[1] between five identical ligands located around the central atom (ten electron silicon shell) corresponds to the TBP in which two axial bonds are longer than three equatorial [385, 386]. The trigonal bipyramid is calculated to be about 8% more stable than the square pyramid, and only slight bond-bending of the TBP structure is necessary to attain the SPY structure. The presence of two unsaturated five-membered rings in spirocyclic derivatives is a general principle in forming a square pyramid for most penta-coordinated main group elements including silicon [387]. The first examples of the SPY structures of penta-coordinate silicon have been obtained only recently by X-ray diffraction for a spirobicyclic derivatives[2] [161–163].

---

1 The repulsive force is proportional to r^{-n}, where $n = 8$–12 [384].

2 Recently, the square pyramidal geometry for anionic germanium [388] and tin [389, 390] species have been detected by X-ray diffraction.

For the TBP molecules of the $MX_{5-n}D_n$ type the following achiral isomers are possible:

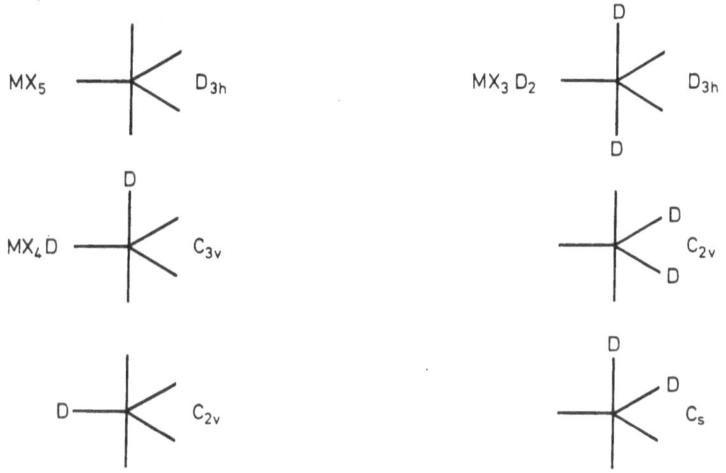

The most electronegative substituents are always located in the axial position. In this respect there is no difference between compounds of penta-coordinate silicon and isoelectronic phosphorus compounds. Thus, in adducts of the type $SiCl_3X \cdot NMe_3$ (X = F, Cl) the substituent X is always in the axial position (C_{3v} symmetry), but if X = H the molecular symmetry changes to C_s since the hydrogen atom is always in the equatorial plane [313]. In solution of $[(C_6H_5)_3P]_2Pt \cdot SiF_4$ the molecules with an equatorial platinum-containing electropositive ligand (C_{2v} symmetry) are predominant [391]. It was believed that the $N(CH_3)_3$ ligand always occupies an axial site in a trigonal bipyramid due to steric interaction [391, 392]. Ab initio MO calculations on $SiF_4 \cdot NH_3$ employing complete geometry optimization and additional polarization functions on Si and N have shown a trigonal bipyramidal geometry with axial NH_3 to be the most stable arrangement and the nitrogen atom to be more negatively charged than F [393].

In the case of SiX_4D_2 complexes, unlike SiX_6 (O_h) and SiX_5D (C_{4v}), there are two possible isomers: trans (D_{4h}) and cis (D_2).

In octahedral acyclic complexes of the $X_4Si \cdot 2D$ type, the D ligands usually occupy the trans position though in some cases geometrical factors can cause cis orientation of the ligands (for instance, $F_4Si \cdot$ bipy [394]). A rule has been formulated [134] according to which in complexes of the $X_4Si \cdot 2D$ type the D ligands occupy the cis position if their special requirements are less than that of X; otherwise, they occupy the trans position. The structure of silicon complexes, however, can be affected by the $(p\text{-}d)_\pi$ bonding in addition to other factors. As quantum-chemical calculations have shown [314], for $X_4Si \cdot 2\,Py$ with X = F, Cl, the cis configuration is prefered with X = F provided that the $(p\text{-}d)_\pi$-bonding was taken into account; and with X = Cl if only electrostatic interactions were considered.

4.2 Stereodynamics

High conformational mobility of the molecules is a specific feature of penta-coordinate species of transition and main group elements including silicon [395–397]. Thus, at room temperature all fluorine atoms of SiF_5^- are equivalent on the NMR time scale which points to inter- or intramolecular exchange processes. Since in ^{19}F NMR spectra the ^{29}Si satellites are observed (148 Hz) [398], the magnetic equivalency of fluorine nuclei cannot be realized at the expense of fast inter-molecular exchange involving the rupture of Si—F bonds. Among the great number of mechanisms proposed for intramolecular rearrangements in penta-coordinate complexes [399–401] the Berry pseudorotation postulated in 1960 [402] is most widely accepted:

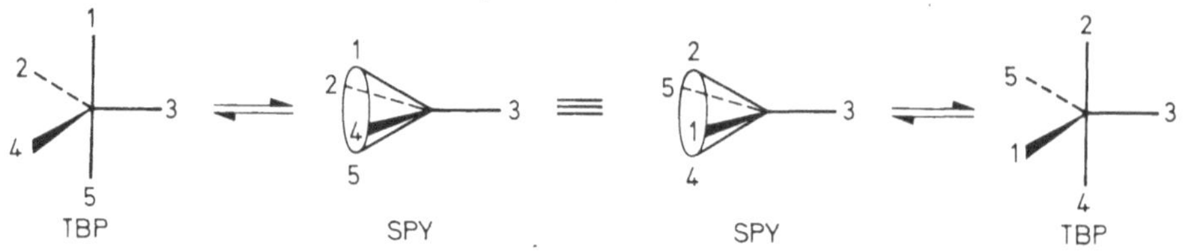

For this mechanism to be realized, it is required that the energies of trigonal bipyramide and square pyramide be relatively low. Indeed. due to weak interaction between ligands in acyclic coordination compounds, the values of energy barriers for such processes do not exceed 6–12 kcal · mol^{-1} [403]. In the process of Berry rotation, one equatorial position (3) is fixed whereas the two others are exchanged with two axial positions. Therefore, for the anions SiX_5^- and $RSiX_4^-$ the pseudo-rotation barriers must be of similar magnitude since the R substituent as pivotal ligand 3 does not participate in this process. Berry rotation, with the point of rest on an electropositive substituent, corresponds to energetically more favorable apical position in the square pyramidal structure [404].

The turnstyle mechanism is another possible way of an intramolecular ligand exchange. In the process of a turnstyle motion one of the axial ligands and one of the equatorial ligands exchange their places as if rotation about the local axis were of second order, the remaining three atoms rotating about the third order axis in the opposite direction [399]:

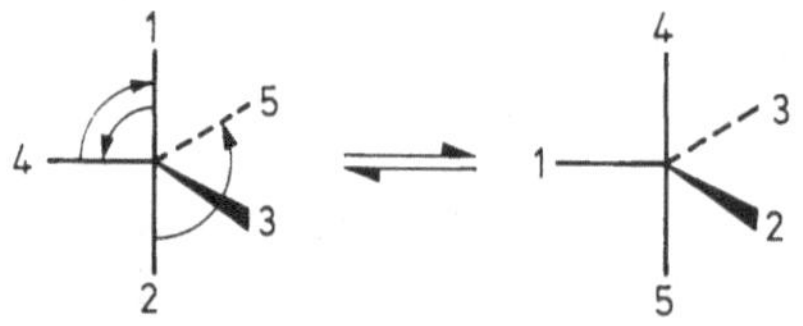

For intramolecular exchange processes in spirocyclic systems this mechanism prevails but it is energetically less favorable than the previous one.

Among other rearrangement mechanisms, mention should be made of the so-called "trigonal twist" and low energy inversion transitions between two structures with close energy values.

Both experimental data and theoretical calculations show that the transition of a trigonal bipyramidal into a square pyramidal structure for compounds of penta-coordinate silicon does not require high energy consumption. Up to now the attempts of experimental separation of diastereomers of organosilicon compounds with penta-coordinate silicon atom were unsuccessful. For spirocyclic bisacetylacetonates, $XYSi(Acac)_2$, containing a hexa-coordinate silicon atom, however, geometric isomerism has been observed [405–409].

4.3 Quantum-Chemical Calculations

As was already mentioned, the silicon coordinate bonds have energies much lower than those of normal covalent bonds (> 50 kcal · mol^{-1}). Therefore, it is more difficult to characterize such bonds (the equilibrium geometry of the complex depends greatly on various factors) [410,411]. Three different contributions considered in quantum chemical calculations are the effects of electrostatic interaction, charge transfer and, to a smaller extent, polarization; these cannot be unambiguosly separated. It was assumed that the main component of the donor acceptor bond is electrostatic interaction rather than charge transfer [412,413].

Some simple rules were supported by empirial evidence, valence shell electron pair repulsion model (VSEPR) and MO calculations, both semiempirical and ab initio. These rules could explain those features of molecular geometry which have been characterized by structural investigations using spectroscopic and diffraction techniques.

Based on a currently popular VSEPR approach[3] the C—F bonds in a hypothetical CF_5^- ion (without steric hindrances) must be about 16% longer than in the tetrahedral CF_4. At the same time, for the anion SiF_6^{2-} to be stable, an increase in the Si—F bond lengths must be as small as 11% [416]. Thus the existence of the silicon atom with an expanded coordination sphere is much facilitated due to its greater size. The repulsion of non-bonded atoms in this latter case does not result in additional hindrances to the increase of the number of atoms to five or six.

The VSEPR model works at its best in rationalizing ground state stereochemistry but does not attempt to indicate a more precise electron distribution. The molecular orbital theory based on 3s and 3p orbitals only is also compatible with a relative weakening of the axial bonds. Use of a simple Hückel MO model, which considers only σ orbitals in the valence shell and totally neglects explicit electron repulsions can be invoked to interpret the same experimental results. It was demonstrated that the electron-rich three-center bonding model could explain the trends observed in five-coordinate species [417]. Various MO models of electronic structure have been proposed to predict the shapes and other properties of non-transition element

[3] In particular, for the siloxane molecules the increase of Si—O—Si bond angles can be explained in terms of the VSEPR method or "one angle non-bonded radii" by the increasing of the repulsive interactions between the shortened Si—O bonds. Recent semi-empirical (CNDO/2)[414] and non-empirical [415] calculations taking into account neither $(p - d)_\pi$-interactions nor additional repulsion of non-bonded atoms attribute the anomalously high oxygen bond angle to a more pronounced ionic character of the Si—O bond than based on the difference in electronegativity between Si and O.

complexes [418–422]. However, the VSEPR approach still provides excellent rules with which to quickly view molecular shapes.

Many quantum chemical calculations were performed for organosilicon compounds in order to study intermediate products of biomolecular S_N2 substitution reactions [393, 404, 423–436]. In some cases the calculated energy values correspond to the experimental data obtained in the gas phase [437].

There is no experimental observation of an SiH_5^- ion, but quantumchemical calculations have shown that SiH_5^- should be more stable with respect to $SiH_4 + H^-$ (for D_{3h}) by 17–20 kcal · mol^{-1} and unstable with respect to $SiH_3^- + H_2$ [404, 423, 424]. Electron correlation has a pronounced effect on this energy, accounting for roughly 6 kcal · mol^{-1} [424]. For the $FSiH_3 + F^-$ and $SiF_4 + F^-$ systems and the complexes $FSiH_3F^-$ and SiF_5^- the energy difference is 50 [423] and 230 kcal · mol^{-1} [425], respectively. The acceptor ability of the penta-coordinate silicon atom is less pronounced than that of tetrahedral atoms. As follows from calculations, the energy of the reactions $SiF_5^- + F^- \rightarrow SiF_6^{2-}$ and $2\,SiF_5^- \rightarrow Si_2F_{10}^{2-}$ is 80 and 120 kcal · mol^{-1}, respectively [425]. For the anion H_4SiX^-, where X = H, NH_2, OH, or F (ab initio calculations using the STO 4LGTO basis set), the energy of the complex is less than the sum of the energics of the unbounded components; whereas for the corresponding carbon analogues H_4CX^- the situation is quite opposite [426]. For a neutral complex $H_4Si \cdot NH_3$, a negative value of the complex formation energy was obtained [427].

An increase in the coordination number of the silicon atom results in lengthening of all bonds (Table 2). In the penta-coordinate SiH_5^- and $F_2SiH_3^-$ anions the axial Si—H and Si—F bonds are by 10% and 6%, respectively, longer than similar bonds in SiH_4 and SiH_3F. In the hexa-coordinate $HOSiF_5^{2-}$ anion the equatorial SiF bond (197.0 pm) is longer than axial (192.5 pm) and the axial Si—O bond (188.4 ppm) is much longer than similar bonds in the tetrahedral molecule of silanols. The lengthening of silicon bonds with an increase of the coordination number from four to five and then to six is accompanied with an enhancement of an additional negative charge on the ligands (Table 3). The concentration of electron density on the axial fragments of a trigonal bipyramid is enhanced and favors the predominant axial arrangement of the most electronegative substituents [441].

Table 2. Calculated Bond Distances (pm) for Tetra-, Penta-, and Hexacoordinate Silicon Compounds

X	Y	SiX_3Y	$[SiX_3Y_2]^-$, D_{3h}		$[SiX_6]^{2-}$	Method	Ref.
			eq	ax			
H	H	147.7 (148.0[a])	153	162		ab-initio	[423]
			154	159			[424]
			162	164		CNDO/2	[404]
H	F	146.7 (147.4[a]) (H) 159.4 (159.4[a]) (F)	152 (H)	168 (F)		ab-initio	[423]
F	F	187.5 (156[a])	189.0	190.0	193.0[b]	CNDO/2	[425]

[a] Experimental [55, 438–440]; [b] See Table 5.

Table 3. Calculated Mulliken Charges on the Atoms of Simple Silicon Complexes

Molecular	Central Atom	Ligand	
	Si	eq	ax
CNDO/2 [425]			
SiF_4	1.214	−0.304	
SiF_5^- (D_{3h})	2.028	−0.399	−0.415
SiF_6^{2-} (O_h)	2.940	−0.490	
SiF_5OH^{2-} (C_{4v})	2.928	−0.502 (F)	−0.482 (F)
Ab initio [423]			
SiH_4	0.63	−0.16	
SiH_5^- (D_{3h})	0.84	−0.29	−0.49
SiH_3F	1.10	−0.15 (H) −0.67 (F)	
$SiH_3F_2^-$ (D_{3h})	1.26	−0.26 (H)	−0.74 (F)
Ab-initio [428]			
SiF_4	1.434	−0.358	
NH_3		−0.509 (N) 0.169 (H)	
$SiF_4 \cdot NH_3$ (C_{3v})	1.470	−0.397 (F)	−0.385 (F) −0.531 (N) 0.205 (H)
$SiF_4 \cdot 2\,NH_3$ (D_{4h})	1.463	−0.463 (F)	−0.533 (N) 0.243 (H)

For the $SiF_4 \cdot NH_3$ adduct formation the electron density transferred from NH_3 to SiF_4 originated on the H atoms and the N in the complex is more negatively charged than either N in NH_3 or F in $SiF_4 \cdot NH_3$ [428]. Viewing Table 3 one finds that in all cases the Si atom becomes more positive in complexes. The hydrogen and fluorine atoms bonded to silicon acquire additional negative charges. Upon forming the SiH_5^- and $F_2SiH_3^-$ complexes the Mulliken population analysis of wavefunctions gives a decrease of charges on the silicon atom by 0.21 and 0.16 e, respectively[4]. These results have been called the "spillover effect" of negative charge at the acceptor atom [446].

The role of 3d AOs in the formation of silane and difluorosilane adducts has been examined. It has been shown that, although these orbitals do not play an essential role, their participation considerably increases the stability of SiH_5^- and $F_2SiH_3^-$. This is especially pronounced in the case of fluorosilane complexes. Participation of 3d AOs decreases the energy of the system with penta-coordinate silicon stronger than that of the system with tetra-coordinate silicon.

The factors determing the arrangement of π-donor or π-acceptor substituents in molecules or penta-coordinate silicon compounds are not yet sufficiently studied.

4 Similar results were calculated for the corresponding carbon compounds by the SCF molecular orbital method in the Gaussian approximation. It was found that the electron density on the carbon atom in CH_5^- and $F_2CH_3^-$ is by 0.32 and 0.12 e lower than that in CH_4 and FCH_3 [442–445].

For π-acceptors the axial position in trigonal bipyramidal complexes is prefered (with neglect of 3d AOs) and for π-donors the equatorial one. Introduction of 3d AOs into calculations complicates the development of certain conclusions since a need arises to take into account such factors as $(p\text{-}d)_\pi$-interaction, additional stabilization of axial and equatorial bonds, and so on. It is widely believed that π-interaction is more effective in axial than in equatorial position, although there are data available which point to the reverse.

In order study the Berry rotation process the energies of square pyramidal and trigonal bipyramidal structures have been calculated. A small difference in the energies confirms that the Berry rotation proceeds readily. For SiH_5^-, for example, the difference in the energies of trigonal bipyramid and square pyramid ranges from 1.5 to 3.2 kcal · mol^{-1} [404,424]. The effect of subsequent substitution of more electronegative ligands for hydrogen atoms in SiH_5^- on the Berry rotation barrier value was considered (neglecting the steric effects and π-interaction)[404]. Introduction of 3d AOs into the basis does not drastically affect the final data but decreases the barrier between the ground state and the transition complex.

4.4 X-Ray and Electronography Data

The most reliable data on the specific intra- and intermolecular interactions are provided by X-ray diffraction. Since the energy of a donor-acceptor bond exceeds that of van der Waals interaction, interatomic distances less than the sum of van der Waals radii forming this bond may serve as a criterion of complex formation. Such a criterion is the most convincing evidence for the presence or absence of the interaction between atoms. These interactions (along with the van der Waals ones) change the geometry and lattice packing of molecules in a distinct manner. Therefore, structural elucidation methods are extremely important for studying the nature of chemical bonds in complexes.

In many silicon systems the molecular structure shows that a molecule defined by several short internuclear distances (bonds) around the central atom often contains several other central atom-ligand distances that are much longer than normal bonded contact but shorter than the sum of the relevant van der Waals radii. The van der Waals radii reported by different authors do not agree well (Table 1). They are, according to Pouling, identical to the anionic radii of these atoms. The data obtained by Glidewell allow an estimate of the minimal intramolecular contacts of not interacting M ... M′ atoms in the MXM′ fragment (one-angle radius). The value of 210 pm for silicon as van der Waals radius (estimated from the density of pure silicon and of liquid SiH_4 by Bondi [205]) is taken to further discussion. It should be noted that the van der Waals radii are usually determined within a limited accuracy and depend on the state of hybridization.

An analysis of intra- and intermolecular contacts indicates that the silicon atom enters into coordinate interaction with not only the atoms of the second (N, O and F) and third (P, S and Cl), but also the fourth (Se) period (see below). The Si←N and Si←O bonds in crystals are longer than normal covalent bonds (176 and 168 pm, respectively) but much shorter than the sum of van der Waals radii of the silicon and nitrogen (365 pm) or oxygen (360 pm)[4] atoms. The greatest known intermolecular Si . . . N and Si . . . O interactions which order the crystal lattice and affect only slightly the intramolecular geometry are 366 pm [207] and 346 pm [447], respectively (Sect. 4.4.5). These values are close to the sum of van der Waals radii[5]. The shortest coordination Si←N and Si←O bonds are found in $(F_3SiNPMe_3)_2$ (185.7 pm) [449] and $(CH_3)_2SiClCH_2NRC(O)CH_3$ (191.8 pm) [450, 451].

Table 4. Bond Distances (pm) for Octahedral Silicon (Neutral Complexes)

Compounds	Positions		Ref.
	ax	eq	
trans-$Cl_4Si \cdot 2\,PMe_3$	226 (2 P)	220 (2 Cl) 230 (2 Cl)	452)
trans-$F_4Si \cdot 2$ Py	193 (2 N)	164 (4 F)	453)
cis-$F_4Si \cdot$ Bipy	165.4 (F) 165.9 (F)	162.9 (F) 163.2 (F) 197.2 (N) 198.2 (N)	394)
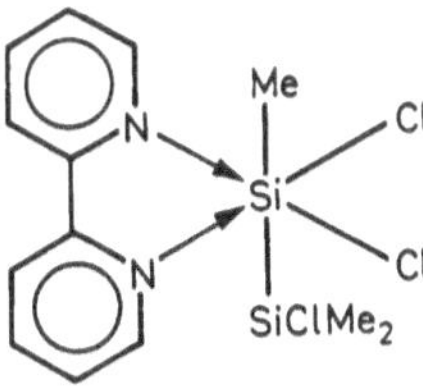 	188.8 (C) 236.7 (Si)	200.7 (N) 202.9 (N) 227.4 (Cl) 239.2 (Cl)	454)
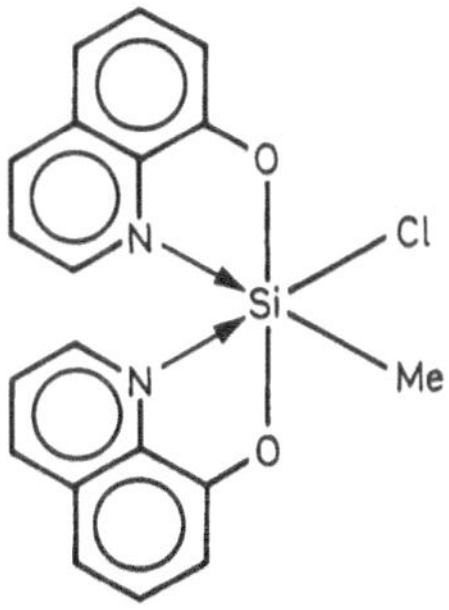 	176.3 (O) 176.6 (O)	201.4 (N) 201.6 (N) 219.9 (Cl) 221.7 (Cl')[a] 194 (C) 183 (C')[a]	455)
$PcSi(OSiMe_3)_2$[b]	167.8 (O) 168.0 (O)	191.5 (N) 191.8 (N) 192.2 (N) 192.4 (N)	456)
$(Me_3SiO)_2MeSiO(PcSiO)_3SiMe(OSiMe_3)_2$	166.2 (20)		457)
$(PcSiO)_n$ polymer[c]	165.9 (20) 166.5 (20)		458) 459)

[a] Correspond to second chiral isomeric form with the same SiO and SiN bond length; [b] Pc = phthalocyanine; [c] From X-ray powder photographs

[5] Minimum non-covalent contacts for nitrogen and oxygen atoms are usually greater than the van der Waals radius [448].

Non-covalent interaction of silicon with an electron-donor atom can result in an intramolecular coordination with the formation of a 3-, 4-, 5-, or 6-membered ring structure. The geometry of the central silicon atom with an expanded valence shell is determined by a compromise between rehybridization of the atom due to coordinate interaction Si←D and optimum non-valence (steric) interaction of ligands in the

Table 5. Bond Distances (pm) for Silicon Compounds of Higher Coordination Number (Ionic Complexes)[a]

Compounds	Position		% (TBP-SPY)[b]	Ref.
	ax	eq		
Anionic Pentacoordinate				
$[SiF_5]^{\ominus}[IrH_2CO(PPh_3)_3]^{\oplus}$	166 (F) 172 (F)	168 (F) 171 (F) 172 (F)		460)
$[SiF_5]^{\ominus}[PhCH_2NMe_3]^{\oplus}$	164.6 (2 F)	160.2 (2 F) 157.9 (F)		461)
	166.8 (F) 167.0 (F)	159.7 (F) 160.6 (F) 187.1 (C)	9	462)
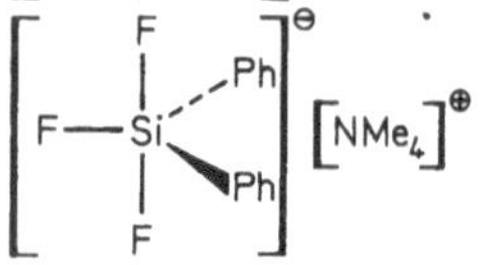	168.8 (2 F)	164.8 (F) 189.3 (2 C)		461)
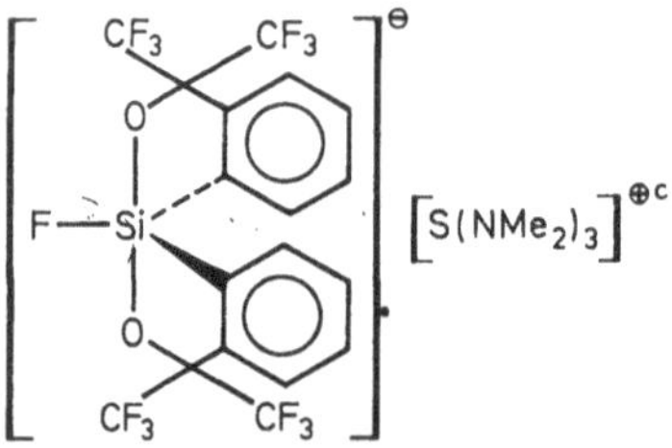	178.7 (O) 179.2 (O) 178.2 (O) 180.6 (O)	163.1 (F) 168.9 (C) 188.3 (C) 163.2 (F) 187.3 (C) 188.9 (C)	28.7 25.4	463)
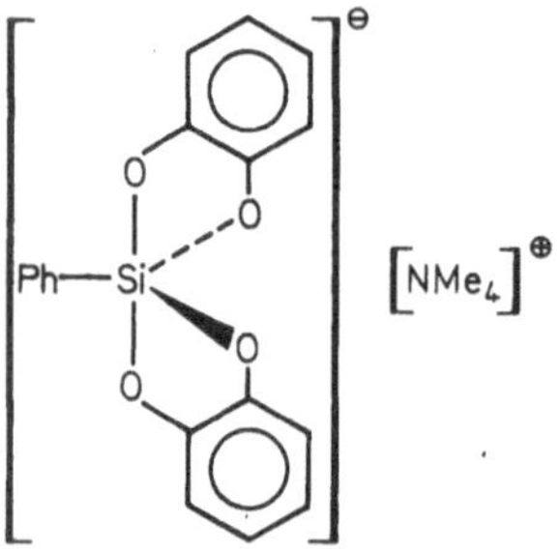	179.4 (2 O)	170.0 (2 O) 188.8 (C)	29.5	464)

Table 5. (Continued)

Compounds	Position ax	Position eq	% (TBP-SPY)[b]	Ref.
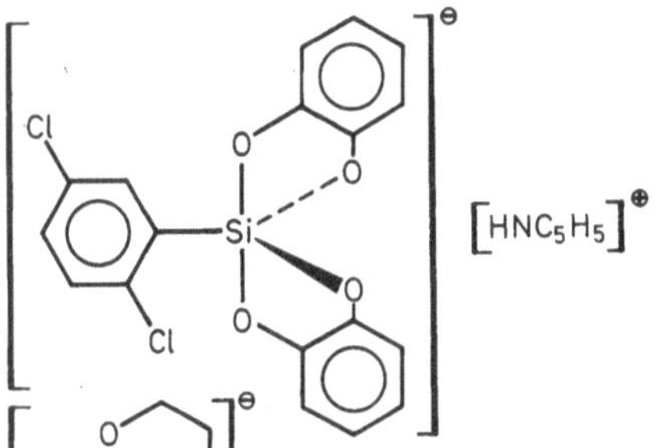	175.1 (O) 182.0 (O)	169.8 (O) 171.0 (O) 188.8 (C)	33.2	163)
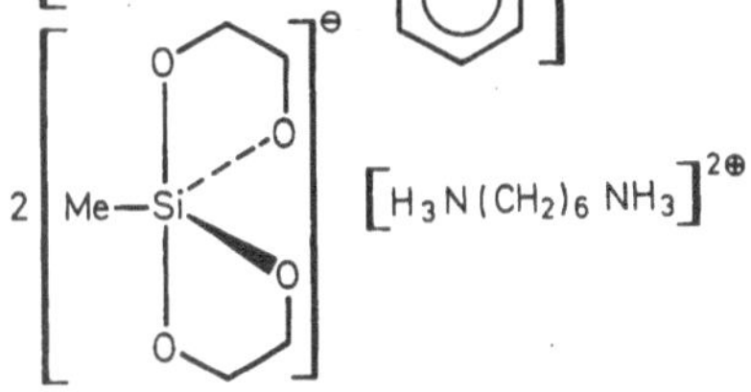	174.9 (O) 176.9 (O)	171.1 (O) 173.5 (O) 187.6 (C)	53.3	465)
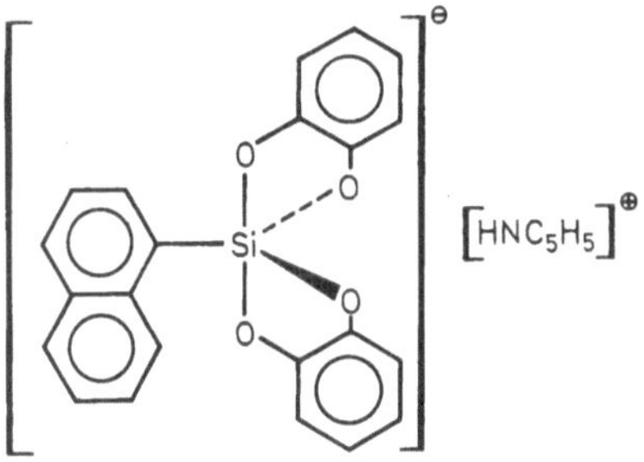	173.4 (O) 177.2 (O)	172.2 (O) 174.2 (O) 187.5 (C)	58.7	163)
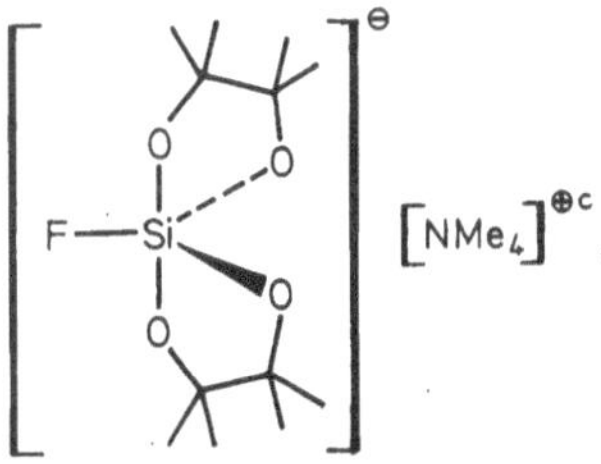	168.9 (O) 173.4 (O)	168.3 (O) 168.4 (O) 164.8 (F)	52.3	466)
	170.7 (O) 173.2 (O)	168.8 (O) 169.2 (O) 164.2 (F)	69.1	
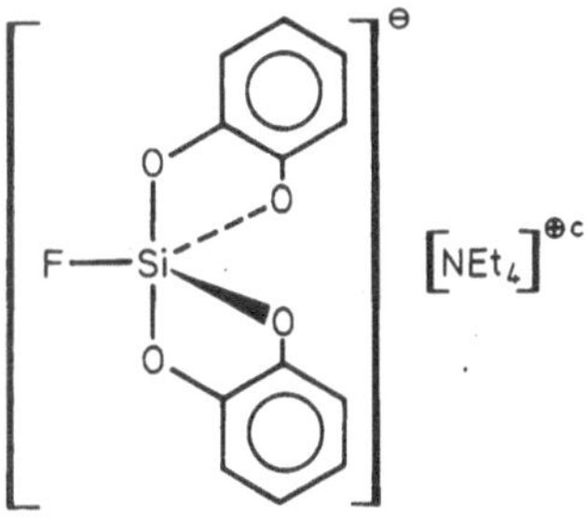	173.8 (O) 174.1 (O)	169.9 (O) 170.1 (O) 159.9 (F)	52.8	161)
	173.6 (O) 174.2 (O)	170.4 (O) 170.6 (O) 160.7 (F)	68.7	

Table 5. (Continued)

Compounds	Position		% (TBP-SPY)[b]	Ref.
	ax	eq		
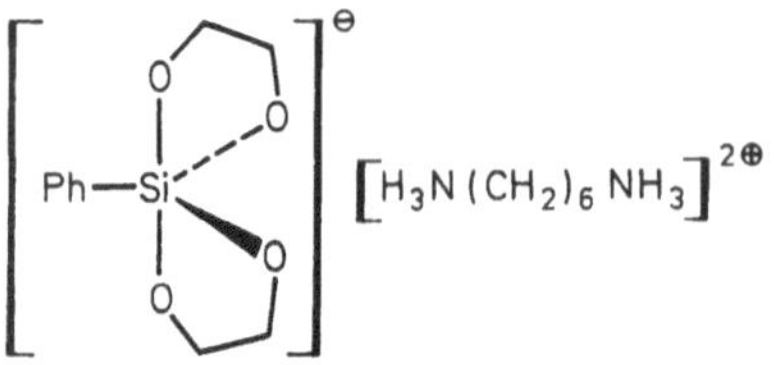	173.8 (O) 175.5 (O)	168.5 (O) 171.7 (O) 188.3 (C)	72.1	162)
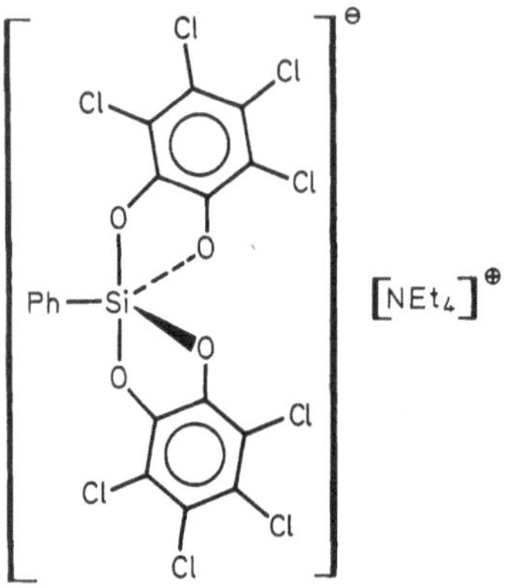	175.3 (O) 175.9 (O)	174.6 (O) 175.7 (O) 185.4 (C)	89.8	162)
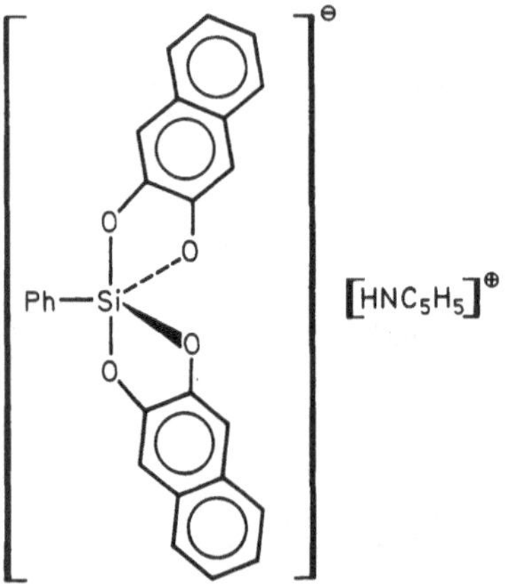	173.4 (O) 175.9 (O)	173.3 (O) 175.0 (O) 187.1 (C)	97.6	163)

Table 5. (Continued)

Compounds	Position ax	Position eq	Ref.
Anionic Hexacoordinate			
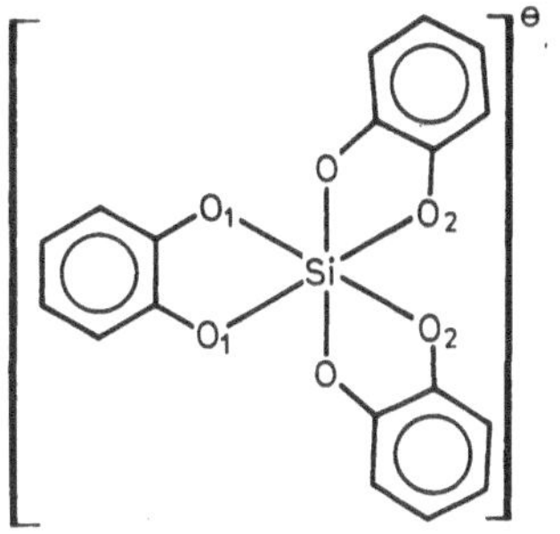 $2[HNC_5H_5]^{\oplus}$	176.5 (2 O)	177.5 (2 O_2) 181.3 (2 O_1)	467)
$[SiP_4O_{13}]^{2\ominus}\ 2[NH_4]^{\oplus}$	177.1 (O) 177.6 (O)	176.2 (O) 176.5 (O) 176.6 (O) 178.8 (O)	468)
SiP_2O_7[d]	173.6 (O) 175.2 (O)	175.9 (O) 177.9 (O) 178.4 (O) 178.6 (O)	469)
$[SiF_6]^{2-}X^{2+}$			
X = 2 K	176 (6 F)		153)
	168.3 (6 F)		470)
2 K · FK	170 (6 F)		471)
2 Na	180 (6 F)		472)
	165 (6 F)		152)
	169.5 (6 F)		473)
Ba	171 (6 F)		474)
2 $[NH_4]$ · FNH_4	171 (6 F)		475)
N_2H_6	166.8 (2 F)	168.5 (2 F) 168.6 (2 F)	476)
	167.1 (2 F)	168.2 (2 F) 168.3 (2 F)	477)
$Co(N\text{-viz})_4$[e]	169.9 (2 F)	166.7 (4 F)	478)
$Co(NH_3)_5Cl$	163.9 (2 F)	166.1 (F) 167.7 (F) 167.9 (2 F)	479)
$Fe(H_2O)_6$	170.6 (6 F)		480)
$Co(H_2O)_6$	164.3 (6 F)		480)
	167.4 (2 F)	167.8 (5 F)	481)
$Ni(H_2O)_6$	167.1 (2 F)	168.3 (4 F)	481)
$Zn(H_2O)_6$	167.1 (2 F)	168.0 (4 F)	481)
$Cu(H_2O)_4$	166.9 (2 F)	168.7 (2 F) 169.9 (2 F)	482, 483)
2 $CuSC(NH_2)_2$	165 (F) 166 (F)	164 (F) 167 (2 F) 168 (F)	484, 485)

Table 5. (Continued)

Compounds	Position		Ref.
	ax	eq	
Cationic Hexacoordinate			
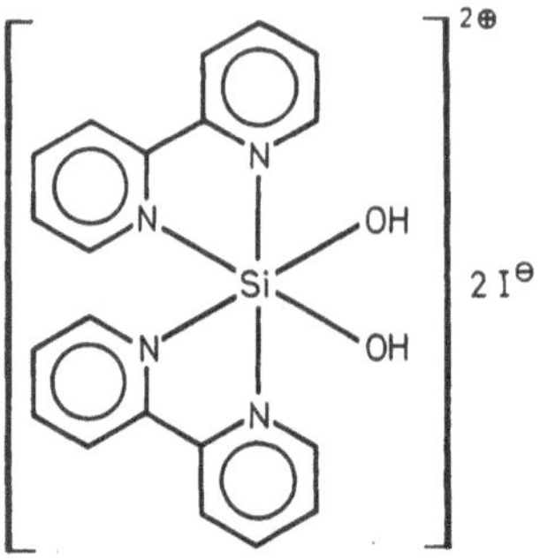	195.3 (2 N)	200.5 (2 N) 164.3 (2 O)	486)

[a] The silicon atom in the crystal of cationic complex $[Ph_3Si(bipy)]^+ \cdot I^-$ is pentacoordinated but the bond lengths are not determined [487]; [b] Geometrical distortion from the TBP to the SPY configuration expresses from the sum of dihedral angle method described in Refs. [488,489]; [c] Two crystallographically independent forms; [d] See the text; [e] viz = vinylimidazole

coordination sphere. The most favorable conditions arise when the central atom Si and the ligand D enter into a five-membered ring (Sect. 2.1).

4.4.1 Neutral Hexa-coordinate Complexes

The question of the intermolecular association in silicon halide has a certain interest. Although the infrared spectra of silicon chloride and fluoride adducts have been reported and interpreted in terms of five- and six-coordinate silicon-atoms [138–143], only the latter adducts have been examined by X-ray diffraction (Table 4). All are based on an octahedral arrangement and involve nitrogen and phosphorus (only one example) atoms attached to silicon. No authenticated penta-coordinate silicon adduct has yet been determined by X-ray diffraction.

The remarkable complexing ability of phthalocyanines (Pc) has permitted the synthesis of neutral silicon complexes with coordination number 6. Based on a X-ray crystal structure study of $PcSi(OSiMe_3)_2$, the central silicon atom is approximately octahedral. The essential planarity of the ring indicates that the silicon atom is small enough to fit into the ring without distorting it. X-Ray diffractometric analyses of $[PcSiO]_n$ polymers [459], the model trimer [457], and an earlier (far more qualitative) study [458] indicate parallel arrangement of metallomacrocycles with inter-planar spacing (Si—Si distance) of 333 pm (Table 4), and also that the Si—O—Si fragments are linear and perpendicular to the PcSi ring planes.

4.4.2 Ionic Complexes

X-Ray parameters indicate that the charge of the central silicon atom affects greatly the length of its bonds (Table 5). In discrete anionic forms the charges on the Si atom are strongly delocalized over the ligands and, therefore, they have the longest Si—O covalent bonds among penta- and hexa-coordinate silicon compounds. In the cis-$[(HO)_2Si \cdot bipy_2]^{2+}$ cation (where no charges are delocalized) the Si—O bonds (164.3 pm) are exceptionally short for octahedral complexes and close to those observed in silanols (164 pm). The length of a less polar Si—C bond depends, to a much lesser extent, on the charge value of the central silicon atom.

Among anionic penta-coordinate silicon derivatives the acyclic silicates are not significantly distorted from a trigonal bipyramid. The structure of pyridinium bis(2,3-naphthalenediolato)phenylsilicate is nearly an ideal square (exactly the rectangular, RP) pyramid. Other five-coordinate anionic species have structures distributed between the two representative geometrics. The distortions from idealized symmetries closely follow the Berry exchange coordinate [402]. This was illustrated by plotting the axial (1.5) and equatorial angles (2.4) at silicon vs. a measure of distortion

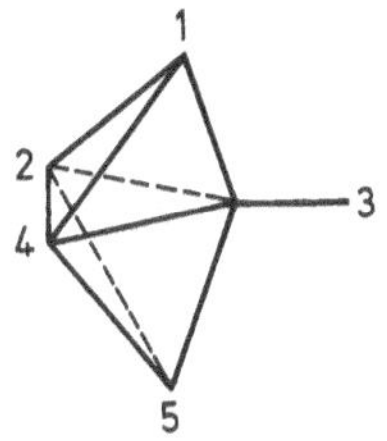

coordinate, for instance, the dihedral angle formed between normals to the TBP faces 124 and 245 that have the common equatorial edge 24 [162].

Excluding ring oxygen atoms that are hydrogen bonded, the difference in bond lengths between the axial and equatorial sets decreases and approaches zero as the RP is approached. From the least-squares equations obtained for the axial and equatorial Si—O bond lengths for penta-coordinate anionic cyclic silicates vs. the distortion coordinate (the dihedral angle) the Si—O bond lengths for the ideal TBP have been calculated to be 178 (axial) and 168 pm (equatorial), and for the RP the basal Si—O bond length to be 173 pm. Changes in Si—O bonds for five-coordinate silicon compounds may imply (with less accuracy) a greater ease of structural distortion from the TBP to the RP [163].

The approach to the rectangular pyramid for spirobicyclic species is consistent with the presence of two unsaturated five-membered rings containing four highly electronegative identical oxygens. The fact that the solid state structures of penta-coordinate silicon compounds fall between the TBP and SPY structures indicates that the energy difference between those two geometries is relatively small. These same features were observed with penta-coordinate phosphoranes [490].

The most accurate parameter values (corrected for thermal motion) of the symmetric $[SiF_6]^{2-} \cdot X^{2+}$ structures are not exactly regular octahedra (O_h). There is a considerable tetragonal distortion (D_{4h}) with four Si—F bonds much longer (shorter) than the other two (Table 5). The distortion can be attributed to the cationic environments of the F atoms [476,478]. An increase in the cation size reduces the density of the cation charge, thus decreasing the distortion of SiF_6^{2-}. Apart from the lattice effect, the Jahn-Teller pseudo-effect is also reported as a possible source for tetragonal distortion of octahedral complexes in terms of ligand-field theory [491–494]. Considering the d orbitals are filled, a non-linear molecular has a degenerate state; then there is at least one vibrational coordinate along which a distortion can occur so as to remove the degeneracy. Thus in an octahedral arrangement the two ligands on the z axis can move towards the silicon, and simultaneously the four ligands in the xy plane move away, so that the d_{z^2} orbital becomes of higher energy than the $d_{x^2-y^2}$ orbital. This distortion continues until the energy gained is just balanced by the energy required to compress two bonds and stretch the other four. Similarly, one can expect the opposite distortion of four short and two long bonds to occur. There is no way of predicting whether a specific system is the more likely, but the 4-long 2-short arrangement seems to take place in octahedral silicon complexes.

Although silicon fairly readily enters into octahedral coordination with fluorine, there is only a limited number of examples in which it is octahedrally coordinated by oxygen. In some phases of silicon dishosphate, SiP_2O_7, the PO_4 tetrahedra are cornerlinked in pairs to form P_2O_7 groups. Each PO_4 tetrahedron shares three corners with three different SiO_6 octahedra forming a three-dimensional framework of P_2O_7 groups and SiO_6 octahedra [469, 495]. Isolated SiO_6 octahedra are found in $(NH_4)_2SiP_4O_{13}$ [468] and thaumasite, $Ca_3[Si(OH)_6](SO_4)(CO_3) \cdot 12\,H_2O$ [496]. Octahedral coordination is also favored by high pressure, the high pressure form of SiO_2 stishovite with rutile structure [497] being one example [333].

There are some data indicating the possible existence of anionic silicon complexes with coordination number 7 and 8 [164-166]. The assumption of hepta-coordination of silicon in the molecule of empirical formula $(NH_4)_3SiF_7$ [164], first proposed by Marignac [498], was later rejected. X-Ray analysis shows that crystals of the above tentative composition do not contain SiF_7^{3-} ions but rather a double salt, $(NH_4)_2SiF_6 \cdot NH_4F$ [475]. Similarly, the structure of K_3SiF_7 consists of an array of K^+, SiF_6^{2-} and F^- ions [471].

4.4.3 Intramolecular Complexes Formed by Five-Membered Ring Closure

As crystal structure analysis shows, unlike some anionic complexes with square pyramid geometry, a trigonal bipyramidal arrangement is usually found for penta-coordinate silicon in neutral Lewis acid-base adducts. There are no structural data of extra-coordinate silicon complexes containing a six-membered ring. This can be explained by the preference of five- over six-membered chelating rings. The five-membered rings connect axial-equatorial sites with the donor-acceptor ring component in an axial position. It is to be expected that the more favorable linear arrangement of tree-center X—Si←D bonding and minimizing of five-membered ring strain, which is achieved by an axial-equatorial rather than an equatorial-equatorial sites, will be the dominating factor in the stereochemistry observed.

Intramolecular complexes of penta-coordinate silicon formed by only one five-membered ring are listed in Table 6. For identical substituents, the axial bound distances are longer than the equatorial ones and both are pronouncedly elongated in comparison with distances in tetravalent silicon compounds. A change in the covalent bond lengths upon the formation of the cycle points to the tendency to acquire an aromatic character.

In eight-membered heterocycles containing atoms of silicon and donor (D) in positions 1 and 5, a coordinate bond Si←D can be formed to yield two condensed five-membered rings. Such diptych structures (largely with transannular interaction

Table 6. Bond Distances (pm) for Pentacoordinate Silicon at the Expense of the Formation of Five-Membered Chelate Ring (Neutral Complexes)

Compounds		Positions ax	Positions eq	Ref.
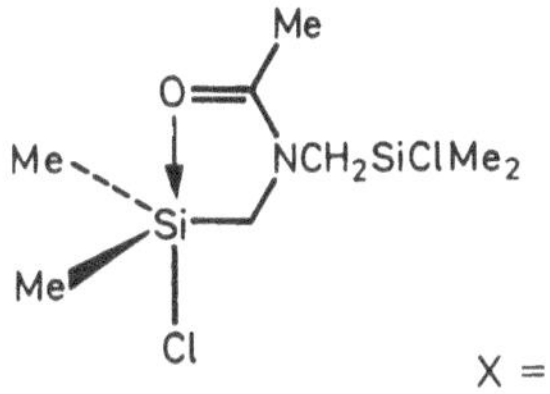		191.8 (O) 234.8 (Cl)	188.8 (CH_2) 185.3 (Me) 185.5 (Me)	450, 451)
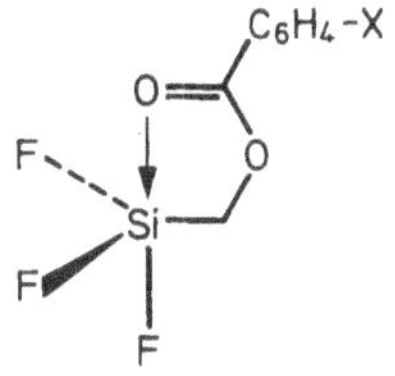	X = 4-Br	194 (O) 169 (F)	181 (C) 168 (F) 169 (F)	499–502)
	H	201 (O)		
	4-F	202.9 (O) 160.2 (F)	186.9 (C) 158.5 (F) 158.2 (F)	
	2-Cl	204.0 (O) 162.0 (F)	181.0 (C) 157.5 (F) 158.1 (F)	
	4-Cl	208 (O) 156 (F)	184 (C) 159 (F) 162 (F)	
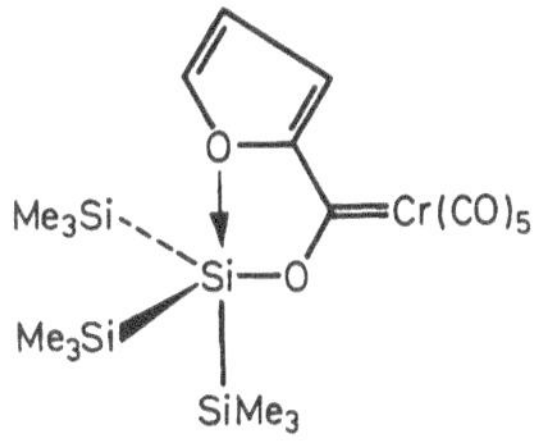		292 (O) 236.1 (Si)	174.2 (O) 236.3 (Si) 236.6 (Si)	503)
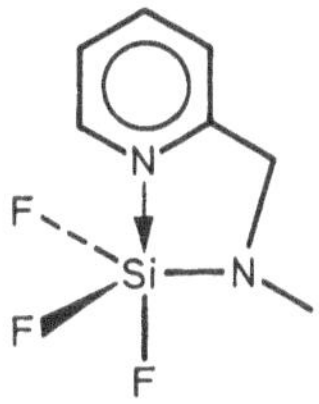		197.4 (N) 162.1 (F)	170.1 (N) 160.3 (2 F)	504)
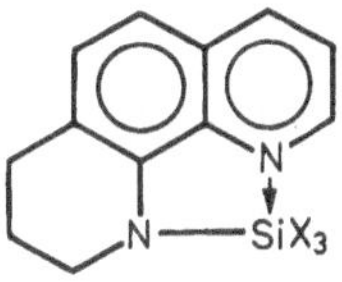	X_3 = Cl_2CH_3	201.8 (N) 221.3 (Cl)	214.4 (Cl) 292 (C)	505–507)
	Cl_3	198.4 (N) 215.0 (Cl)	173.7 (N) 209.4 (2 Cl)	
	F_3	196.9 (N) 162.1 (F)	173.2 (N) 159.0 (2 F)	
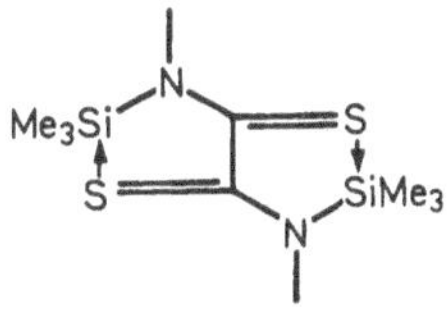		371.8 (S) 185.3 (C)	181.3 (N) 185.3 (C) 184.7 (C)	508)
		371.9 (S) 186.9 (C)	181.7 (N)[a] 184.9 (C) 186.0 (C)	

[a] Two crystallographically independent molecules

Table 7. Bond Distances (pm) for Pentacoordinate Silicon at the Expense of Transannular Interaction

Compounds		Positions			Ref.
		ax		eq	
$RR'Si(OCH_2CH_2)_2NR''$					
RR'Si	R″				
$\overline{OCH_2CH_2OSi}$	H	200.4 (N)	171.0 (O)	166.5 (O) 166.7 (O) 168.3 (O)	509)
$\overline{OC(CF_3)_2C(CF_3)_2OSi}$	CH_3	203.2 (N)	173.1 (O)	163.2 (O) 164.0 (O) 169.0 (O)	510–512)
$\overline{OC(CH_3)_2CH_2C(CH_3)_2OSi}$	CH_3	224.7 (N)	167.1 (O)	164.3 (O) 165.1 (O) 165.5 (O)	513)
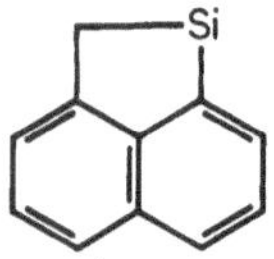	CH_3	226.3 (N)	190.1 (C)	164.6 (O) 165.8 (O) 188.1 (CH_2)	514)
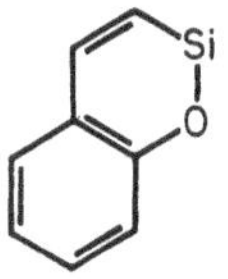	CH_3	229.7 (N)	169.8 (O)	163.6 (O) 164.3 (O) 185.1 (C)	515)
$(C_6H)_2Si$	H	230.1 (N)	190.1 (C)	164.4 (O) 165.9 (O) 188.6 (C)	516)
$(C_6H_5)_2Si$	CH_3	268 (N)	188 (C)	162 (O) 164 (O) 185 (C)	517)
$(C_6H_5)_2Si$	C_6H_5	308 (N)	187 (C)	163 (O) 165 (O) 186 (C)	517)
$(C_6H_5)_2Si$	$C(CH_3)_3$	316 (N)	195 (C)	163 (O) 166 (O) 195 (C)	517)
$(CH_3)_2Si$	C_6H_5	319 (N)	185 (C)	163 (O) 163 (O) 183 (C)	517)
$C_6H_5N(CH_2CH_2O)_2S$	C_6H_5	328 (N)			510)
$(CH_3)_2Si[(o\text{-}C_6H_4)CH_2]_2NC(CH_3)_3$		295.1 (N)	188.4 (CH_3)	186.2 (CH_3) 188.5 (C) 189.3 (C)	518)
$(CH_3)_2Si[(o\text{-}C_6H_4)CH_2S\text{—}]_2$		343.8 (S)	187.2 (CH_3)	185.6 (CH_3) 188.0 (C) 189.0 (C)	519)

Si←N) are given in Table 7. As one should expect the shortest Si←N bond lengths have been observed in spirocyclic molecules in which a stimulated decrease of the silicon bond angles by an additional five-membered ring favors the expansion of its coordination sphere. The silicon axial bond length is significantly longer than the equatorial distance as expected for trigonal bipyramidal coordination.

Using these heterocyclic compounds one can follow the effect of different factors on intramolecular interaction. A rise in the bulk of the substituent near the nitrogen atom weakens the Si←N interaction. Thus, when R = R' = C_6H_5, a change of R'' from H to CH_3 and then to $C(CH_3)_3$ increases the interatomic Si←N distance (230, 268 and 316 pm, respectively). Conjugation of a nitrogen long electron pair with a phenyl group (R'' = C_6H_5) also considerably decreases the Si←N distance. With an increase in the Si . . . N distance in these compounds the conformation of the eight-membered ring passes from "boat-boat" to "chair-chair" and the silicon atom becomes tetrahedral [517].

At present the molecular structures of about 40 silatranes containing three condensed rings with a common Si←N bond have been determined (Table 8). The Si . . . N distance in these molecules varies within wide range (from 200 to 290 pm) but always remains less than the sum of van der Waals radii of silicon and nitrogen. The length of the Si←N bond in Si-substituted silatranes $XSi(OCH_2CH_2)_3N$ is related linearly to the Taft inductive constant σ^* of the substituent X: [534] $l_{Si \leftarrow N} = 220 - 6.3\,\sigma^*$ (r = 0.947) and has a more complicated relationship with its electronegativity [547, 548]: $l_{Si \leftarrow N} = 302.1 \cdot \chi^{-0,257}$.

A change in the Si←N distance can be characterized by a displacement of the silicon (Δ_{Si}) or nitrogen (d_N) atom from the equatorial plane defined by three oxygen atoms [534]:

$$\Delta_{Si} = -114 + 0.62 \cdot l_{Si \leftarrow N} \qquad (r = 0.991)$$
$$d_N = 115 + 0.38 \cdot l_{Si \leftarrow N} \qquad (r = 0.978)$$

The equations show that a change in $l_{Si \leftarrow N}$ is more dependent on Δ_{Si} than on d_N values. The longer the Si←N bond, the more planar the configuration of the NC_3 grouping and the structure of the $XSiO_3$ fragments approaches tetrahedral. Changes in the equatorial fragment are much weaker and do not correlate with those in the axial part. Thus, the degree of the intramolecular interaction in silatranes depends greatly on the electronic effects of substituents at the silicon atom.

The substitution of a methylene group for an oxygen atom in silatrane framework is accompanied with a great lengthening of the Si←N bond. Whereas in 1-methylsilatrane this value equals 217.5 pm, in 1-methyl-2-carbasilatrane it is 233.6 pm. This is due to the fact that the acceptor ability of the silicon atom and, con-consequently, the degree of the transannular Si←N interaction depends on the total electron effect of the neighboring substituents. Therefore, in 1-ethylsilatrane and 1-methoxy-2-carbasilatranes, where the sums of electronegativity of substituents at the silicon atom are similar, the lengths of the Si←N bonds are almost identical (221 and 222 pm).

The real value of the Si←N distance is determined by two factors:

i) the tendency of the Si and N atoms to approach each other due to attractive interaction, and

Table 8. Axial Bond Distances (pm) for Silicon in Silatranes

Compounds	d(Si–N)	d(Si–X)	Ref.
$N(CH_2CH_2O)_3SiX$			
$X = PtCl[P(C_6H_5)(CH_3)_2]_2$	289	229.2	294, 295)
CH_3	217.5	187.0	520)
CH_3 (gas)	245.3	185.3	297)
$CH_2CH_2Si(CH_3)_2OC_4H_3$[a]	223		521)
CH_2CH_3	221	188	522)
$CH_2CH_2CH_2Cl$	218	188	523)
$CH_2OCOC_6H_5$	212.2	195.4	
CH_2Cl	212.0	191	524)
$CH_2P^+(C_6H_5)_3 \cdot I^-$	207.5[b]		525)
	212.3[b]		
$CH_2N^+(CH_3)_3 \cdot I^-$	208.3		525)
$CH_2S^+(CH_3)_2 \cdot I^-$	204.6	193.0	526)
$C_6H_5-\alpha$	219.3	188.2	527)
$C_6H_5-\beta$	215.6	190.8	528)
$C_6H_5-\gamma$	213.2	189.4	529)
$C_6H_4-NO_2-3$	211.6	190.4	530)
OC_4H_3[c]	215		531)
OC_4H_3[a]	211		521)
OC_6H_4-Cl-3	207.9	169.0	532)
SC_4H_3[a]	213.5		521)
F	204.2	162.2	533)
Cl	202.3	215	534)
$N(CH_2CH_2O)_2(CH_2CH_2CH_2)SiX$[d]			
$X = OCH_3$	222	167	535)
CH_3	233.6	187.7	536)
$N(CH_2CH_2O)_2(CH_2COO)SiX$[e]			
$X = C_6H_4-F-4$	212.9	188.5	537)
$C_6H_4-CF_3-3$	210.6	188.4	537)
$N(CH_2CH_2O)(CH_2COO)_2SiCH_3$[f]	215	184	538)
$N(o\text{-}C_6H_4O)_3SiC_6H_5$	234.4	185.3	539)
$N(CH_2CH_2O)(CH_2CHCH_3O)_2SiCH_2Cl$	212	188	540)
$N(CH_2CHCH_3O)_3SiX$			541, 542)
$X = F$	215		
H	214.6		
CH_2Cl	212.2		
$N(CH_2CH_2O)_2[CH(CH_2CH_3)CH_2O]SiH$	216		543)
$N(CH_2CH_2O)_2(CH_2CH_2CH_2O)SiCH_2Cl$[g]	225	189	544)
$N(CH_2CHCH_3O)_2(CH_2CH_2CH_2O)SiC_6H_5$	242		545)
$N(CH_2CH_2O)_2[CH_2CH_2OSi(CH_3)_2]SiCH_3$	276.8	186.8	546)

[a] OC_4H_3 = 2-furyl, SC_4H_3 = 2-thienyl; [b] Two crystallographically independent molecules; [c] 3-furyl; [d] 2-carba; [e] 3-on; [f] 3,7-dion; [g] 3-homo

ii) the tendency of the silatrane skeleton as a whole to retain an equilibrium configuration.

Therefore, introduction of additional units into the framework immediately affects the Si←N bond length (225 pm in 1-chloromethylhomosilatrane and 212 pm in 1-chloromethylsilatrane). Since in these molecules the displacements of silicon atoms from the plane of three oxygen atoms (Δ_{Si}) are almost the same

(16.3 and 16.7 pm, respectively) the difference between the bond lengths may be due to the fact that intramolecular interaction is more favorable for a five-membered $\overset{\downarrow}{Si}OCH_2CH_2N$ ring than for a six-membered $\overset{\downarrow}{Si}OCH_2CH_2CH_2N$ ring.

Substitution of a carbonyl group for a methylene group in five-membered rings is accompanied, as a rule, with a rise in the ring strength and coplanarity [549]. This is typical for the structure of atrane rings as well. The carbonyl-containing heterocyclic ring $SiOC(O)CH_2N$ in silatrane-3-ones is essentially planar, and the Si—O bond is by 7 pm longer and the C—O bond is by 21 pm shorter than in silatrane. At the same time the Si←N distance decreases no more than 3 pm.[6]

The crystalline structure of 3,710-methylsubstituted silatranes containing asymmetric carbon atoms is a mixture of diastereomers [552–555] whose composition cannot be refined by X-ray structural analysis [540, 542]. The presence of the C-methyl group in the silatrane framework shortens the neighboring C—C bond from 154 pm to 143 pm; the length of the Si←N bond, however, remains unchanged.

X-Ray diffraction analysis was used for studying the spatial structure of other metalloatranes and their analogs (intracomplex of triethanolamine derivatives) containing heteroatoms Na [556, 557], Sr [558], Ba [559], B [560–563], Al [564], Ge [565, 566], Sn [567–569], P [570, 571], Zn [572], Mo [573], V [574].

4.4.4 Intramolecular Complexes Formed by Four- or Three-Membered Ring Closure

The strain in four- and three-membered rings (contrary to five-membered systems) hinders the silicon and donor atoms to approach each other when they occupy the β or α position in a linear chain. The observed distortions of tetrahedral silicon environment, however, and intraatomic distances shorter than the sums of van der Waals radii point to the possibility of intramolecular Si←D coordination in these systems (Table 9).

Thus, in the molecule of 2,2,4,4-tetramethyl-3-benzoyl-6-phenyl-2,4-disila-1,3,5-oxadiazine the coordination polyhedron of one of the silicon atoms is a distorted trigonal bipyramid. Both oxygen atoms are in the axial position (the O—Si—O angle is 161°), and the Si . . . O coordination distance (261 pm) is much less than the sum of the Si and O van der Waals radii (360 pm).

Based on vibrational spectroscopy and conductometry [602] data, the acetoxysilanes exhibit intramolecular Si←O=C interaction. In the gas phase the configuration of $H_3SiOCOR$ molecules (R = H, CH_3, CF_3) is close to the cis form (with respect to the O—C bond). The deviation of the Si—O and C=O bonds from the planar cis configuration does not exceed 21°. The Si . . . O=C distance varies from 280 to 290 ppm, considerably shorter than the sum of van der Waals radii. The distinctly shorter C . . . Si contact in silyl formate (261 pm) than that observed in methoxysilane (266 pm [603]) again suggests an interaction attracting the silicon atom to the carbonyl group. In molecules of isostructural organic compounds, i.e., CH_3OCOR (R = H, CH_3) [604, 605] with a smaller positive charge on the carbon atom interacting with oxygen, the C . . . O=C distances are also less than the sum

---

6 Similar features are revealed in the molecules of 1-phenylgermatraneone [550] and boratranetrione [551] as well.

Table 9. Intramolecular Contacts and Important Bond Lengths (pm) for Silicon at the Expense of Four- or Three-Membered Ring

Compounds		Intramolecular Contact	Covalent Bond	Ref.
Four-Membered Ring				
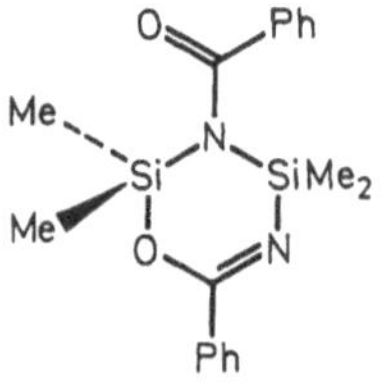		261.3 (O)	167.0 (O) 177.4 (N) 185.8 (C) 186.4 (C)	575)
$H_3SiOCOH$	gas[a]	286.5 (O)	169.5 (O)	576)
$H_3SiOCOCH_3$	gas[a]	279.5 (O)	168.5 (O)	447)
	crystal[b]	283.2 (O)	169.6 (O)	447)
$H_3SiOCOCF_3$		285 (O)	169.8 (O)	175)
$H_3SiOCSCH_3$	gas[a]	314.3 (S)	171.7 (O)	577)
	crystal[b]	318.5 (S)	169.9 (O)	577)
$Si(OCOCH_3)_4$		292.7 (4 O)	162.5 (4 C)	578)
$Me_3SiOCONHPh$		293 (O)	170.7 (O)	579)
$Me_3SiOCONHSiMe_3$		290.7 (NSi · · · OSi) 298.1 (OSi · · · O=C)	175.1 (N) 168.5 (O)	580)
$Me_3SiNHCONH(C_6H_{11})$		299 (O)	173 (N)	81)
t-$BuMe_2SiON(O)=CPh$		283.2 (O)	171.6 (O)	582)
t-$BuMe_2SiON(O)=CPh_2$		280.0 (O)	170.9 (O)	582)
Three-Membered Ring[c]				
$ClMeSi(ON=CMe_2)_2$		250 (2 N)	165 (2 O)	583)
$Si(ON=CMe_2)_4$		250.4 (4 N)	165.6 (4 O)	584)
$PhSi(ON=CMe_2)_3$		251.0 (N) 251.2 (N) 255.6 (N)	164.7 (O) 164.5 (O) 164.0 (O)	585)
$Me_3SiOOSiMe_3$		253.8 (O)	168.1 (O)	586)
$Ph_3SiOOGePh_3$[d]		249.1 (O)	174.2 (O)	587)
$(PhCH_2)Me_2SiOOSiMe_2(CH_2Ph)$		249.1 (O)	168.7 (O)	588)
$(Me_2SiOO)_3$		251.3 (2 O)	167.4 (2 O)	589, 590)
$Me_3SiCH_2N(CH_2)_3CH_2$ (ring)		286.4 (N)	188.2 (CN)	591)
H_3SiN_3	gas[a]	267.4 (N)	171.9 (N)	592)
H_3SiNSO	gas[a]	297.4 (S)	176.2 (N)	593)
$(H_3Si)_2N_2$	gas[a]	269.7 (N)	173.1 (N)	594)
$ClCH_2SiCl_3$	gas[a]	300.8 (Cl)	202.8 (3 Cl) 185.1 (C)	595)
$Cl_2CHSiCl_3$	gas[a]	302.4 (Cl)	202.4 (3 Cl) 191.0 (C)	596)
Cl_3CSiCl_3	gas[a]	300.7 (Cl)	201.1 (3 Cl) 193.1 (C)	597)
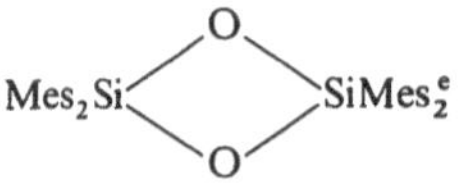		231 (Si)	166 (O) 172 (O)	598)

Table 9. (Continued)

Compounds	Intramolecular Contact	Covalent Bond	Ref.
$(LiSiMe_3)_6$[f]	265 (2 Li)	188.5 (C)	599)
	277 (Li)	189.5 (2 C)	
$Li_2Hg(SiMe_2Ph)_4$[f]	290–300 (Li)		600)
1-Br-μ-$Me_3SiB_5H_7$[f]	232 (2 B)	187 (3 C)	601)

[a] Electron diffraction determinations; [b] Intermolecular contacts in Table 11; [c] Silicon-transition-metal compounds in Table 10; [d] Averaged two-bond contacts between oxygen and silicon or germanium; [e] Mes = 2,4,6-trimethylphenyl; [f] See the text

of van der Waals radii (300 pm). This has been viewed to indicate a positive (probably electrostatic) attraction between these groups [604]. The silicon atom in crystalline $Si(OCOCH_3)_4$ is tetrahedrally coordinated. The Si—O bond length is 162.5 pm while the second oxygen atom of the same acetate is 292.7 pm apart from the silicon atom [578]. The observed Si . . . D (D = O, S) intramolecular three-bond distances are longer than the sum of hard-sphere radii suggested by Bartell [606] and Glidewell [416] for two-bond contacts between Si and D.

On the basis of NMR [607–610], dipole moment [610, 611], NQR [612–614], IR [615, 616], and basicity data [617, 618] of organosilicon compounds containing Si—C—X (X = F, Cl, O, S, N) or Si—M—X (M ≠ C) groups, it was demonstrated that a number of anomalous physicochemical properties exist as compared to carbon analogs (α-effect). This is due to the fact that the interaction of Si and X atoms follows not only the inductive and resonance mechanisms but can also occur through space [614, 617, 619]. In the latter case the silicon atom can interact either directly with X atom or with the C—X bond and the X atom with the Si—C bond. The importance of electrostatic contribution to geminal interaction was stressed in some cases [620]. The α-effect grows with increasing the electronegativity of the atom of the Group IVA element, i.e., in the order Si < Ge < Sn, which is opposite to the ability of these elements to undergo $(p\text{-}d)_\pi$ conjugation [345, 346].

Doublets in the SiH stretching region of H_2SiHCH_2Cl and $R_2SiHCHCl_2$ species were assigned by Egorochkin et al. to "free" and intramolecularly bonded silicon [621]. Enthalpy of geminal interaction in the molecule of dimethyl(chloromethyl)silane, for example, is rather low (<1 kcal · mol^{-1}) and the interaction can be disturbed by even low-basic solvents [621]. The asymmetry parameters of the electric field gradient at Cl in $ClCH_2SiCl_n(CH_3)_{3-n}$ were very small indicating interaction of chlorine with silicon [622]. As follows from the electron diffraction data [595–597], the intramolecular Cl . . . Si contact in the gaseous phase $Cl_nCH_{3-n}SiCl_3$ (n = 1–3) is close to the sum of the non-bonded radii 300 pm [416] (the van der Waals value = 390 pm).

However, typical bond splitting for the SiH stretching vibration in the IR and Raman spectra of $(XCH_2)R_2SiH$ (X = Cl, Br, I) were attributed to rotational isomers [623–625]. Gas-phase basicities [626], Fourier analysis of potential curves [624, 627] and analysis of orbital correlation diagrams [628] of internal rotation around the C—X bond in a series of silyl alcohols and amines, R_3SiCH_2X (X = NH_2, OH), was used to study by CNDO/2 calculations of the character of intramolecular interactions in relations to the mechanism of α-effect. In some cases the experimental results gave no indication for geminal Si . . . X interaction [629,630].

The α-effect is clearly pronounced in molecules of acetoximoxysilanes $R_{4-n}Si[ON=C(CH_3)_2]_n$. The Si . . . N distance is 250 pm with n = 2 and 4 which is much less than the sum of van der Waals radii of silicon and nitrogen (365 pm). With n = 3 (R = C_6H_5), two interatomic Si . . . N distances are 250 pm and the third is 256 pm. Consequently, for compounds with n = 3 and 4 the existence of seven- and eight-coordinate silicon atom is possible. Thus, another four atoms can be placed in the coordination sphere of a silicon atom in addition to four covalently bonded substituents.

An electron diffraction investigation of bis(trimethylsilyl)peroxide [586] and X-ray studies of bis(dimethylbenzylsilyl)peroxide [588] and triphenylsilyl(triphenylgermyl)-peroxide [587] showed the trans conformation for the peroxy ring. In organosilicon peroxides the expansion of the coordination sphere of the silicon atom in the Si—O—O group is performed due to electrostatic interaction of the silicon atom with a remote oxygen atom. In a cyclic $(Me_2SiOO)_3$ molecule the O—Si—C angles (102.4°) are contracted because of an interaction between each silicon atom and two non-adjacent oxygen atoms.

The increase in thermal stability of $(Me_3SiO)_nMe_{3-n}SiOOCMe_3$ as n increases was attributed to increasing intramolecular Si←O coordination [631]. These two-bond contacts are also remarkably similar to non-bonded distances in other silicon compounds: for example, Si . . . C is 275.5 pm (in H_3SiNMe_2 [632]), 280.9 pm (in $H_3SiNCNSiH_3$ [592]) and 283,3 pm (in H_3SiNCO [633]) compared with the hard-sphere sum of 280 pm [206].

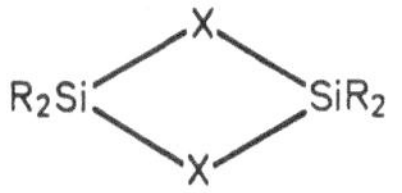

A large number of ring systems of the above type where X = NR [634-639], S [690], CH_2 [641,642], SiR_2 [643] suggest the presence of a silicon-silicon bond. In tetra-mesitylcyclodisiloxane (X = O), however, the Si—O—Si angle is very small (86°) giving the Si . . . Si distance of 231 pm which is shorter not only than those found for the cited four-membered rings (260–270 pm), but also somewhat shorter than the normal Si—Si single-bond length of 234 pm. The molecular structure was described as penta-coordination about the silicon with distorted TBP geometry. Each silicon occupies an equatorial position with respect to the other and the oxygens axial ones for both Si atoms [598]. A CNDO calculation for R = H, Me and X = O, NH has shown significant Si—Si bond indices of 0.3 (X = O) to 0.6 (X = NH), as well as a N . . . N bonding interaction in comparison with O . . . O to a considerable extent more important [644].

In some cases five close contacts for a Si atom can be attributed to extra-coordinated species due to the closure of a formally three-membered ring. Crystal structure data show the participation of silicon in an electron-deficient structure. It is found that each Me_3Si group in hexameric trimethylsilyllithium $(LiSiMe_3)_6$ is located centrally above a triangular face on the six-membered lithium ring (chair form) forming three lithium—silicon bonds. The two average Li—Si distances are 265 and 277 pm. The bonding is suggested to consist of fourcentered electron-deficient Si—Li bonds [599, 645]. The fact is that there is no Li—Si—Li bridge bonding in the complexed dimer $(LiSiMe_3)_2(Me_2NCH_2CH_2NMe_2)_3$ and the silicon atom is tetra-coordinated [646]. In the structure of lithium tetrakis-(dimethylphenylsilyl)mercurate(II), $Li_2Hg(SiMe_2Ph)_4$, two silicon atom interact with two lithium cations, whereas the others only interact with a single lithium cation [600, 647].

The silicon atom in 1-bromo-μ-trimethylsilylpentaborane (Table 9) is penta-coordinated. Structural parameters, however, point to a tetrahedral configuration of the fragment $(Si(CH_3)_3$, i.e., to sp^3 hybridization of Si atom. The B_2—Si fragment can be described in terms of three-center two-electron bonds involving quasitetrahedral silicon orbitals and orbitals of each boron atom [601].

Close contact between a hydride ligand an silicon atom is consistently observed in η-cyclopentadienyl transition-metal complexes, containing a $HMSiR_3$ fragment (Table 10); these species play an important role in hydrosilylation and related reactions. Though X-ray structure determinations of hydrogen positions have a low degree of accuracy, an interpretation of strong deviation of the substituents at silicon from a tetrahedral arrangement suggests not only different steric influences [654, 657, 658, 662, 663] but also the Si . . . H attractive interaction [648, 664]. The coordination polyhedron at silicon can best be described as a distorted trigonal bipyramid with the hydrogen atom in the apical position. Platinum, manganese and iron hydrides seem to be limiting cases, the former showing a short silicon-hydrogen distance (170–180 pm close to the sigma bond 150 pm), the latter

Table 10. Intramolecular Si—H Distances in Molecules of Silicon-Metal Compounds

Compounds	d(Si—H), pm	Ref.
$\{(\mu\text{-}SiMe_2)PtH[P(C_6H_{11})_3]\}_2$	172	648)
$Ph_3SiMnH(CO)_2(Cp)^a$	176	649)
$Cl_3SiMnH(CO)_2(MeCp)^a$	179	650)
$Cl_2PhSiMnH(CO)_2(Cp)$	179	651)
$FPh_2SiMnH(CO)_2(MeCp)$	180.2	652)
$Ph_2HSiMnH(CO)_2(MeCp)$	180	653)
$Ph_3SiReH(CO)_2(Cp)$	219	654)
$[(\mu\text{-}SiEt_2)ReH_2(CO)_3]_2$	b	655)
$[(\mu\text{-}SiPh_2)ReH(CO)_4]_2$	c	156)
$(\mu\text{-}SiEt_2)_2Re_2H_2(CO)_7$	b	657)
$(F_2MeSi)_2FeH(CO)(Cp)$	206	658)
$(Cl_3Si)_2FeH(CO)(Cp)$	210	659)
$(Me_2PhSi)_2FeH(CO)(Cp)$	210	651)
cos-$Ph_3SiFeH(CO)_4$	273	657)
$Me_2SiOSiMe_2IrH(CO)(PPh_3)_2$ (ring: Si—Ir)	b	660, 661)

[a] Cp = $(\eta^5\text{-}C_5H_5)$; [b] No distinction between terminal hydride ligand with or without weak interactions with the silicon atom is established; [c] Location of the hydrogen atom is uncertain. Close contact between the hydride ligand and the silicon atom has been postulated from the evidence of bond lengths and angles involving the silicon and metal atoms

showing weak (or no) interaction between silicon and hydrogen. The covalent Mn—Si bond is considerably longer than the Re—Si distance when the sizes of the metal atoms are taken into account. Various valence — bond representations assume an incipient three-center two-electron bond in the M—H—Si triangle [665, 666].

Thus, the silicon atom can be involved in electron-deficient bonds typicyl of bridge systems.

4.4.5 Intermolecular Association of Silicon Compounds

Intermolecular coordinate interactions involving an expansion of the silicon coordination number were revealed by X-ray diffraction mainly for monosubstituted silanes of the H_3SiX type, in whose molecules the central Si atom is sterically accessible. Since all the compounds of this type are gases or volatile liquids, their association was observed only at low temperatures. There are only few exceptions (Table 11).

The crystal and molecular structure for the N-(trifluorosilyl)trimethylphosphinimine dimer demonstrates that the molecule is dimeric with a planar four-membered $(SiN)_2$ ring containing trigonal bipyramidal pentacoordinate silicons and trigonal nitrogens. The axial Si—F (166.8 pm) and Si—N (185.7 pm) bond distances are significantly longer than the equatorial bond length 160.6 (2F) and 173.6 pm (N). The non-bonded Si . . . Si distance is 276 pm [449].

In the solid dimethylaminosilane molecules, species are grouped into cyclic pentamers $[H_3SiN(CH_3)_2]_5$ with alternating silicon and nitrogen atoms. The nitrogen atoms occupy the corners of a regular pentagon; the silicon atoms are located symmetrically between the adjacent nitrogen atoms and have a trigonal bipyramidal structure. The mean Si—N distances 197.6 pm are greater by 26 pm than that found for the isolated molecule in the gas phase (171.2 pm [319]).

Table 11. Shortest Intermolecular Contact Distances (pm) in Crystals of Associated Silicon Compounds

Compounds	d(Si ... D)	Ref.
$F_3SiNP(CH_3)_3$ dimer	185.7 (N)	449)
H_3SiX		
X = $N(CH_3)_2$ pentamer	198 (N)	154)
	197.6 (N)	156)
CN	280	667)
NCO	330.3 (O)	321)
	331.1 (N)	
OCOH	286.5 (O)	576)
$OCOCH_3$	272.1 (O)	447)
	346.4 (O)	
$OCSCH_3$	338.2 (S)	577)
$OSiH_3$	311.5 (O)	320)
$SSiH_3$	355 (S)	668)
	356 (S)	
$SeSiH_3$	358 (Se)	668)
	362 (Se)	
$(CH_3)_3SiCN$	366 (N)	207, 669)
$(CH_3)_2Si(CN)_2$	348 (N)	669, 670)

Silyl cyanide H_3SiCN forms a linear ... Si—C—N ... Si chain [671,672] with an unusually short N ... Si distance of 280 pm [667]. Crystalline $(CH_3)_3SiCN$ shows the molecules to be aligned head-to-tail with an intermolecular N ... Si contact of 366 pm [207]. In $(CH_3)_2Si(CN)_2$ the N ... Si interaction is stronger (353 pm), and the molecular shape is somewhat distorted from tetrahedral (C—Si—C = 120°). There are also very weak N ... Si interaction 397 pm [669,671].

There are significant differences between the intramolecular geometries of silyl and methyl acetates in the solid phase. The molecular parameters for silyl acetate corresponds closely with the carboxyl geometry for a COOH group in the crystal structures of carboxylic acids [673,674]. This emphasizes the parallels between the $O-SiH_3$ ... O interaction and the O—H ... O hydrogen bonds [447].

In the crystal lattice of disiloxane, $H_3SiOSiH_3$, the intermolecular Si ... O contact is 312 pm which is by 50 pm less than the sum of van der Waals radii of silicon and oxygen. This interaction is strong enough to effect alignment of the molecules but too weak to affect the intramolecular geometry to any significant extent. In crystalline hexamethyldisiloxane there are no short intermolecular contacts. Methylation at the Si atom reduces its ability to participate in donor-acceptor interactions [320]. In the series $H_3Si—X—SiH_3$ (X = O, S, Se), intermolecular contact for Si and X atoms is less than the sum of van der Waals radii by 35–50 pm. With X = O, only one silicon atom interacts with the oxygen atom of the neighboring molecule; if X = S or Se, the association is performed via interaction of both silicon atoms with the X atom of the neighboring molecules. The coordination number of the oxygen is 3 and that of sulfur and selenium 4 [668].

4.4.6 A General Change Pattern of Structural Parameters with Increasing the Silicon Coordination Number

An analysis of the experimental data on the structure of silicon extra-coordinate compounds allows one to follow certain trends in the change of molecular parameters as compared with similar tetrahedral molecules. When the silicon atom passes from tetrahedral to the penta- and then hexa-coordinate state, its bond angles and bond lengths vary drastically. In all penta- and hexa-coordinate silicon compounds, the silicon atom occupies the center of a trigonal or, correspondingly, hexagonal bipyramid whose structure can be more or less distorted. A sharp difference in equatorial bond angles between penta- and hexa-coordinate silicon atom attracts attention. In the first case they are equal or close to 120° which makes the central atom sterically accessible. In the second case they are equal or close to 90° and this, along with an insufficient atomic radius of silicon impairs a further increase in the silicon coordination number. The silicon bond angles formed by an axial or equatorial substituent and by two axial substituents are identical (90° and 180°, respectively) for penta- and hexa-coordination.

Along with the change in silicon bond angles, the expansion of silicon coordination sphere is often accompanied with a distortion of bond angles of the atoms attached to silicon. Silatranyl groups with strong electron-donating effect produce deformations that are consistently larger than the error of structure determinations. For the benzene ring bonded to silicon, the ipso C—C(Si)—C angle α increases and ortho angles β decrease with a decrease in the Si←N bond distances in silatranes. For α–$C_6H_5Si(OCH_2CH_2)_3N$ (219 pm) and $C_6H_5Si(OC_6H_4)_3N$ (234 pm) these angles are 115.9° and 118.0° for α and 121.8° and 120.8° for β, respectively. In tetrahedral molecules of $C_6H_5Si(OR)_3$ and in the absence of coordinate interaction the mean angles are 120° [675]. The angular deformation changes indicate [676, 677] that the electronegativity of the silatranyl group decreases with strengthening the coordination Si←N bond.

With increasing coordination number of the silicon atom, not only bond angles but other structural parameters of the molecules change. The Si—X bonds (X=F, Cl, O, N, C, etc.) both in ionic and neutral complexes of penta- and hexa-coordinate silicon are always longer than the corresponding silicon bonds in compounds with normal coordination. For example, the Si—F bonds in SiF_6^- are longer than in SiF_4 (170 and 156 pm, respectively). The axial Si—Cl bonds in 1-chlorosilatrane (215 pm) and in $(Me_2ClSiCH_2)_2NCOMe$ (234.8 pm) and equatorial bonds in trans-$Cl_4Si \cdot 2P(CH_3)_3$ (220 and 230 pm) are much longer than in molecules of the type $R_{4-n}SiCl_n$ with n = 1–4 (202 pm [596]). When the silicon valence sphere expands, the lengthening of the Si—C(sp^3) bonds is less than that of the Si—C(sp^2) bonds. Thus, in 1-methylsilatrane and 1-methyl-2-carbasilatrane the axial Si—C(sp^3) bonds are by 3 and 2 pm, respectively, longer than in molecules of $CH_3Si(OCH_3)_3$ [678] and $(CH_3)_2Si(OCH_3)_2$ [679]. At the same time, the lengthening of the Si—C(sp^2) bond in 1-phenylsilatrane, as compared to a similar tetrahedral bond in $C_6H_5Si(OR)_3$ is 5 pm [675].

In molecules of 1-chloromethylhomosilatrane and 1-chloromethylsilatrane the length of the Si—C bond is 189 and 191 pm, and of the Si←N bond 225 and 212 pm, respectively. This means that strengthening of the interaction between silicon and nitrogen atoms (i.e., shortening of the Si←N bond) results in the lengthening of the axial covalent Si—C bond. The difference between the lengths of axial and equatorial Si—C bonds in molecules of $(C_6H_5)_2Si(OCH_2CH_2)_2NR$ increases with strengthening of transannular Si←N interaction (Table 7). Thus, the stronger the Si←N interaction in compounds of penta-coordinate silicon, the longer the covalent X—Si bonds. This is more pronounced in the axial X—Si←D fragment than in the equatorial one.

Such a relationship between bond lengths points to the enhanced ionic character of the covalent X—Si bond on complexation. Bong longthening means increase in bond polarity, i.e., increase in net negative charge at the more electronegative atom X and increase in net positive charge at the less electronegative atom Si. The conventional view is that the polarity of a Si←D bond is in the sense Si^-—D^+, but there is now sufficient experimental [680] as well as calculated evidence (Sect. 4.3), that the gain in negative charge at the acceptor atom is passed on the other parts in the acceptor unit, including possibly a portion of the negative charge that originally resided at the acceptor atom in the free acceptor unit.

Attractive interactions may be viewed as incipient valence shell expansion and as the stages of bimolecular nucleophilic displacement reactions [681]. Crystallographic data for the reaction pathways first described for cadmium complexes by Burgi [682] reveal a correlation between the two X—Si and Si←N distances in the linear

X—Si←N fragment. Use of the Pouling relationship between bond distance increment and bond number $\Delta d = -C \times \log n$ and the Burgi assumption between the bond number and the out-of-plane displacement of the Si atom or valent angles about silicon [683–686] in penta-coordinate species including silatranes, a value of C = 111 pm [321] and 117 pm [687] was obtained. On the basis of experimental data and results of quantum chemical calculations, Burgi has estimated [688] the constants C = 36–46 pm (for SiH_4/SiH_5^-), 43 pm (SiF_4/SiF_5^-) and 60 pm ($R_3NSiH_3/(R_3N)_2SiH_3$), significantly smaller than 80 pm (for germanium) [689, 690], 120 (tin) and 166 pm (lead) [689]. This reflects a smaller tendency to increase the silicon coordination number above four as compared to tin and lead. Besides, this confirmed the concepts of the structure of a transition state during numerous bimolecular nucleophilic reactions and inter- and intramolecular rearrangements of organosilicon compounds.

4.5 NMR-Spectroscopy Data

While complete X-ray analysis will establish the structure in the solid state, it is useful to have NMR data on the solution state that illustrate the increase of the coordination number of silicon. It would seem that NMR spectroscopy of nuclei participating directly in donor-acceptor interaction is especially important in investigating silicon compounds with an expanded coordination sphere. This requires the use of ^{29}Si NMR spectroscopy since the electron shell of the silicon atom, the bond angles and lenghts are strongly affected upon complexation. Valuable information could also be obtained with by ^{14}N, ^{15}N, ^{17}O, ^{19}F NMR data since these elements act as donors. Chemical shifts of nuclei other than hydrogen are determined by various factors and not yet understood well anough to provide easily applied correlations of other physical properties of the molecules.

4.5.1 ^1H-NMR Data

It is known that proton chemical shifts are largely determined by electronic effects [691]. The coordinate Si←D interaction, weakening the covalent silicon bonds (increasing their ionic character), enhances the electron density on the atoms of a hydrocarbon substituent which results, as a rule, in an increase of proton shielding. This is usually accompanied by a deshielding of protons of the donor fragment [692]. This chemical shift contribution, however, is often insignificant and masked by other effects, in particular, when the silicon atom has a π-donor substituent.

greater shielding of protons of the substituent X in silatranes $XSi(OCH_2CH_2)_3N$ as compared with that in model tetrahedral molecules (Table 12) was attributed to the influence of the intramolecular Si←N bond (i.e., to the transfer of electron density to a silicon atom and then to the substituent X) [697–706]. On strengthening of the Si←N interaction, the $\delta^1H(X)$ value should be decrease, for example, in the order

$CH_3Si(OCH_2CH_3)_3 > CH_3Si\begin{matrix}OCH_2CH_2CH_2\\(OCH_2CH_2)_2\end{matrix}N > CH_3Si(OCH_2CH_2)_3N$ and the

coordination shift $\Delta\delta^1H$ decrease in the order silatrane > homosilatrane > carbasilatrane, based on the X-ray diffraction data (Table 8). In practice, however, this is

Table 12. ^{1}H Chemical Shifts (ppm) for Si-Substituent of Silatranes in $CDCl_3$ [693–696)]

Compounds		δ	Δδ[a]
$XSi(OCH_2CH_2)_3N$			
X = H		3.94	−0.35[h]
CH_3		−0.07	−0.20
$ClCH_2$		2.66	−0.15
Cl_2CH		5.15	−0.10
$CH_2{=}CH$	trans	5.73	−0.38
	cis	5.78	−0.22
	gem	5.96	0.07
$Cl_2C{=}CH$		5.96	0.11
C_6H_5	ortho	7.72	0.07
$CH_3Si(OCH_2CH_2CH_2)(OCH_2CH_2)_2N$		−0.08	−0.21[b]
$CH_3Si(CH_2CH_2CH_2)(OCH_2CH_2)_2N$		−0.12	−0.18[c]
$CH_3Si(NHCH_2CH_2)_3N$		−0.37	−0.38[d]

[a] Coordination shifts are the difference between the model tetracoordinate molecule, $\Delta\delta = \delta - \delta_m$; [b] Relative $XSi(OCH_2CH_3)_3$; [c] Relative $(CH_3)_2Si(OCH_2CH_3)_2$; [d] Relative $CH_3Si[N(CH_3)_2]_3$

not the case. The shielding of methyl protons in 1-methylhomosilatrane (—0.08 ppm) is even higher than in 1-methylsilatrane (—0.07 ppm) and the $\Delta\delta^1$H values for 1-methyl-2-carbasilatrane and 1-methylsilatrane are almost identical. Besides, the order of change in $\Delta\delta^1$H(X) in the series of 1-substituted silatrane is not related to the electronegativity of X and changes even when the solvent is replaced [696)]. Consequently, the chemical shifts in SiCH protons in silatranes are determined by several factors including, possibly, an anisotropic contribution of both a bicyclic structure and the solvent.

On the other hand, the shielding of silatrane skeleton protons is linearly related to the electronic effects (δ_I, δ_R°) of substituent X at the silicon atom ($CHCl_3$, r = 0.989) [696)]:

$$\delta OCH_2 + 3.829 + 0.465\sigma_I + 0.261\sigma_R^\circ$$
$$\delta CH_2N = 2.866 + 0.493\sigma_I + 0.342\sigma_R^\circ$$

The value of $\delta OCH_2 - \delta CH_2N$ remains constant over a wide range of changes in chemical shifts. Formally, this means that the Si←N bond transmits the effect of the substituent X more effectively than each Si—O bond[7]. There is a linear relationship

[7] Similar trends are observed in germatranes [707)].

between the shielding of the CH_2N protons and the length of the Si←N bond [696], i.e., $\delta CH_2N = -1.21\ l(Si \leftarrow N) + 5.46\ (r = 0.96)$.

A detailed analysis of the ^1H NMR spectra of silatranes with general formula $X\overset{\downarrow}{Si}(OCH_2CH_2)_2(YCH_2)N$ was carried out [287]. The nonequivalency of methylene protons in atran-3-one (Y = OCO) and 2-carbatrane (Y = CH_2CH_2) frameworks as well as changes of their ^1H chemical shifts relative to the model symmetric molecules (Y = OCH_2) provided information concerning intraannular nonbonded interactions. Interactions of the following atom groups occur in each eight-membered ring of atranes:

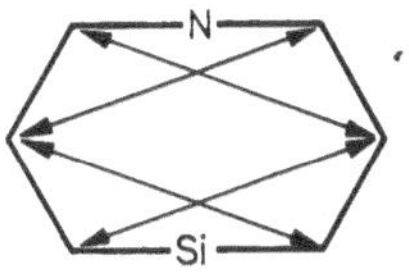

This intramolecular hydrogen-hydrogen interaction of methylene group in atranes is implied by NMR data and is similar to that described for 1-azabicyclo[3.3.3.]undecane ("manxine") [290, 708]. It is the reason that determines the pronounced physico-chemical features of medium-ring bicyclic compounds.

The ^1H and ^{19}F NMR data for C-substituted silatranes, $X\overset{\downarrow}{Si}(OCH_2CH_2)_2(OCHRCH_2)N$, where R = CH_3 or CF_3, revealed a C_3-symmetry of theses molecules, i.e., any of three eight-membered rings in atrane exhibits a boat-chair (BC) conformation [552, 553, 709]. As reported recently [710] chair-chair (CC) and boat-boat (BB) conformations are detected for the unsubstituted eight-membered ring from the ^1H NMR spectra of atran-3-ones.

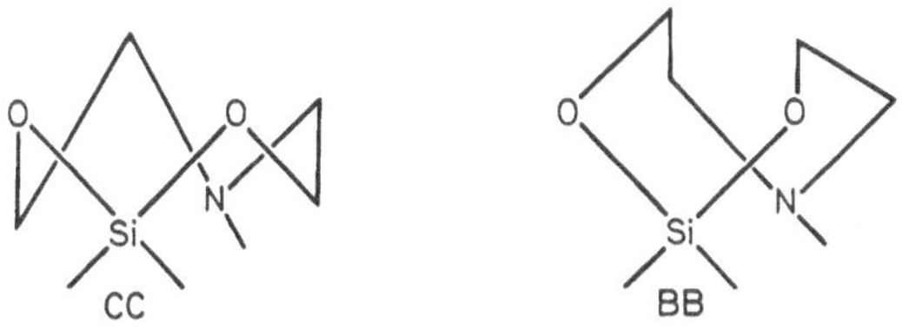

The CC conformation has also been suggested for 1,3-dioxa-6-aza-2-silacylooctanes, $XPhSi(OCH_2CH_2)_2NR$ when different substituents at the silicon atom (X ≠ Ph) are involved [266]. These conclusions are based, however, on an erroneous suggestion about the equivalence of vicinal ^1H—^1H spin-spin coupling constants in OCH_2CH_2N fragments of the investigated compounds.

A change in the chemical shifts of HSi proton of triethoxysilane in different solvents and a decrease of coupling constants $^1J_{SiH}$ from 289.4 Hz in CCl_4 to 285.0 Hz in DMSO [711] indicates that solvation of a $HSi(OCH_2CH_3)_3$ molecule is

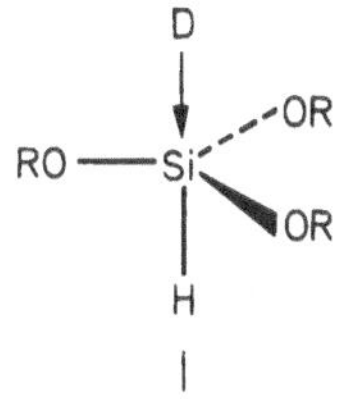

accompanied by the formation of a complex with the solvent [329,711]. The donor atom of the solvent and the H—Si bond are arranged axially in the complex. On the other hand, an increase in $^1J_{SiH}$ by 23 Hz upon complexation of 1,1,2,2-tetra-fluoroethylsilane with triethylamine suggests a possible formation of a structure with an enhanced s-character of the H—Si bond, i.e., with an equatorial arrangement of hydrogen atoms [712]:

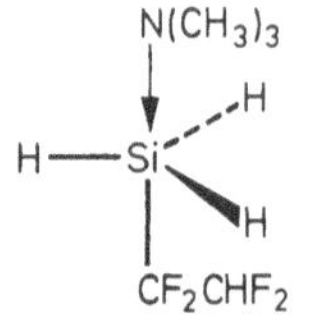

The $^1J_{SiH}$ value of SiH_4 is slightly (0.5 Hz) changed upon addition of varying amounts of pyridine, due to a weak coordinative interaction of silicon atom [740].

It was assumed that the silicon atom is tetrahedral in N,N-bis(chloromethyl-dimethylsilyl)acetamide [713]. In the NMR spectrum of this compound enriched with ^{15}N, however, the coupling constant with protons is observed only for the one NCH_2 group. This points to a rigid orientation of the protons of another NCH_2 group since its hydrogen atoms are fixed due to the formation of a five-membered ring with a penta-coordinate silicon atom:

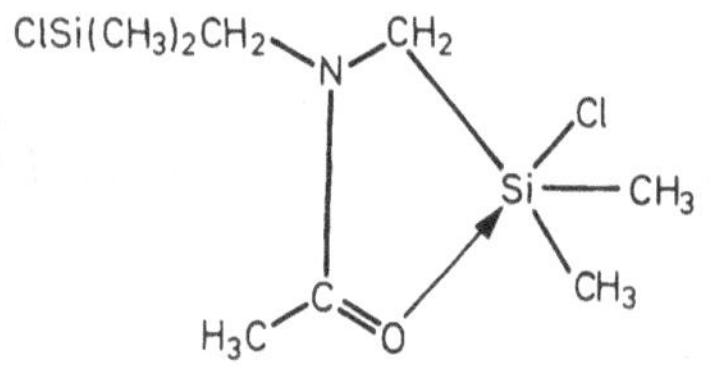

Such a structure is confirmed by five-bond coupling between the $SiCH_2$ protons of the ring and the CCH_3 protons. This corresponds to a trans orientation of these groups relative to the N—C bond [115,451,714]. In the crystalline state of this molecule the silicon atom is also penta-coordinated [450].

The ^1H NMR spectra of anionic and neutral phenyl-substituted penta-coordinate silicon complexes exhibited two distinct sets of aromatic multiplets arising from ortho-protons at low field and meta and para protons at high field. Thus, in the series $RSi[(o\text{-}C_6H_4)C(CF_3)_2O]_2^-$ the multiplet separation is 0.73 (R = CH_3) and 0.66 ppm (H) relative to 0.35 ppm in $Si[(o\text{-}C_6H_4)C(CF_3)_2O]_2$ [238,239]. For 1-phenylsilatrane and $PhSi[o\text{-}C_6H_4)O]_2^-$ the relevant values are 0.48 (Table 12) and 0.40 ppm [715] whereas for phenyltrimethoxysilane only 0.23 ppm. An increase in the magnitude of the multiplet separation when passing from tetravalent silicon compounds to its derivatives with a higher valency is consistent with a decrease in electronegativity of the central Si atom [716]. The enhanced deshielding of the ortho-protons of aryl systems may reflect a spectral feature of penta-coordinate silicon species.

The methyl and ring proton resonances of cationic acetylacetonato complexes of silicon are shifted downfield relative to the resonance of neutral silicon diketonates, whereas the resonances of the anionic complexes are shifted upfield [405,406,717]. For $[Si(acac)_3](HCl_2)$ the CH_3 and CH proton resonances are shifted by 0.28 and 0.86 ppm, respectively, as compared to the enol form of free acetylacetone. The low-field shifts can be explained in terms of either benzenoid resonance [718] or the positive charge on the ion without the necessity of invoking ring currents in the chelated six-membered rings [719]. The latter assumption is confirmed by calculation of the contribution to the shifts from the electric field [720]. The ^1H NMR spectral data of cis and trans configurations in octahedral silicon complexes involving unsymmetrical β-diketonates and isomerization process have been summarized [721,722].

The ^1H chemical shifts of phthalocyanine and porphyrine complexes with hexacoordinate silicon [723–729] as well as theoretical calculations of the contribution to ^1H magnetic shielding [730,731] show a high degree of aromatic nature of the macrocycle; compounds of this type can even be used as diamagnetic shift reagents [732].

In compounds $(CH_3)_2XSiCH_2NCH_3(CH_2)_nNR_2$ (X = Cl, Br; R = CH_3, C_2H_5; n = 2, 3) the shielding of α-protons of the NR_2 group is by ~1 ppm less than in model compounds when a lone electron pair of nitrogen is non-bonded [116]. A rise in the ^1H chemical shift for the donor fragment is observed for other silicon complexes as well [733–735]. In complexes containing an unsaturated ligand, however, the resonance of the protons is able to shift to low fields [736]. An anomalous high-field shift of the ^1H signals of Si substituents may be caused by the intramolecular Si←D interaction due to anisotropy of the acceptor fragment [262]:

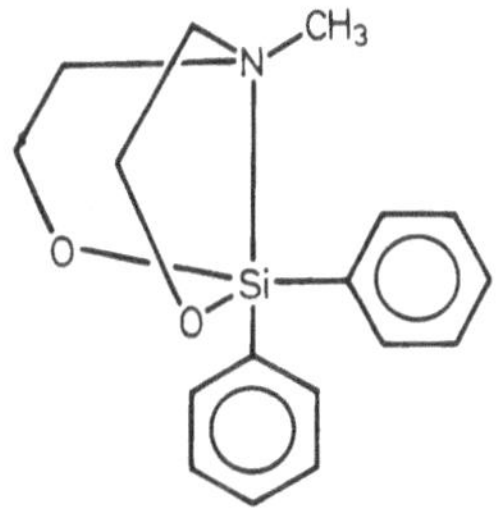

Intramolecular interactions are sometimes responsible for different NMR spectra of chemically equivalent atoms or groups. The magnetic non-equivalency of two aryl groups in the ^1H (and ^{13}C) NMR spectra of 2,2-di(2-furyl)-1,3-dioxa-6-aza-2-silacyclooctane at −60° points to their equatorial and axial arrangement in silicon trigonal bipyramidal environment [263]. In the molecule of $(Me_3Si)_3SiOC(C_4H_3O)$ $[=Cr(CO)_5]$ the non-equivalence of trimethylsilyl groups means that there is an intramolecular interaction of the furyl ring oxygen with the silicon atom [503]. The absence of nitrogen inversion in silicon oxymates $X_{4-n}Si(ON{=}CMe_2)_n$ upon intramolecular Si ... N interaction may result in non-equivalence of methyl groups in the NMR spectra [737]. In stannyl derivatives of this type, however, the tin atom is pentacoordinated due to an intermolecular Sn←O coordination and the methyl groups NMe_2 are equivalent [738]. In the ^1H NMR spectrum of $H_3SiOCON(CH_3)_2$ the methyl groups are magnetically equivalent even at −40 °C. In a similar compound,

$H_3SiOCSN(CH_3)_2$ the free rotation barrier differs only slightly from that of $H_3SiOCON(CH_3)_2$ and no magnetic equivalence is observed even at +60 °C. Therefore, an intramolecular interaction between the silicon atom and the carbonyl oxygen in the first compound has been suggested [739]. The fact that no broadened signals of the CH_2N moiety are observed in the ^1H NMR spectra of $(CH_3)_3SiCH_2NC$ (unlike alkylisonitrils) was related to a poorer symmetry of the nitrogen atom due to intramolecular Si ... N interaction [740].

The ^1H (and ^{13}C) NMR spectra of compounds with a chiral silicon atom point unambiguously to intramolecular coordinate interaction. For example, in molecules of the type $(CH_3)_2NCH_2(o\text{-}C_6H_4)SiXYR$ the nitrogen atom undergoes a slow inversion and methyl groups are non-equivalent at low temperatures. Since at low temperatures the singlet signal in the ^1H NMR spectra corresponds to a free ligand, the hindered rotation about the C—C and C—N bonds is ruled out. Intramolecular Si←N interaction is responsible for non-equivalence of $N(CH_3)_2$ methyl groups in the ^1H and ^{13}C NMR spectra. The formation of the intramolecular Si←N bond induced diastereotopy of the two Me groups[8] [741]:

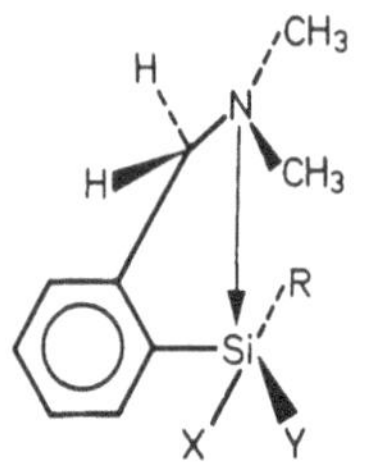

Activation parameters at coalescence temperature show that the coordinate interaction in these compounds is not a function of the electronegativity of X but is controlled by the ability of the nitrogen atom to stretch the Si—X bond. The tendency of the silicon atom to increase its valency decreases in the order: X = OCOR, Br, Cl > SR; F > OR, H [741]. This sequence corresponds directly to the rate of racemization of halosilanes and to the substitution of R_3SiX with inversion of configuration [129]. Although no intramolecular coordination was observed in solutions of acetoxysilanes $(CH_3)_nSi(OCOR)_{4-n}$ by the ^{29}Si NMR method [747], the shape of the ^1H NMR spectra of these compounds with chiral silicon atom points to Si←O interaction [748].

^1H NMR spectra have also been studied for other silicon compounds containing an expanded coordination sphere [749–773].

4.5.2 ^{13}C-NMR Data

Among the silicon compounds with an expanded coordination sphere, the silatranes are best studied by ^{13}C NMR spectroscopy. Independently of sp^3-, sp^2-, or sp-hybridization, the ^{13}C chemical shift of the carbon atom attached to penta-

[8] Similar non-equivalence is apparent in the ^1H NMR spectra of isostructural germanium [742] and tin compounds [743–746].

Table 13. ^{13}C Chemical Shifts (ppm) for Si-Substituent of Silatranes in $CDCl_3$ [693-695, 774-778]

Compounds		δ	Δδ[a]
$XSi(OCH_2CH_2)_3N$			
X = CH_3		−0.72	7.0[b]
$CH_2{=}CH$	α	140.61	10.4
	β	129.60	− 6.5
$(CH_3)_3SiCH{=}CH$	α	150.23	10.5
	β	147.73	− 9.6
C_6H_5	ipso	142.42	11.0
	para	127.57	− 2.8
$(CH_3)_3SiC{\equiv}C$	α	116.42	12.1
	β	102.90	−11.0
$XSi(OCH_2CH_2CH_2)(OCH_2CH_2)_2N$			
X = CH_3		−1.48	6.21[b]
C_6H_5	ipso	141.24	9.29
	para	128.50	− 2.34
$XSi(CH_2CH_2CH_2)(OCH_2CH_2)_2N$			
X = CH_3		2.47	5.25[c]
C_6H_5	ipso	146.24	10.85
	para	128.11	−2.28
$CH_3Si(NHCH_2CH_2)_3N$		2.46	8.82[d]

See the footnotes of Table 12

coordinate silicon is always greater than in the case of tetracovalent silicon compounds (Table 13). If the α-carbon atom of the substituent is sp- or sp^2-hybridized, its coordinated chemical shift is almost doubled as compared to that of the sp^3-hybridized atom [778]. Unfortunately, different contributions to the ^{13}C shielding (even in the case of organic derivatives of tetracovalent silicon) cannot presently be considered quantitatively [779,780]. A down-field shift of the ^{13}C signal of the C—M bond is also observed when the coordination sphere of other IVA Group elements (M = Ge [775], Sn [781]) is expanded. This may be due to an increase in the paramagnetic term of the shielding constant under the σ-electron density redistribution caused by the inductive effect.

The shielding of the carbon atoms of π-polarizable substituent may increase at a distance of two and more covalent bonds from silicon participating in the coordinate interaction. This is observed for vinyl (C_β), phenyl (C_{para}) and ethynyl (C_β) derivatives. Such an increment in chemical shifts of β-carbon atoms of double and triple bonds in silatranes relative to the model derivatives of triethoxysilane has a similar magnitude but opposite sign (Table 13). Thus, intramolecular interaction in silatranes inverts the ^{13}C chemical shift values of unsaturated bonds likely due to a change in the π-electron density of these bonds. A considerable deshielding of carbon at the silicon atom [782-786] as well as smaller coupling constants $^1J_{cc}$ [786-788] for double and triple bonds as compared to those of carbon analogs are most often attributed to $(p-d)_\pi$ interaction. The contributions of σ and π electron densities of the double bond to the C_α and C_β shielding are different. In first approximation, however, an internal chemical shift $\delta C_{\alpha\beta} = \delta C_\alpha - \delta C_\beta$ was proposed [784] as a measure of the π-system polarization. For example, in trimethylvinylsilane and vinyltriethoxysilane ($\delta C_{\alpha\beta}$ = 9.1 [784] and −5.9 [695] ppm, respectively) polarization of the π-system is determined by electronegativity of substituents at the silicon atom [784]:

$$(CH_3)_3Si-\overset{\delta+}{C}H=\overset{\delta-}{C}H_2 \qquad \overset{\delta+}{C}H_2=\overset{\delta-}{C}H-Si(OCH_2CH_3)_3$$

Apparently, for (trimethylsilylvinyl)triethoxysilane an essential contribution to the ground state can be due to a polarity in the sense:

$$(CH_3)_3Si-\overset{\delta+}{C}H=\overset{\delta-}{C}H-Si(OCH_2CH_3)_3.$$

For silatrane, however, containing a penta-coordinate silicon atom, such a resonance structure is improbable. Both the enhanced σ-donor (an increase in δC_α) and lowered π-acceptor (a decrease in δC_β) effects of the silatranyl grouping result in not only a decrease of double bond "polarization" ($\delta C_{\alpha\beta}$) but also in a chang of the sign. The $\delta C_{\alpha\beta}$ values for trimethylsilylvinylsilatrane and the model triethoxysilane as calculated by an additive scheme with the use of $\delta C_{\alpha\beta}$ = 11.0 ppm for 1-vinylsilatrane [695]

Table 14. Comparison of Inductive and Resonance Constants for Tetra- and Pentacoordinate Organosilicon Groups[a]

Groups	δ_I	δ_R°
$Si(OCH_2CH_3)_3$	−0.23	0.122
$Si(OCH_2CH_2)_3N$ (N→Si)	−0.56	0.026
$Si(OCH_2CH_2CH_2)(OCH_2CH_2)_2N$ (N→Si)	−0.61	0.042
$Si(OCHCH_3CH_2)(OCH_2CH_2)_2N$ (N→Si)	−0.59	0.026
$Si(CH_2CH_2CH_3)(OCH_2CH_3)_2$	−0.27	0.104
$Si(CH_2CH_2CH_2)(OCH_2CH_2)_2N$	−0.56	0.016

[a] Calculated using the correlation equations $\delta_R^\circ = 0.05(\delta C_{para} - \delta C_{meta})$ and $\delta_I = 0.424(\delta C_{meta} - \delta C_{C_6H_6}) + 0.032(\delta C_{para} - \delta C_{C_6H_6})$ [775] for the substituent chemical shifts in monosubstituted benzenes. Benzene taken as 129.05 ppm in $CDCl_3$

are 1.9 and −15.0 ppm, respectively. This agrees well with experimental values of 2.3 and −17.8 ppm [778].

The ^{13}C chemical shifts of meta and para (but not ortho [789]) carbon atoms in monosubstituted benzenes correlate well with the inductive and resonance constants of the substituents [790–792]. Examination of σ_I and σ_R° constants obtained from ^{13}C NMR data (Table 14) shows that for the groups containing a penta-coordinate silicon atom the σ-donor component is enhanced and the π-acceptor one is lowered. Similar conclusions have been reached by other measurements [793–796], e.g., the induction constant σ_I for the silatranylmethyl group is −0.36 ($\sigma_R^\circ = 0.21$) whereas for trimethoxysilylmethyl group it is −0.10 ($\sigma_R^\circ = 0.17$) [794]. Nevertheless, attempts to find a direct relationship between these constants and the degree of Si←N interaction in molecules of silatranes, homosilatranes, and 2-carbasilatranes have failed. Thus, in spite of the fact that the interatomic Si←N distance in silatranes is about 10 pm shorter than that in 2-carbasilatranes (Table 8), the difference between σ_I values for $C_6H_5Si(OCH_2CH_3)_3$ and $C_6H_5\overset{\downarrow}{S}i(OCH_2CH_2)_3N$ on one hand and $C_6H_5(C_3H_7)Si(OCH_2CH_3)_2$ and $C_6H_5\overset{\downarrow}{S}i(OCH_2CH_2)_2(CH_2CH_2CH_2)N$ on the other hand is small with 0.33 and 0.29, respectively[9]. Besides, the calculated $\sigma_I = -0.61$ for the homosilatranyl group, in which the Si←N distance is about 10 pm larger than that in the silatranyl group, exceeds the value of −0.56 for the latter. Therefore, no direct relationship between electronic characteristics of the silatranyl group and the Si←N interaction was observed. Only for Si-substituted 2,2-diphenyl-1,3-dioxy-6-aza-2-silacyclooctane, containing an intramolecular coordinate Si←N bond, a linear relationship between the calculated para-carbon chemical shift $\Delta\delta C_p$ for a coordinate form (the model compound is $Ph_2Si(OCH_2CH_3)_2$) and the displacement of silicon from the equatorial plane OOC: $\delta^{13}C = -10.1 + 17.3\,\Delta Si$ ($r = 0.999$), have been reported [263].

For the silatranes of the general formula $(CH_3)_3SiZ\overset{\downarrow}{S}i(OCH_2CH_2)_3N$, with Z = $CH_2{-}CH_2$, CH=CH, C≡C, and model triethoxysilanes, the absolute value of the one-bond $^{13}C{-}^{29}Si$ coupling constants of both penta- and tetra-coordinate silicon increases with a rise of the bond order: 109.2 and 99.0 Hz (for $CH_2{-}CH_2$), 126.0 and 117.2 Hz (CH=CH), and 146.7 and 145.9 Hz (C≡C) [778]. A comparison of other carbon—silicon coupling constants for the penta- and tetra-coordinate silicon compounds shows that the positive inductive effect of the silatranyl group exceeds that of the triethoxysilyl one. Thus, the $^1J_{SiC}$ values of the trimethylsilyl group are always greater in the silatrane series than in the corresponding triethoxysilane derivatives, consistent with dependence of the coupling constants for trimethylsilyl compounds Me_3SiX on the substituent electronegativity [798]. In the $(CH_3)_3SiC{\equiv}CY$ series, $^1J_{SiC_\alpha}$ depends linearly on the inductive (σ_I) and resonance (σ_I) and resonance (σ_R) effects of the substituent Y [799]. A greater $^1J_{SiC}$ value for silatrane with Z = C≡C than that for the corresponding derivative of triethoxysilane (80.2 and 74.4 Hz, respectively) and its closeness to 79.0 Hz found for

[9] Similar differences were observed in germatranes and 2-carbagermatranes [797].

$(CH_3)_3SiC{\equiv}CSi(CH_3)_3$ [800]) shows that the σ-acceptor ability of the silatranyl group with respect to the triple bond is lowered and the total electron effect is close to that of the $Si(CH_3)_3$ group.

^{13}C chemical shifts for the $(OCH_2CH_2)_3N$ moiety for Si-substituted silatranes in $CDCl_3$ changes over a narrow range of 57.5 ± 0.5 ppm (CO) and 51.2 ± 0.3 ppm (CN) [801]). In general, protonation of tertiary amines [802]) or complexation with Lewis acids [803]) induces a diamagnetic shift to 3 ppm for α-carbons. The upfield shift of the ^{13}C resonance on protonation has been noted for N-methyl piperidine (equatorial position) [804]) and for diazabicyclo[2.2.2.]octane [805]). Nevertheless, the differences in ^{13}C shifts of Si-substituted silatranes and tris(2-trimethylsiloxyethyl)amine ($\delta^{13}C$ = 61.9 (CO) and 58.8 ppm (CN) [256])) are too large to be dependent only on the Si←N bonding. This may be explained by steric 1,4-interactions of methylene groups in the silatrane framework (see Sect. 4.5.1). Upfield shifts are produced by polarization of electrons from interacting hydrogens along C—H bonds [806, 807]).

The ^{13}C NMR spectra of $Si(acac)_2X_2$ indicate the presence of a nonexchanging mixture of the cis and trans isomers for X = acetato, and the trans structure for X = Cl. The latter disproportionates in solution to $Si(acac)_3Cl$ and $SiCl_4$.

Coordinative interaction of the type Si←O=C produces a low field shift of the ^{13}C resonance of the carbonyl carbon in 1-(trimethylsilyl)-2-propanone [808]) and in (aryloxymethyl)trifluorosilanes $XC_6H_4COOCH_2SiF_3$ [809]). In the latter cases a correlation between ^{13}C chemical shifts and substituent constants has been established (r = 0.974): $\delta^{13}C = 175.3 - 2.8 \cdot \sigma_I - 1.0 \cdot \sigma_R^-$.

The NMR resonance signal of the isocyano carbon in hexa-coordinate silicon complex $\{Si[NCCr(CO)_5]_6\}(NEt_4)_2$, 166.3 ppm, shifts to downfield by 19.5 ppm relative to $[Cr(CN)(CO)_5]^-$ [810]).

Small downfield shifts of the C_{meta} and C_{para} resonance for aryltrichlorosilane in acetonitrile as compared with those of neat liquid apparently cannot be considered as convincing proof of complexation with the solvent [811]).

4.5.3 ^{19}F-NMR Data

The ^{19}F NMR spectra of penta- and hexa-coordinate silicon compounds containing a Si—F bonds generally exhibit a singlet due to a fast intramolecular (and often intermolecular) exchange [174, 812]). When passing from SiF_4 to SiF_5^- and SiF_6^{2-}, the ^{19}F resonance shifts downfield from −160 to −136 and −128 ppm (Table 15). Since the latter two values are averaged over the values of $\delta^{19}F$ for equatorial and axial atoms, the shielding of the fluorine atoms participating in the formation of hypervalent bonds is less than that of covalent bonded by about 35 ppm. For the $Ph_2SiF_3^-$ anion (in the case of no pseudorotation) the shielding of two axial fluorine nuclei is by 36 ppm less that of equatorial [398]).

The ^{19}F resonance of $(Ph_3P)_2Pt \cdot SiF_4$ is observed at lower field (−137.3 ppm) than the signal for SiF_4. A great broadening of the first ^{19}F signal (to 0.6 Hz) and the absence of coupling with ^{31}P and ^{195}Pt nuclei indicate that an intermolecular exchange takes place in the solution [391]). Accordingly, complexation of SiF_4 with anions (for instance, with BH_4^-) is accompanied by a downfield shift of the ^{19}F resonance (−137.7 ppm [819])) with respect to the SiF_4 signal.[10] The donor-acceptor Si←N

[10] Simultaneously the ^1H signal of BH_4^- (−0.34 ppm) shifts highfield with respect to a free BH_4^- anion (−0.16 ppm).

Table 15. ^{19}F NMR Data for Tetra-, Penta-, and Hexa-coordinate Silicon with Fast Intramolecular Rearrangement

Compounds	Solvent (°)	δ($CFCl_3$), ppm	$^1J_{SiF}$, Hz	Ref.
Tetra-coordinate				
SiF_4	SO_2	−158.8	172	813)
		−160.3	178	814)
	CH_2Cl_2	−156.8		398)
			170–178[a]	815)
CH_3SiF_3	CH_2Cl_2	−139.2	260	398)
$ClCH_2SiF_3$	CH_2Cl_2	−144	267	398)
$C_6H_5SiF_3$	CH_2Cl_2	−143.0	266	398)
$(C_6H_5)_2SiF_3$	CH_2Cl_2	−144.1	290	398)
Penta-coordinate				
SiF_5^-	$CHCl_3$	−137.5		816)
	CH_2Cl_2	−136.4		460)
		−136.7	140	817)
		−136.0	148	398)
$CH_3SiF_4^-$	CH_2Cl_2 (−60)	−110.9	218	398)
$C_6H_5SiF_4^-$	CH_2Cl_2	−118.7		818)
$(C_6H_5)_2SiF_3^-$	CH_2Cl_2	−110	229	398)
$BH_4SiF_4^-$	CH_2Cl_2	−137.7		819)
$SiF_4NO_2^-$	CH_2Cl_2	−136.5	150	813)
SiF_4CN^-	SO_2	−130.1	105	813)
$SiF_4(OCOCH_3)^-$	SO_2	−120.8	155	813)
	SO_2 (−70)	−130.6		820)
$F(CH_3)(C_6H_5)\overline{Si(o\text{-}C_6H_4)C(CF_3)_2CH_2^-}$		−97.6		463)
$[Ph_3P]_2PtSiF_4$	$(CH_3)_2CO$	−137.3		391)
$F_3SiCH_2CH_2S(O)CH_3$	CH_2Cl_2	−124.7	226.4	821)
$F_3SiCH_2OC(O)C_6H_5$	CH_2Cl_2	−136.56	225.0	822)
$F_3Si(o\text{-}C_6H_4)CH_2N(CH_3)_2$		−128		742)
Hexa-coordinate				
SiF_6^{2-}	H_2O	−118.6	108.1	823, 824)
	SO_2	−125.0	110	813)
	CH_3OH	−131.0	109.8	825)
	CH_2Cl_2	−127.4		818)
$SiF_5(NH_3)^-$	CH_2Cl_2	−123.6	120	817)
$SiF_2(Bipy)_2^{2+}$		−126.9	197	826)

[a] The values have been obtained for 25 solvents

interaction causes an upfield shift of the ^{19}F resonance for the fluorine separated from silicon by one carbon atom [712].

On increasing the coordination number of silicon bonded to three and more fluorine atoms, the $^1J_{SiF}$ value decreases regularly (Table 15). The ratio between coupling constants in the spectra of SiF_4, SiF_5^-, and SiF_6^{2-} (1.6:1.35:1) corresponds to a regular decrease of s- and p-character of the bonding hybride silicon AOs in molecules of these compounds (1.5:1.2:1).

In the series of $RSiF_5^{2-}$ the one-bond Si—F constant is a minimum for R = F

Table 16. ^{19}F NMR Data for Penta- and Hexacoordinate Silicon With Slow Intramolecular Rearrangement

Compounds	Solvent (°C)	$\delta^{19}F(CFCl_3)$, ppm	$^{1}J_{Si}$, Hz	$^{n}J_{FF}$, Hz	Ref.
Penta-coordinate					
$(C_6H_5)_2SiF_3^-$	CH_2Cl_2 (−100)	− 98.0 (2 F_a) −138.0 (F_e)	254 212	5	398)
$[(AcO)F_3SiF'SiF_3(OAc)]^-$	SO_2 (−70)	−128.1 (F_a) −123.9 (2 F_e) − 84.6 (F′)	120 (3 F)	18 (F_aF_e) 52 (F_eF') 0 (F_aF')	813)
$FSi[(o\text{-}C_6H_4)C(CF_3)_2CH_2O]_2^-$	CD_3CN	−130.1 (F_e)	227		463)
$F_3Si(o\text{-}C_6H_4)CH_2N(CH_3)_2$	CD_2Cl_2 (−40)	−128 (F_a) −148 (2 F_e)			742)
$F_2(CH_3)Si(o\text{-}C_6H_4)CH_2N(CH_3)_2$	CD_2Cl_2 (−95)	−111 (F_a) −154 (F_e)			742)
$F_3SiCH(CH_3)NC(O)CH_2CH_2CH_2$	$CH_2Cl_2:CDCl_3:CCl_4$ 13:27:60 (−102)	−123.49 (F_a) −135.53 (F_e) −139.71 ($F_{e'}$)	243 (F_a) 220 (2 F_e)	15.0 (F_eF_e) 41.5 (F_aF_e)	830)
$F_3SiCH_2OC(O)C_6H_5$		−127.43 (F_a) −140.69 (2 F_e)			822)
$F_3SiCH_2CH_2S(O)C_2H_5$		−113.1 (F_a) −131.4 (F_e)		19.5 (F_eF_e) 36.6 (F_aF_e)	821)
$XF(CH_3)Si(o\text{-}C_6H_4)CH_2N(CH_3)_2$	CH_2Cl_2 (−90)				
X = H $OC(CH_3)_3$ $N(C_2H_5)_2$ Cl		−122.5 (F_a) −119 (F_a) −120 (F_a) −142 (F_e)			831)

Hexa-coordinate					
SiF_6^{2-}			108 (2 F_a)		824)
			108 (4 F_e)		
$(C_6H_5)SiF_5^{2-}$	CH_2Cl_2 (—28)	—114.5 (4 F_e)	181.8 (4 F_e)	11.0	832)
		—121.3 (F_a)			
$[Si(C_2O_4)F_4]^{2-}$	CH_3OH	—126.4 (2 F_{cis})	137.7	2.4	825)
		—138.0 (2F_{trans})	118.5		
$[Si(C_2H_4)_2F_2]^{2-}$	CH_3OH	—133.5 (2 F_a)	142.9		825)
$SiF_5 \cdot NH_3^-$	CH_2Cl_2 (—83)	—136.6 (F_a)	120 (F_a)		817)
		—118.6 (4 F_e)	125 (4 F_e)		
$SiF_5 \cdot (C_2H_5)_2NH^-$	CH_2Cl_2 (—60)	—136.9 (F_a)			817)
		—127.5 (4 F_e)			
$SiF_4 \cdot$ Bipy	$ClCH_2CN$	—121.5 (2 F_a)		13	833)
		—143.9 (2 F_e)			

and rises with increasing the electron-donor ability of the substituent R. Nevertheless, there is no simple relationship between a change in the strength of the Si—F bond and $^1J_{SiF}$ values [827]. One may only state that downfield ^{19}F NMR shifts and smaller $^1J_{SiF}$ values correspond to longer Si—F bonds in fluorosilane complexes.

A fast exchange of the fluorine atoms in $CH_3SiF_4^-$ was established by dynamic NMR spectroscopy. The equilibrium of the process $CH_3SiF_4^- \rightleftarrows CH_3SiF_3 + F^-$ is shifted to the left. A kinetic study of this process has shown two parallel mechanisms of the fluorine exchange. One of them prevails within the temperature region above —25 °C (the activation energy being 10.2–12.3 kcal · mol^{-1}). The process follows the scheme $RSiF_4^- + HF \rightleftarrows RSiF_3 + FHF^-$, which is most likely due to the presence of catalytic HF impurities. The other mechanism, dominating within the temperature region below —25 °C (the activation energy being 0.4–2.0 kcal ×mol^{-1}), involves an intermolecular exchange with methyltrifluorosilane, which is present in the solution, according to [828]:

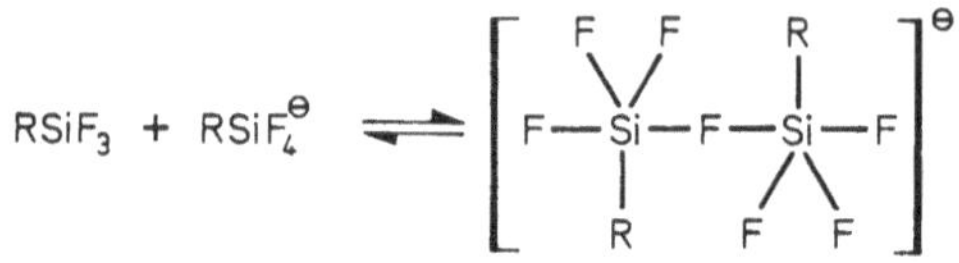

It was, however, impossible to establish a correlation between the effect of any HF impurity and bimolecular exchange by NMR spectroscopic data. The activation energy of the exchange reaction between SiF_6^{2-} and HF as calculated from the shape of the ^{19}F resonance line is 8.5 ± 1.0 kcal · mol^{-1} [829]. One should bear in mind that HF and other compounds which interact with $RSiF_4^-$ to form HF, that is H_2O, CH_3OH, etc., favor fluorine exchange. The acceptors Pr_2NH, $(Me_3Si)_2NH$, etc. bond HF in the solution and inhibit this process [817].

A fast intermolecular exchange of the fluorine atoms proceeds also in solutions of mixtures of silicon fluorides with different coordination number of the Si atom. If the silicon coordination number in the mixture of fluorosubstituted compounds is the same, fluorine exchange proceeds very slowly. In solutions of methyltrifluorosilane no intermolecular fluorine exchange is observed; however, addition of diethylamine initiates an exchange due to the formation of $CH_3SiF_3 \cdot HN(C_2H_5)_2$, in which silicon is penta-coordinate. When ammonia is added to a solution containing SiF_5^- and SiF_6^{2-}, the exchange is slowed down due to the formation of $[NH_3 \cdot SiF_5]^-$ which is characterized by a slow exchange with SiF_6^{2-} [818].

Careful purification of solutions of fluorosilane and fluorosilicate anions and a decrease in concentration and temperature result in a fine structure of ^{19}F NMR spectra due to a non-equivalent arrangement of ligands (Table 16). For example, in the ^{19}F NMR spectra of octahedral complexes of the type $RSiF_5^{2-}$ a doublet (equatorial atoms) and a quintuplet are observed whose mutual arrangement depends on the nature of the ligand [834]. In fluorine derivatives of penta- or hexa-coordinate silicon with slow exchange, the $^1J_{SiF}$ values (if they can be determined) are greater for the fluorine atoms with chemical shifts in low field.

Replacement of fluorine in SiF_6^{2-} by phenyl causes a downfield shift of the ^{19}F resonance signals with the quintuplet in low field. For substituted octahedral anionic complexes, $^1J_{SiF}$ decreases in the order cis > trans and both values are larger than that for SiF_6^{2-} (Table 16). These changes are in accord with the mutual influence of ligands in non-transition element compounds applied to NMR chemical shifts [835] and coupling constants [827, 836].

Replacement of a fluorine atom in $F_3Si(o\text{-}C_6H_4)CH_2N(CH_3)_2$ with a methyl group in the equatorial plane ($CSiF_2$) shifts the ^{19}F resonance of the axial fluorine to low field and that of the equatorial atom to high field [831]. For the adducts F_5SiD^- (D = NH_3, $HN(C_2H_5)_2$) the shileding of the fluorine nuclei in the trans position remains constant with −137 ppm and equal to that for the SiF_5^- anion. For the remaining equatorial fluorine atoms this value decreases, the decrease being greater for D = NH_3 (−118.6 ppm) than for D = $HN(C_2H_5)_2$ (−127.5 ppm) [817]. In the series of (aroyloxymethyl)trifluorosilanes the axial Si—F bond is most sensitive to the degree of coordinate Si←O interaction due to the trans-effect [837]. In hexa-coordinate chelate silicon anions $[F_4SiDD]^{2-}$ the shielding of ^{19}F nuclei in the trans-position with respect to bidentate ligand is by ~4.5 ppm less for five-membered rings (oxalate) than for six-membered species (malonate), whereas it does not change in the cis-position [825].

The geminal coupling constant $^2J_{FF}$ = 13 Hz for $F_4Si\cdot$ bipy is considerably smaller as compared with the value of 58 Hz for analogous germanium complexes [833]. The first value is close to $^2J_{FSiF}$ for the SiF_2 fragment in tetrahedral organisilicon compounds [838, 839].

At lower temperatures, the ^{19}F signals of $XFMeSi(o\text{-}C_6H_4)CH_2NMe_2$ shift downfield for X = H, O-t-Bu and NEt_2 whereas they shift upfield for X = Cl. This has been attributed to axial F in the former cases and equatorial in the latter case. However, for X = $OCOC_6H_5$ a mixture of two penta-coordinate conformers with axial and equatorial fluorine arrangements was obtained below the coalescence temperature [831].

If an asymmetric atom is located in the chelate ring plane upon intramolecular interaction this gives rise, apart from axial-equatorial non-equivalence, to diastereotopy of two equatorial substituents at the silicon stom. This is observed in $F_3SiCH(CH_3)NC(O)CH_2CH_2CH_2$ [837] and $F_3SiCH_2CH_2S(O)R$ (R = CH_3, C_2H_5, $C_6H_5CH_2$) [821] with asymmetric atoms of carbon and sulfur. According to the ^{29}Si chemical shifts and ^{19}F coalescence temperature the Si←O interaction in the latter case weakens in the given sequence.

Non-equivalence of the CF_3 groups in the ^{19}F NMR spectrum of $FSi[(o\text{-}C_6H_4)C(CF_3)_2O]_2^-$ supports the trigonal bipyramidal structure of silicon [238, 840]. The appearance of the ^{19}F—^{31}P splitting in the NMR spectra of $(F_3SiCH_2CH_2PMe_2)_2PdCl_2$ allowed the silicon atom to be penta-coordinated due to the Si←Pd interaction [841]. The ^{19}F chemical shifts of anionic fluorosilicates correspond to the values for isolated ion only at inifinite dilution. The ion-pairs interaction between the hexafluorosilicates and various cations have been studied [842−846].

4.5.4 ^{29}Si-NMR Data

A number of reviews concerning all aspects of ^{29}Si NMR is available [847−849]; they include but a brief note about silicon complexes. All experimental data for compounds in which the coordination number of silicon rises above four display the

Table 17. ^{29}Si Chemical Shifts (ppm) of Some Penta- and Hexa-coordinate Silicon Complexes

N	Compounds	Solvent	δ	Ref.
Penta-coordinate				
Cationic				
1.	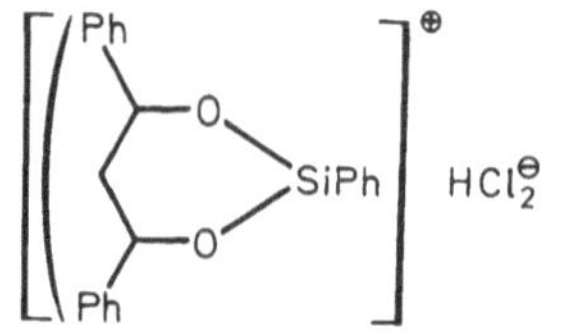	DMSO	−175.8	850)
2.	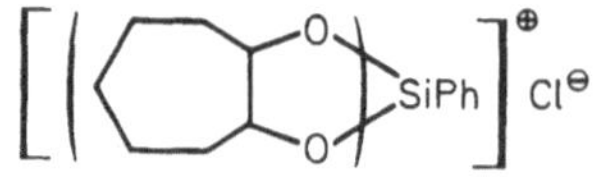	DMSO	−141.3	848)
Neutral		See Table 18		
Anionic				
3.	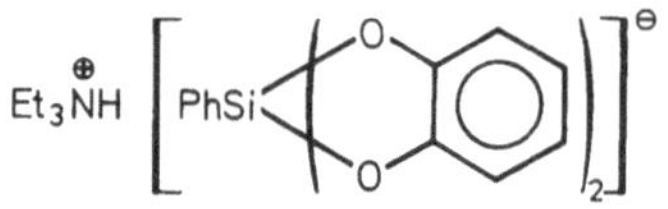	$Cl_2DCCDCl_2$	− 87.0	850)
4.	$(Me_2N)_3\overset{+}{S}\{FSi[(o\text{-}C_6H_4)C(CF_3)_2O]_2\}^-$	CD_3CN	− 76.6	463, 840)
Hexa-coordinate				
Cationic				
5.	$(Acac)_3Si^+ ZnCl_3^-$	DMSO	−193.7	850)
6.	$(Acac)_3Si^+ HCl_2^-$	DMSO	−192.4	850)
		$CHCl_3$	−194.4	824)
7.	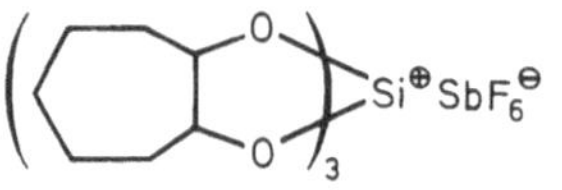	CD_3OD	−139.4	850)
8.	$[Bipy_2SiCl_2]^{2+}$ 2 Cl^-	$CH_3OH/(CH_3)_2CO$ (1:1)	−161.1	826)
Neutral				
9.	$(Acac)_2Si(OAc)_2$	DMSO	−196.8	850)
10.	$(Acac)_2SiClCH_3$	$CHCl_3$	−149.5	848)
11.	$CH_3Si(OCOCH_2)_3N$ [a]	DMSO	−135.8	710)
12.	$C_6H_5Si(OCOCH_2)_3N$ [a]	DMSO	−146.5	710)
Anionic				
13.	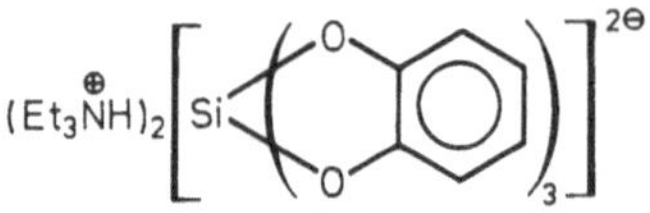	DMSO	−139.3	850)
14.	$(n\text{-}Bu_4N^+)_2[Si(C_2O_4)_3]^{2-}$	C_6F_6	−173.3	824)

[a] Intermolecular coordinated by solvent

Table 18. ^{29}Si Chemical Shifts (ppm) of Some Neutral Penta-coordinate Silicon Complexes

Compounds	δ	Δδ[a]	Ref.
$XSi(OCH_2CH_2)_3N$			695, 851, 852)
X = H	−83.6	24.1	
CH_3	−64.0	20.6	
$ClCH_2$	−77.2	21.1	
Cl_2CH	−83.2	14.8	
$CH_2{=}CH$	−81.6	23.2	
C_6H_5	−81.7	23.3	
$4\text{-}ClC_6H_4O$	−98.6	11.9	
$XSi(OCH_2CH_2CH_2)(OCH_2CH_2)_2N$			776)
X = CH_3	−61.5	17.3	
$ClCH_2$	−85.1	26.8	
$XSi(CH_2CH_2CH_2)(OCH_2CH_2)_2N$			776)
X = CH_3	−29.1	24.8	
C_6H_5	−45.6	26.0	
$Me_2ClSiCH_2NC(O)CH_2CH_2CH_2$	− 9.7	32.6[b]	854)
$Me_2ClSiCH_2NC(O)CH_2CH_2CH_2CH_2$	−34.7	57.6[b]	854)
$MeFHSi(o\text{-}C_6H_4)CH_2NMe_2$	−36.1	44.7	332)
$Me_2ClSiCH_2NMeCH_2CH_2NMe_2$	− 4.5	27.8	116)
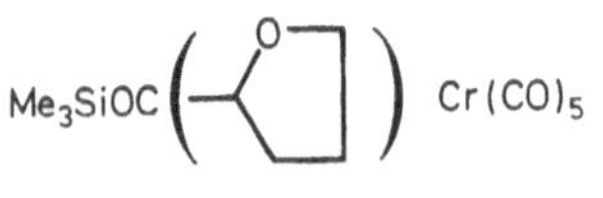	1.8	11.7	503)
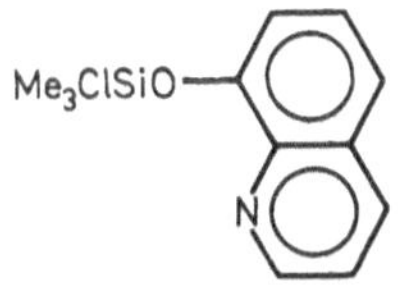	−21.0	35.1	855)
$F_3SiCH_2OC(O)C_6H_5$	−94.8	23.5	822)
$F_3SiCH_2CH_2S(O)CH_2C_6H_5$	−83.6	31.8	837)

[a] Coordination shifts are the difference between the model tetracoordinate molecule, $\Delta\delta = \delta - \delta_m$;
[b] Relative chloromethyl derivatives

^{29}Si signals at much higher field as compared with model tetra-coordinate analogs. For the penta-coordinate $CH_3SiF_4^-$ anion, the δ^{29}Si value of 110.7 ppm [849] is 55.5 ppm upfield of CH_3SiF_3; the difference between the δ^{29}Si for SiF_6^{2-} (−185.3 [824])–184.2 ppm [823]) and SiF_4 is 72 ppm. The ^{29}Si chemical shift for the $[F(C_6H_5)(CH_3)Si(o\text{-}C_6H_4)C(CF_3)_2O]^-$ anion is 99.4 ppm less than that for

$(C_6H_5)(CH_3)\overset{\downarrow}{Si}(o\text{-}C_6H_4)C(CF_3)_2O$[463]. Donor-acceptor interactions during the formation of neutral complexes usually cause smaller changes in $\delta^{29}Si$ than in the case of ionic complexes and do not exceed 70 ppm (Tables 17 and 18).

Structurally similar cationic (5) and neutral (9) compounds as well as cationic (7) and anionic (13) complexes containing a hexa-coordinate silicon atom bonded only to oxygen atoms have identical ^{29}Si chemical shifts. This may be due to a symmetric arrangement of the ligands in a coordination octahedron which effectively quenches any charge on the central atom [850]. This is not observed in penta-coordinate complexes. For example, the difference between the ^{29}Si chemical shifts in spectra of compounds (2) and (3) is more than 50 ppm.

Most semiempirical interpretations of silicon chemical shifts are based on the variations of the paramagnetic term only [856–858]. According to Engelhardt et al. [856] the value of paramagnetic contribution to the ^{29}Si shielding constant is parabolically related to the total electron charge on a tetrahedral silicon atom[11]. When the silicon atom is surrounded by electronegative substituents, i.e., when it has a pronounced ability to increase its coordination number, the ^{29}Si resonance is displaced upfield with a decrease in the negative charge. This is applicable to penta-coordinate silicon compounds only under the assumption that the positive charge on the silicon nucleus increases when coordinate interactions take place. The electron changes at the acceptor atom itself lead to an increase in positive net charge because the original gain in electron density at the acceptor atom may be transferred to other parts of the acceptor component [337, 860, 861]. Recently, it has been shown by X-ray spectroscopy that the silicon atom in silatranes has an enhanced positive charge. A linear dependence was observed between the ^{29}Si chemical shift and the position of SiK_α lines in the X-ray fluorescence spectra: $\Delta E_e(SiK_\alpha) \cdot 10^2 = 29 - 0.36\delta^{29}Si$ ($r = 0.95$) [680].

According to Ernst et al. [862] empirical parabolic dependence of ^{29}Si chemical shifts on the sum of electronegativities of substituents R_i at the tetra-coordinate central silicon atom, $\delta^{29}Si = F\left[\sum_i (\chi_{R_i} - \chi_{Si})\right]$, coordinate interaction (for instance, Si←N), decreasing the electronegativity of silicon by $\Delta\chi_{Si}$, increases the shielding of a silicon nucleus by $\Delta\delta^{29}Si = \Delta F_i(4\,\Delta\chi_{Si})$. The calculated decrease in electronegativity of the silicon atom of 1-methylsilatrane under Si←N interaction in the molecule is 0.2. This agrees with an enhanced σ-donor ability of the silatrane skeleton $\overset{\downarrow}{Si}(OCH_2CH_2)_3N$ as compared to that of the triethoxysilyl group $Si(OCH_2CH_3)_3$.

The concepts developed for compounds of tetrahedral silicon do not explain many experimental peculiarities of ^{29}Si NMR spectroscopy in the case of penta-coordination. Thus, for example although the interatomic Si ... N distance in 1-chloromethylhomosilatrane is much greater than in 1-chloromethylsilatrane, the shielding of silicon in the former is greater (−85.1 ppm) than in the latter (−77.2 ppm). Even in a narrow series of 1-substituted silatranes the coordination shifts $\Delta\delta_{Si}$

[11] The dependence of ^{29}Si chemical shifts on the silicon charge is approximated for some compounds by the fifth-order polynom [859].

Table 19. Coordination Chemical Shifts $\delta^{29}Si$ (ppm) for Structurally Studied Silatranes

Si-substituent		$\Delta\delta$	$l_{Si \leftarrow N}$, pm	Ref.
CH_3	2-carba[a]	24.8	234	[776]
$ClCH_2$	3-homo[a]	26.8	225	[776]
CH_3O	2-carba[a]	20.5	222	[776]
C_2H_5		21.2	221	[776]
α-C_6H_5		23.3	219	[695]
CH_3		21.7	217.5	[553]
		20.6		[695]
CH_3	3,7-dion[a]	34.0	215	[864]
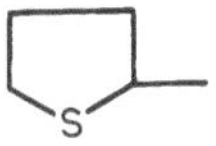		17.1	213	[863]
$ClCH_2$		19.2	212	[553]
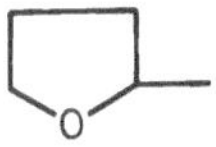		22.1	211	[863]
$CH_2S^+(CH_3)_2 \cdot I^-$		23.4	204.6	[553]

[a] See Table 8 and the Text

(Table 19) are related neither to Si←N distances in the solid (which vary over a wide range) nor to the chemical shifts of CH_2N protons which properly reflect the degree of Si←N bonding in solutions [865].

Thus, ^{29}Si shielding in silatranes is caused not only by the degree of Si←N interaction. An additional effect for the coordination shift that may be mentioned is a sensitivity of the ^{29}Si chemical shift to O—Si—O bond angles. An empirical correlation correlating ^{31}P chemical shifts and bond angles of phosphate esters is known [866]. The importance of this is supported by a few factors. For compounds of hexa-coordinate silicon there are great differences between ^{29}Si resonance of five- ($<OSiO \approx 86°$) and six-membered ($\sim 93°$) chelate rings (Table 17 compounds (1) and (2) as well as (5) and (7)). Besides, there is a dependence of the chemical shift on the arrangement of silicon covalent bonds among non-equivalent positions of a trigonal bipyramide. Molecules of 1-ethylsilatrane ($\delta^{29}Si = -67.1$ ppm) and 1-methoxy-2-carbasilatrane (—62.6 ppm), which have almost identical Si←N distances (222 and 221 pm, respectively) and, consequently, close trigonal bipyramidal configuration, differ only by that the Si—C bond is axial in the former and equatorial in the latter compound. The difference in shielding is 4.5 ppm.

Molecules of N-substituted 2,2-diaryl-1,3-dioxa-6-aza-2-silacyclooctanes exist in solutions in two equilibrium forms: non-coordinate (without a Si←N bond) and coordinate [263]. An increase in the donor ability of the substituents at the silicon atom and in size of the substituents at the nitrogen atom shifts the equilibrium to a conformation lacking the Si←N bond. The activation energy of < 11 kcal · mol^{-1} for this process is much less than that of similar tin derivatives [867]. A calculated change in the ^{29}Si chemical shift for a completely coordinated boat-boat shape as compared with a model molecule of diphenyldiethoxysilane depends linearly on the shift of the silicon atom from the equatorial plane of three adjacent atoms OOC [263]: $\Delta\delta^{29}Si = -77.7 + 142.0 \cdot \Delta Si$ ($r = 0.99$). Consequently, for an ideal trigonal bipyramid ($\Delta Si = O$) the silicon shielding due to coordinative Si←N interaction rises by ~78 ppm.

Due to Si←N interaction in N-(dimethylchlorosilylmethyl)-lactames, $Me_2ClSiCH_2NC(O)(CH_2)_nCH_2$ with n = 1–3, the ^{29}Si shielding is much greater than in the corresponding chloromethyl derivatives of tetracovalent silicon (Table 18). The intramolecular coordinative Si←O interaction increases with the nucleophilicity of oxygen in the following order of n: 1 > 3 > 2 [868]. No interaction was observed, however, in the corresponding derivative of succinimide, which is a pyrrolidone-2 analog with two C=O groups located equivalently ($\delta^{29}Si$ = 24.2 ppm, $\Delta\delta^{29}Si$ = 1.3 ppm [854]).

The shielding of the silicon atom in dimethyl(trifluoroacetoxymethyl)acetoxysilane ($\delta^{29}Si$ = −4.5 ppm) is by 26.6 ppm greater than in trimethylacetoxysilane [869]. Examination of temperature and concentration dependences of $\delta^{29}Si$ in solutions of $(CH_3)_2(CH_3COO)SiCH_2OCOCF_3$ points to intramolecular Si←O coordination [869]. Methyltris(trifluoroacetoxy)silane forms penta-coordinate ionic complexes with RCOOM salts ($R = CH_3, CF_3$; $M = Na, NR_4$) and 2:1 and 1:1 neutral adducts with pyridine and HMPTA, respectively [869, 870]. Titration of chloromethyltrifluorosilane with HMPTA by ^{29}Si NMR spectroscopy provides evidence for the formation of a 1:1 complex. In the case of $PhCOOCH_2SiF_3$, 1:1 and 1:2 complexes are observed, indicating an intramolecular Si←O interaction in the 1:1 complex [837].

Introduction of a methyl group into position 3 of the four-membered ring of 1,1-dimethyl-1-silacyclobutane shifts considerably the ^{29}Si resonance highfield; this may be due to transannular 1,3-interaction [871]. The ^{29}Si shielding in a series of 1,1-dimethylsilacycloalkanes $(CH_3)_2Si(CH_2)_nCO(CH_2)_{m-1}CH_2$ (n, m = 2–6) at first decreases with increasing n and m and then, beginning with m + n = 8, rises. This was attributed to transannular Si←O interaction in their molecules [872].

Thus, the increase in the coordination number of the silicon atom is accompanied by a highfield ^{29}Si chemical shift. An increase in shielding upon complexation is also observed for other heavy nuclei of the Group IVA elements, ^{73}Ge [873], ^{119}Sn [874–878], ^{207}Pb [875]. The high sensitivity of the ^{29}Si resonance to a change in the coordination number of the silicon atom is a simple and reliable way of establishing intra- and intermolecular coordinative interaction in molecules of organosilicon compounds.

4.5.5 ^{14}N-, ^{15}N-, ^{17}O- and ^{31}P-NMR Data

Nitrogen NMR spectroscopy would appear to be ideal for the study of donor-acceptor interaction of nitrogen ligands, but quadrupole broadening for ^{14}N and experimental difficulties for ^{15}N, even with enriched samples, have limited such studies. Tertiary amines undergo downfield shifts in polar or acid solvents possibly due to a decrease in the electron density caused by hydrogen bonding to the nitrogen lone pair or protonation [879–881]. Early NMR experiments have shown that the ^{14}N resonance of 1-methylsilatrane ($\delta^{14}N = -346.2 \pm 3.6$ ppm) and 1-hydrosilatrane (-354.7 ± 4.4 ppm) are at higher field relative to that of triethylamine and differ greatly from boratrane (-325.0 ± 2.0 ppm) [882]. Hence, there is an influence of both deviation of the $N(CH_2)_3$ valence angles (diamagnetic cyclization shifts [883]) and the acceptor atom, the latter being not predictable. Therefore, the analysis of nitrogen chemical shifts may lead to contradictory results [884, 885].

A lower ^{14}N resonance half-high band width for 1-hydrosilatrane 638 ± 68 Hz as compared to 2211 ± 103 Hz in 1-methylsilatrane shows that a stronger Si←N interaction in the former leads to a smaller electric fields gradient and/or a smaller distortion towards the planar nitrogen results in a marked increase in the quadrupolar coupling constant.

The ^{15}N shifts of Si-substituted silatranes were correlated with the Taft constants of X and Si←N bond distances (pm) [881, 887]

$$\delta^{15}N = -358.12 + 3.29\sigma^* \qquad r = 0.98$$

$$\delta^{15}N = -236.8 - 0.544 l_{Si \leftarrow N} \qquad r = 0.99$$

The same relationships were found for $R^1R^2Si(OCH_2CH_2)_2NCH_3$ [885]. The ^{15}N and ^{29}Si NMR data obtained for the silatranones indicate that an increase in the number of carbonyl groups in the atrane framework enhances charge transfer by the donor-acceptor Si←N bond [710].

Silicon-nitrogen coupling constants, $^1J(^{29}Si-^{15}N)$, measuring the "s" electron density in the Si←N bond, form a regular trend in reasonable agreement with the known order of the donor-acceptor bond strengths (Table 20).

The dependence on the aggregate state is a specific feature of labile intramolecular bond in silatranes and 1,3-dioxa-6-aza-2-silacyclooctanes [889-892]. For example, the Si←N bond distances and ^{15}N and ^{29}Si chemical shifts for 1-methylsilatrane vary greatly:

	Crystal state	Solution state				Gas
		D_2O	DMSO	CCl_4	C_6H_{12}	
$\delta^{15}N$, ppm	−355.6	−353.5	−356.4	−363.4	−364.8	−370.7
$\delta^{29}Si$, ppm	− 70.8		− 69.1	− 59.3		
l_{Si-N}, ppm	217.5					245.3

The empirical relationship $\delta^{15}N = -407.8 - 0.75\delta^{29}Si$ ($r = 0.99$) was found by least-square procedure for 1-methylsilatrane in various solvents [889]. The changes in $\delta^{29}Si$ linearly reflect smaller changes in $\delta^{15}N$ for the silatrane. Thus it is clear that the chemical shifts of both silicon and nitrogen, i.e., the acceptor and donor nuclei, are subjected to joint influences. On strengthening of the Si←N interaction the ^{15}N resonance shifts downfield whereas the ^{29}Si shift is displaced in the opposite direction. Other things being equal, the $\delta^{15}N$ and $\delta^{29}Si$ range decreases with strengthening of the Si←N bond in silatranes and weakening in silacyclooctanes [890, 891].

Solvent effects on silatranes are found to be better described by Koppel-Palm [893, 894] and Taft-Kamlet [895] solvent parameters. A good relationship has been established for ^{15}N and ^{29}Si shifts for 1-methylsilatrane:

Table 20. Silicon-Nitrogen Coupling Constants Through Dative Bond in Silatranes [710, 888]

Compounds	Solvent	$^1J(^{15}N-^{29}Si)$, Hz
$XSi(OCH_2CH_2)_3N$ (Si←N)		
X = CH_3	$CDCl_3$	0.2
	DMSO	0.70
C_6H_5	DMSO	1.49
$ClCH_2$	$CDCl_3$	1.5
	CH_3CN	2.4
	DMSO	2.64
CH_3O	DMSO	1.72
Cl_2CH	CH_3CN	3.3
Cl	$CDCl_3$	1.98
$XSi(OCOCH_2)_3N$ (Si←N)	DMSO	3.37
X = CH_3	DMSO	8.2
$ClCH_2$	DMSO	10.9

$$\delta^{15}N = -372.25 + 20.74Y + 11.34P + 0.229E \qquad r = 0.976 \quad n = 21$$ [889]

$$\delta^{15}N = -364.7 + 8.1(\pi^* - 0.20\delta) + 2.9\alpha \qquad r = 0.990 \quad n = 16$$ [896]

$$\delta^{15}N = -364.23 + 7.48(\pi^* - 0.31\delta) + 2.74\alpha \qquad r = 0.99 \quad n = 21$$ [897]

Three solvent parameters are shown to contribute to these variations: the polarity Y, π^*, the polarizability P, δ, and the acidity E, α.

The ^{17}O chemical shifts of equatorial oxygen atoms in silatranes do not differ much from those in triethoxysilanes. This suggests that the greater ability of the silatrane groups for complexing with electrophilic reagents as compared to appropriate model compounds is due to better steric availability of cyclic oxygens rather than to a change in the electronic density on the latter. In accordance with the trans-effect, the axial oxygen shielding in the molecule of 1-ethoxysilatrane is by 14 ppm lower than in tetraethoxysilane [898, 899].

^{17}O NMR spectroscopy presents a fairly informative method in the case of coordinate Si←O interaction involving the oxygen atom of the carbonyl group. In the series of (aroyloxymethyl)trifluorosilanes the ^{17}O chemical shifts depend linearly on the π- and σ-withdrawing properties of substituent X: $\delta^{17}O_{CO} = 277.3 + 28.0 \cdot \sigma_p$ $(r = 0.97)$. In analogous equations for model molecules $XC_6H_4COOCH_2CF_3$ the coefficient of the σ_p constant is considerably lower, i.e., 17.1. With silicon compounds, the carbonyl ^{17}O resonance is highfield shifted by 52–65 ppm whereas the ether ^{17}O resonance is shifted downfield by 18–25 ppm. This as well as the difference in chemical shifts of carbonyl oxygens in the two series of compounds $\delta^{17}O(Si) - \delta^{17}O(C) = -60.6 + 10.9\sigma_p$ $(r = 0.98)$ results from Si←O interaction [900]. This conclusion is also supported by the correlation dependence of ^{29}Si and ^{19}F chemical shifts on the substituent X electronic effects [822].

In the series considered, the ^{17}O shielding of the carbonyl group increases in the following order of changing X in the para-position: Cl (283.3 ppm) > F (280.2 ppm) > H (274.2 ppm) with decreasing the intramolecular Si←O distance (Table 6).

Strengthening the coordination Si←O=C interaction shifts the ^{17}O resonance to highfield and the ^{13}C resonance to downfield[900]. In lactame derivatives, $Me_2ClSiCH_2\overline{NC(O)CH_2(CH_2)_nCH_2}$ (n = 1 and 2), the ^{17}O resonance for the five-membered ring is more downfield relative to that for the six-membered ring, i.e., 273 and 256 ppm, respectively, although the Si←O interaction is stronger in the former, according to the ^{29}Si NMR data[837].

Ph_3PO solutions of various halosilanes (e.g., $SiCl_4$, $MeSiCl_3$, Me_2SiCl_2, Me_3SiCl) show significant downfield ^{31}P shifts relative to the free base. The coordination shift of approximately 8–9 ppm indicates bonding interaction between phosphoryl oxygen and silicon. An effect of added MeCN on the chemical shift of Ph_3PO—Me_3SiCl solutions has been interpreted in terms of a coordination interaction between the nitrile group and silicon[323, 324].

4.5.6 Typical NMR Properties of Coordinate Silicon Compounds

The results of ^1H, ^{13}C, ^{15}N, ^{19}F and ^{29}Si NMR spectroscopic investigations provide unambiguous evidence for the existence penta- and hexa-coordinate silicon in the liquid phase. When the coordination number of silicon is expanded, some NMR parameters change in a fairly predictable manner:

1. An increase of the coordination number of the silicon atom is indicated by a highfield shift of the ^{29}Si signal. However, there exists no relationship between the increasing chemical shifts and the increase in the coordinate interactions of the silicon atom. The substituents rather than any simple donor-acceptor inductive effects dominate the ^{29}Si chemical coordination shift.
2. In comparison with ligands which have an uncoordinated nitrogen atom, in compounds with Si←N coordination the ^1H signals of the protons adjacent to the nitrogen atom are downfield shifted. The differences of the chemical shifts in related compounds correlate with increasing the Si←N interaction.
3. The enhanced separation of the ortho-proton resonance from the meta- and para-signals can be employed as a diagnostic tool to indicate the presence of the penta-coordinate state for phenyl-substituted silicon compounds in solution.
4. Non-equivalence of substituents attached to the silicon atom, usually observed at low temperatures, is caused by the difference between equatorial and axial bonds in a trigonal-bipyramidal configuration of penta-coordinate silicon.
5. There exists a correlation between the NMR spectral parameters of the donor fragment (or silicon-containing group) and electronic effects of substituents at the silicon atom (or at the donor fragment) which cannot be explained in terms of transmission through covalent bonds.
6. Diastereotopic groups or atoms in donor fragment are observed upon intra-molecular interaction; they result from slow inversion of the nitrogen atom and chirality of the silicon atom.
7. There exists a considerable dependence of chemical shifts of silicon (or donor atoms) on temperature and nucleophilicity of the solvent.

A reliable evidence for the silicon coordinative interaction is the observation of the both chemical shifts of more peripheral nuclei (^{1}H, ^{13}C, ^{19}F), as well as the spin-spin coupling constants between the silicon and nearly atoms (largely ^{1}H and ^{19}F).

It is clear that NMR spectroscopy is a suitable analytical tool to ascertain the silicon coordination number in the solution state.

5 Conclusion

The most important recent developments in the structural chemistry of silicon have involved non-tetracoordinate derivatives. In the last 20 years there was an ever increasing interest to penta- and hexa-coordinate silicon compounds. New classes of these compounds were discovered and studied in detail. This progress was stimulated by the development of new experimental techniques and theoretical approaches as well as the specific biological activity finding for some of silicon compounds with expanded coordination sphere.

Silicon donor-acceptor interactions may be a significant factor in determining the minimum energy arrangements in silicon compounds containing oxygen, nitrogen and fluorine atoms. The majority of short silicon contacts can reasonably be described as attractive interactions. This conclusion is based on a survey of many reported crystal structures determined very precisely by neutron diffraction. Comprehensive results obtained by X-ray and electron diffraction data for coordinated silicon compounds ranging from weak (van der Waals type) through medium to strong interaction in a variety of ionic complexes have been tabulated. A meaningful set of structural principles that will aid in the understanding of the penta- and hexa-coordinate silicon formation has been discussed.

Theoretical and experimental studies on penta-coordinate silicon derivatives demonstrate that their existence is determined by a combination of factors: electronegativity of the substituents and steric interactions between substituents. One should also emphasize the role and significance of polydentate ligand (chelate effect), the size and number of chelate rings involving the silicon atom, and strain reduction for five-membered ring system, wich can stabilize unusual structures, as well as to the role of medium effect. This is consistent with the results of NMR spectroscopic studies.

The practical application of NMR spectroscopy to different aspects of coordination chemistry of silicon has been demonstrated.

6 Acknowledgements

The authors are indebted to Profs. Yu. T. Struchkov and Yu. A. Ustynyuk for critically reading and commenting through the manuscript.

7 References

1. Gay-Lussac, J. L., Thenard, L. J.: Mem. Phys. Chem. Soc., Arcueil *2*, 317 (1809)
2. Davy, J.: Phil. Trans. *1*, 352 (1812); Liebigs Ann. Chim. *86*, 178 (1813)
3. Berzelius, J. J.: Ann. Chim. Phys. *27*, 291 (1824)
4. Dolgov, B. N.: Khimija Kremneorganicheskikh soedinenij. Leningrad: Goskhimtekhizdat 1933

5. Dilthey, W.: Ber. dtsch. Chem. Ges. Ber. *36*, 923, 1595, 3207 (1903); Liebigs Ann. Chem. *344*, 300 (1906)
6. Rosenheim, A. V. A.: Ber. dtsch. Chem. Ges. *36*, 1833 (1903)
7. Iler, R. K.: The Chemistry of Silica. New York: Wiley-Interscience 1979
8. Voronkov, M. G., Zelchan, G. I., Lukevits, E. J.: Kremni i Zhizn (Silicon and Life). Riga: Zinatne 1971; 2nd Edition 1977; Silicium and Leben. Berlin: Akademie Verlag 1975
9. Bendz, G., Lindquist, I. (eds.): Biochemistry of Silicon and Related Problems. New York: Plenum Press 1978
10. Voronkow, M. G.: Top. Curr. Chem. *84*, 77 (1979)
11. Voronkov, M. G., D'yakov, V. M.: Silatranes. Novosibirsk: Nauka 1978
12. Fessenden, R., Fessenden, J.: Adv. Organometal. Chem. *18*, 275 (1980)
13. Gusel'nikov, L. E., Nametkin, N. S., Vdovin, V. M.: Acc. Chem. Res. *8*, 18 (1975)
14. Jutzi, P.: Angew. Chem., Int. Ed. Engl. *14*, 232 (1975)
15. Gusel'nikov, L. E., Nametkin, N. S.: Chem. Rev. *79*, 529 (1979)
16. Coleman, B., Jones, M., Jr.: Revs. Chem. Intermediates *4*, 297 (1980)
17. Bertrand, G., Trinquier, G., Mazerolles, P.: Organometal. Chem. Revs. *12*, 1 (1981)
18. Schaefer, H. F.: Acc. Chem. Res. *15*, 283 (1982)
19. Armitage, D. A.: Organosilanes in Comprehensive Organometallic Chemistry, Vol. 2, Part 9. 1, Oxford—New York: Pergamon 1982
20. Barton, T. J., Hoekman, S. K., Burns, S. A.: Organometallics *1*, 721 (1982)
21. Aiule, Z. H., Chojnowski, J., Eaborn, C., Stanezyk, W. A.: J. Chem. Soc., Chem. Commun. 493 (1983)
22. Arrington, C. A., West, R., Michl, J.: J. Am. Chem. Soc. *105*, 6176 (1983)
23. Kudo, T., Nagase, S.: J. Organometal. Chem. *253*, C23 (1983)
24. Gordon, M. S., George, C.: J. Am. Chem. Soc. *106*, 609 (1984)
25. Guimon, C., Pfister-Guillouzo, G., Lavayssiere, H., Dousse, H. G., Barrau, J., Satge, J.: J. Organometal. Chem. *249*, C17 (1983)
26. Gusel'nikov, L. E., Volkova, V. V., Avakyan, V. G., Volnina, E. A., Zaikin, V. G., Nametkin, N. S., Polyakova, A. A., Tokarev, M. I.: J. Organometal. Chem. *271*, 191 (1984)
27. Gordon, M. S., Pople, J. A.: J. Am. Chem. Soc. *103*, 2945 (1981)
28. Gordon, M. S.: J. Am. Chem. Soc. *104*, 4352 (1982)
29. Lischka, H., Köhler, H.-J.: J. Am. Chem. Soc. *105*, 6646 (1983)
30. Binkley, J. S.: J. Am. Chem. Soc. *106*, 603 (1984)
31. Barton, T. J., Vuper, M.: J. Am. Chem. Soc. *103*, 6788 (1981)
32. Reyes, L. M., Canuto, S.: J. Mol. Struct. *89*, 77 (1982)
33. Markl, G., Rudnick, D., Schulz, R., Schweig, A.: Angew. Chem., Int. Ed. Engl. *21*, 221 (1982); Angew. Chem. Suppl. 523 (1982)
34. Gordon, M. S., Boudjouk, P., Anwari, F.: J. Am. Chem. Soc. *105*, 4972 (1983)
35. Chandrasekhar, J., Schleyer, R., Baumgartner, R. O. W., Reetz, M. T.: J. Org. Chem. *48*, 3453 (1983)
36. Markl, G., Hollriegl, H., Schlosser, W.: J. Organometal. Chem. *260*, 129 (1984)
37. Baldridge, K. K., Gordon, M. S.: J. Organometal. Chem. *271*, 369 (1984)
38. Brook, A. G., Abdesaken, F., Gutekunst, B., Gutekunst, G., Kallury, R. K.: J. Chem. Soc., Chem. Commun. 191 (1981)
39. West, R., Fink, M. J., Michl, J.: Science *214*, 1343 (1981)
40. Brook, A. G., Abdesaken, F., Gutekunst, G., Plavac, N.: Organometallics *1*, 994 (1982)
41. Boudjouk, P., Han, B.-H., Anderson, K. R.: J. Am. Chem. Soc. *104*, 4992 (1982)
42. Brook, A. G., Nyburg, S. C., Abdesaken, F., Gutekunst, B., Gutekunst, G., Kallury, R. K. M. B., Poon, Y. C., Chang, Y.-M., Wong-Ng, W.: J. Am. Chem. Soc. *104*, 5667 (1982)
43. Wiberg, N., Wagner, G.: Angew. Chem., Int. Ed. Engl. *22*, 1005 (1983)
44. Fink, M. J., Michalezyk, M. J., Haller, K. J., West, R., Michl, J.: J. Chem. Soc., Chem. Commun. 1010 (1983)
45. Masamune, S., Murakami, S., Snow, J. T., Tobita, H., Williams, D. J.: Organometallics *3*, 333 (1984)
46. Michalczyk, M. J., West, R., Michl, J.: J. Am. Chem. Soc. *106*, 821 (1984)
47. West, R.: Pure Appl. Chem. *56*, 163 (1984)
48. Smit, C. N., Lock, F. M., Bickelhaupt, E.: Tetrahedron Lett. *25*, 3011 (1984)

49. Cowley, A. H.: Polyhedron *3*, 389 (1984)
50. Fink, M. J., Michalczyk, M. J., Haller, K. J., West, R., Michl, J.: Organometallics *3*, 793 (1984)
51. Wiberg, N., Wagner, G., Muller, G., Riede, J.: J. Organometal. Chem. *271*, 381 (1984)
52. Nakatsuji, H., Ushio, J., Yonezawa, T.: J. Organometal. Chem. *258*, C1 (1983)
53. Voronkov, M. G., Lavrent'yev, V. I.: Top. Curr. Chem. *102*, 199 (1982)
54. Burger, H.: Fortschr. Chem. Forsch. *9*, 1 (1967)
55. Ebsworth, E. A. V., in: Organometallic Compounds of the Group IV Elements, Vol. 1, Pt. 1, Ch. 1 (ed. A. G. MacDiarmid). New York: Dekker 1968
56. Drake, J. E., Riddle, Ch.: Quart. Rev. *24*, 263 (1970)
57. Voronkov, M. G., Mileshkevich, V. P., Yuzhelevskii, Y. A.: Siloxane Bonding. Novosibirsk: Nauka 1976; (Studies in Soviet Science) New York 1978
58. Csakvari B., Golubinszkij, A. V., Gomory, P., Hargittai, I., Masztrjukov, V. Sz., Mijlhoff, F. C., Rozsondai, B., Vilkov, L. V., Wagner, Z.: Kem. Kozl. *46*, 473 (1976)
59. Hargittai, I., Rozsondai, B.: Kem. Kozl. *50*, 427 (1978)
60. Glidewell, C., Liles, D. C.: J. Organometal. Chem. *212*, 291 (1981)
61. Glidewell, C., Holden, H. D.: Acta Crystallogr. *B37*, 754 (1981)
62. Ebsworth, E. A. V., Murray, E. K., Rankin, D. W. H., Robertson, H. E.: J. Chem. Soc., Dalton Trans. 1501 (1981)
63. Robertson, D. W. H. H. E.: J. Chem. Soc., Dalton Trans. 265 (1983)
64. Shen, Q.: J. Mol. Struct. *102*, 325 (1983)
65. Barrow, M. J., Ebsworth, E. A. V.: J. Chem. Soc., Dalton Trans. 563 (1984)
66. Schomaker, V., Stevenson, D. P.: J. Am. Chem. Soc. *63*, 37 (1941)
67. Harrison, P. G.: Coord. Chem. Revs. *20*, 1 (1976)
68. Connolly, J. W., Hoff, C.: Adv. Organometal. Chem. *19*, 123 (1981)
69. Shirryaev, V. I., Mironov, V. F.: Uspekhi Khim. *52*, 321 (1983)
70. Krmse, W.: Carbene Chemistry. New York: Academic Press 1971
71. Nefedov, O. M., Kolesnikov, S. P., Ioffe, A. I.: Organometal. Chem. Revs. *5*, 181 (1977)
72. Ioffe, A. I., Nefedov, O. M.: Zh. Vses. Khim. O-va *24*, 475 (1979)
73. Bürger, H., Eujen, R.: Top. Curr. Chem. *50*, 1 (1974)
74. Chernyshev, E. A., Komalenkova, N. G., Bashkirova, S. A.: Uspekni Khim. *45*, 1782 (1976)
75. Caspar, P. R., in: Reactive Intermediates. (eds. Jones, M., Moss, R. A.). New York: Wiley *1*, 229 (1978); *2*, 335 (1981)
76. Gmelin Handbock, No 15, Silicon (Supplement Volume B 1), Berlin: Springer 1982
77. Herzberg, G.: Molecular Spectra and Molecular Structure I, Spectra of Diatomic Molecules. Princeton: Van Nostrand 1950
78. Ring, M. A., Jenkins, R. L., Zanganeh, R., Brown, H. C.: J. Am. Chem. Soc. *93*, 265 (1971)
79. Lambert, J. B., Urdaneta-Perez, M., Sun, H.-H.: J. Chem. Soc., Chem. Commun. 806 (1976)
80. Olah, G. A., Hunadi, R. J.: J. Am. Chem. Soc. *102*, 6989 (1980)
81. Sakurai, H., in: Free Radials, Vol. 2 (ed. K. Kochi). New York: Wiley-Interscience 1973
82. Schleyer, P. R., Apeloig, Y., Arad, D., Luke, B. T., Pople, J. A.: Chem. Phys. Lett. *95*, 477 (1983)
83. Gordon, M. S., Gano, D. R., Bootz, J. A.: J. Am. Chem. Soc. *105*, 5771 (1983)
84. Power, D., Brint, P., Spalding, T.: J. Mol. Struct. *108*, 81 (1984)
85. Eyler, J. R., Silverman, G., Battiste, M. A.: Organometallics *1*, 477 (1982)
86. Hopkinson, A. C., Lien, M. H.: J. Mol. Struct. *104*. 303 (1983)
87. Corriu, R. J. P., Henner, M.: J. Organometal. Chem. *74*, 1 (1974)
88. Boe, B.: J. Organometal. Chem. *107*, 139 (1976)
89. Cowley, A. H., Cushner, M. C., Riley, P. E.: J. Am. Chem. Soc. *102*, 624 (1980)
90. Weber, W. P., Felix, R. A., Willard, A. K.: Tetrahedron Lett. 907 (1970)
91. Groenewold, G. S., Gross, M. L., Bursey, M. M., Jones, P. R.: J. Organometal. Chem. *235*, 165 (1982)
92. Pietro, W. J., Hehre, W. J.: J. Am. Chem. Soc. *104*, 4329 (1982)
93. Murphy, M. K., Beauchamp, J. L.: J. Am. Chem. Soc. *98*, 5781 (1976); *99*, 2085, 4992 (1977)
94. Corey, J. Y.: J. Am. Chem. Soc. *97*, 3237 (1975)

95. Corey, J. Y., Gust, D., Mislow, K.: J. Organometal. Chem. *101*, C7 (1975)
96. Barton, T. J., Hovland, A. K., Tully, C. R.: J. Am. Chem. Soc. *98*, 5695 (1976)
97. Apeloig, Y., Godleski, S. A., Heacock, D. J., McKelvey, J. M.: Tetrahedron Lett. *22*, 3297 (1981)
98. Godleski, S. A., Heacock, D. J., McKelvey, J. M.: Tetrahedron Lett. *23* 4453 (1982)
99. Lambert, J. B., Schulz, W. J., Jr.: J. Am. Chem. Soc. *105*, 1671 (1983)
100. Truong, T., Gordon, M. S., Boudjouk, P.: Organometallics *3*, 484 (1984)
101. Sooriyakumaran, R., Bondjouk, P.: J. Organometal. Chem. *271*, 289 (1984)
102. Tasjo, L.: Acta Chem. Scand. *18*, 456 (1964)
103. Seifert, J.: Dissertation Univ. Karlsruhe 1982
104. West, R.: Pure Appl. Chem. *13*, 1 (1966)
105. MacDiarmid, A. G.: Intra-Sci. Chem. Rep. 7, 83 (1973)
106. Chojnowski, J., Cypryk, M., Michalski, J.: J. Organometal. Chem. *161*, C31 (1978)
107. Corrin, R. J. P., Dabosi, G., Martineau, M.: J. Organometal. Chem. *186*, 25 (1980)
108. Hensen, K., Zengerly, T., Pickel, P., Klebe, G.: Angew. Chem. Suppl. 973 (1983); Angew. Chem., Int. Ed. Engl. *22*, 725 (1983)
109. Albanow, A. I., Baukov, Yu. I., Voronkov, M. G., Kramarova, E. P., Larin, M. F., Pestunovich, V. A.: II Vsesoyuzn. Konferents. Spektroskop. YaMR tyazhelykh yader elemento-organich. soed., Tezisy dokl., p. 51, Irkutsk 1983
110. Pestunovich, V. A., Larin, M. F., Albanov, A. I., Voronkov, M. G., Baukov, Yu. I.: 7th Intern. Symp. Organosilicon Chem., Kyoto 1984, Abstracts of Papers, p. 115
111. Hensen, K., Busch, R.: Z. Naturforsch. *B27*, 1174 (1982)
112. Borbely-Kuszmann, A., Zimonyl-Hegedus, E., Nagy, J.: Period. Polytechn. Chem. Eng. *20*, 255 (1976)
113. Jolibois, H., Doucet, A., Dubry, J. L.: Inorg. Nucl. Chem. Lett. *12*, 759 (1976)
114. Graddon, D. P., Rana, B. A.: J. Organometal. Chem. *140*, 21 (1977)
115. Koacher, J. K., Tandon, J. P., Mehrotra, R. C.: J. Inorg. Nucl. Chem. *41*, 1409 (1979)
116. Yoder, C. H., Cullinane, J. A., Martin, G. F.: J. Organometal. Chem. *210*, 289 (1981)
117. Bassindale, A. R., Stout, T.: J. Organometal. Chem. *238*, C41 (1982)
118. Lal, K.: Monatsh. Chem. *113*, 33 (1983)
119. Kettrup, A., Ohrbach, K.-H.: Thermochim. Acta *29*, 273 (1979)
120. Huggett, P. G., Manning, K., Wade, K.: J. Inorg. Nucl. Chem. *42*, 665 (1980)
121. Thornton, T. J., Dhar, S. K.: J. Inorg. Nucl. Chem. *42*, 1202 (1980)
122. Aylett, B. J.: Adv. Inorg. Chem. Radiochem. *25*, 1 (1982)
123. Schafer, H., MacDiarmid, A. G.: Inorg. Chem. *15*, 848 (1976)
124. Sommer, L. H.: Stereochemistra and Mechanisms in Silicon. New York: McGraw-Hill 1965
125. O'Brien, D. H., Hairston, T. J.: Organometal. Chem. Rev. A, *7*, 95 (1971)
126. Gielen, M., Dehouck, C., Mokhtar-Jami, H., Topart, J.: Revs. Silicon, Germanium, Tin, Lead Compounds *1*, 9 (1972)
127. Sommer, L. H.: Intra-Sci. Chem. Rep. *7*, 1 (1973)
128. Kwart, H., King, K.: d-Orbitals in the Chemistry of Silicon, Phosphorus and Sulfur, Ch. 5E, Berlin: Springer 1977
129. Corriu, R. J. P., Guerin, C.: J. Organometal. Chem. *198*, 231 (1980)
130. Corriu, R.: Organometal. Chem. Rev. *9*, 357 (1980)
131. West, R., Barton, T. J.: J. Chem. Educ. *57*, 334 (1980)
132. Corriu, R. J. P., Guerin, C.: Adv. Organometal. Chem. *20*, 265 (1982)
133. Kwart, H.: Phosphorus and Sulfur *15*, 293 (1983)
134. Beattie, I. R.: Quart. Rev. *17*, 382 (1963)
135. Alpatora, N. M., Kessler, M. Yu.: Zh. Strukt. Khim. *5*, 332 (1964)
136. Gielen, M., Sprecher, N.: Organometal. Chem. Revs. *1*, 455 (1966)
137. Dyke, C. H., in: Organometallic Compounds of the Group IV Elements (ed. A. G. MacDiarmid). Vol. 2, Part 1. New York: Dekker 1972
138. Ennan, A. A., Kats, B. M.: Russ. Chem. Rev. *43*, 539 (1974)
139. Ennan, A. A., Nikitin, V. I., Petrosyan, V. P., Plakida, O. A.: Deposited Doc. 1976, VINITI 3338–76
140. Campbell-Ferguson, H. J., Ebsworth, E. A. V.: J. Chem. Soc. A *1966*, 1508; *1967*, 705
141. Beattie, I. R., Ozin, G. A.: J. Chem. Soc. A *1970*, 370

142. Jolibois, H.: Ann. Sci. Univ. Besançon Chim. 3 (1976)
143. Janier-Dubry, J. L.: Ann. Sci. Univ. Besançon Chim. 29 (1976)
144. Porritt, C. J.: Chem. Ind. (London) 215 (1979)
145. Ault, B. S.: Inorg. Chem. *20*, 2817 (1981)
146. Lorenz, T. J., Ault, B. S.: Inorg. Chem. *21*, 1758 (1982)
147. Guertin, J. P., Onyszchuk, M.: Can. J. Chem. *46*, 987 (1968)
148. Aylett, B. J., Ellis, I. A., Porritt, C. J.: J. Chem. Soc., Dalton Trans. 1953 (1972); 83 (1973)
149. Aylett, B. J., Ellis, I. A., Porritt, C. J.: Chem. Ind. (London) 499 (1970)
150. Groth, P.: Chemische Krystallographie. Leipzig: Wilhelm Engelmann 1906
151. Hassel, O., Salvesen, J. R.: Z. Physik. Chem. *128*, 345 (1927)
152. Gossner, B., Kraus, O.: Z. Kristallogr. *88*, 223 (1934)
153. Ketelaar, J. A. A.: Kristallogr. *92*, 155 (1935)
154. Rudman, R., Hamilton, W. C., Novick, S., Goldfarb, T. D.: J. Am. Chem. Soc. *89*, 5157 (1967)
155. Turley, J., Boer, F. P., Frye, C.: Chem. Eng. News *45*, No 42, 46 (1967)
156. Blake, A. J., Ebsworth, E. A. V., Welch, A. J.: Acta Crystallogr. *C40*, 895 (1984)
157. Voronkov, M. G.: Pure Appl. Chem. *13*, 35 (1966)
158. Voronkov, M. G., D'yakov, V. M., Kirpichenko, S. V.: J. Organometal. Chem. *233*, 1 (1982)
159. Sidorkin, V. F., Pestunovich, V. A., Voronkov, M. G.: Uspekhi Khim. *49*, 789 (1980)
160. Ault, B. S.: J. Am. Chem. Soc. *105*, 5742 (1983)
161. Harland, J. J., Day, R. O., Vollano, J. F., Sau, A. C., Holmes, R. R.: J. Am. Chem. Soc. *103*, 5269 (1981)
162. Holmes, R. R., Day, R. O., Harland, J. J., Sau, A. C., Holmes, J. M.: Organometallics *3*, 341 (1984)
163. Holmes, R. R., Day, R. O., Harland, J. J., Holmes, J. M.: Organometallics *3*, 347 (1984)
164. Hofmann, B., Hoppe, R.: Z. Anorg. Chem. *458*, 151 (1979)
165. Prokof'eva, T. I., Prokof'ev, A. I., Lavrukhin, B. D., Belostotskaja, I. S., Ershov, V. V., Kabachnik, M. I.: Dokl. Akad. Nauk SSSR *257*, 1412 (1981)
166. Voronkov, M. G., Deriglazov, N. M., Brodskaya, E. I.. Kalistratova, E. F., Gubanova, L. I.: J. Fluorine Chem. *19*, 299 (1982)
167. Aylett, B. J.: Progr. Stereochem. *4*, 213 (1969)
168. Chvalovsky, V., in: Organosilicon Compounds, Vol. 3, p. 129 (ed. V. Bazant). Prague: Institute of Chemical Process Fundamentals 1973
169. Bleidelis, J. J., Kemme, A. A., Zelchan, G. I., Voronkov, M. G.: Khim. Geterotsikl. Soed. 617 (1973)
170. Pike, R. M.: Coord. Chem. Rev. *2*, 163 (1967)
171. Bokii, N. G., Shklover, V. E., Struchkov, Yu. T., in: Kristallokhimija. VINITI, p. 94, Moskow 1974
172. Prokof'ev, A. K.: Uspekhi Khim. *45*, 1028 (1976)
173. Weiss, A., Herzog, A.: in Ref. 9, p. 109
174. Burger, H.: Angew. Chem., Int. Ed. Engl. *12*, 474 (1973)
175. Glidewell, G.: Inorg. Chim. Acta Rev. *7*, 69 (1973)
176. Solomennikova, I. I., Zelchan, G. I., Lukevics, E. J.: Khim. Geterotsykl. Soed. 1299 (1977)
177. Voronkov, M. G., Mileshkevich, V. P., Yuzhelevskii, Yu. A.: Uspekhi Khim. *45*, 2253 (1976)
178. Gel'mbol'dt, V. O., Ennan, A. A.: Koord. Khim. *8*, 1176 (1982); *9*, 579 (1983)
179. Kumada, M., Tamao, K., Yoshida, J.-I.: J. Organometal. Chem. *239*, 115 (1982)
180. Commission on Atomic Weights and Isotopic Abundances (Holden, N. E., Martin, R. L.), Pure Appl. Chem. *55*, 1101 (1983); *56*, 653 (1984)
181. Waber, J. T., Cromer, D. T.: J. Chem. Phys. *42*, 4116 (1965)
182. Slater, J. C.: J. Chem. Phys. *41*, 3199 (1964)
183. Sanderson, R. T.: J. Am. Chem. Soc. *105*, 2259 (1983)
184. Shannon, R. D.: Acta Crystallogr. *A32*, 751 (1976)
185. Negishi, E.-I.: Organometallics in Organic Synthesis, Vol. 1, New York; Wiley 1980
186. Moore, C. E.: Ionization Potentials and Ionization Limits Derived from the Analysis of Optical Spectra. NSRDS-NBS 34, Washington: National Bureau of Standard 1970
187. Day, M. C., Jr., Selbin, J.: Theoretical Inorganic Chemistry. New York: Reinhold 1969

188. Pauling, L.: J. Am. Chem. Soc. *54*, 3570 (1932)
189. Pauling, L.: The Nature of the Chemical Bond. 3rd Edition, Ithaca, N.Y.: Cornell University 1960
190. Matcha, R. L.: J. Am. Chem. Soc. *105*, 4859 (1983)
191. Mulliken, R. S.: J. Chem. Phys. *2*, 782 (1934)
192. Pritchard, H. O., Skinner, H. A.: Chem. Rev. *55*, 745 (1955)
193. Batsanov, S. S.: Uspekhi Khim. *51*, 1201 (1982)
194. Gordy, W.: Phys. Rev. *69*, 604 (1946)
195. Allred, A. L., Rochow, E. G.: J. Inorg. Nucl. Chem. *5*, 269 (1958)
196. Allred, A. L.: J. Inorg. Nucl. Chem. *17*, 215 (1961)
197. Voronkov, M. G., Kovalev, I. F.: Izv. Akad. Nauk Latv. SSR, Ser. Khim. 158 (1965)
198. Sanderson, R. T.: J. Inorg. Nucl. Chem. *28*, 1553 (1966)
199. Mande, C., Deshmukh, P., Deshmukh, P.: J. Phys. *B10*, 2293 (1977)
200. Blustin, P. H., Raynes, W. T.: J. Chem. Soc., Dalton Trans. 1237 (1981)
201. Ohwada, K.: Polyhedron *2*, 423 (1983)
202. Hargittai, I., Bliefert, C.: Z. Naturforsch. *B38*, 1304 (1983)
203. Parr, R. G., Pearson, R. G.: J. Am. Chem. Soc. *105*, 7512 (1983)
204. Robles, J., Bartolotti, L. J.: J. Am. Chem. Soc. *106*, 3723 (1984)
205. Bondi, A.: J. Phys. Chem. *68*, 441 (1964)
206. Glidewell, C.: Inorg. Chim. Acta *36*, 135 (1979)
207. Barrow, M. J.: Acta Crystallogr. *B38*, 150 (1982)
208. Schlemper, E. O., Britton, D.: Inorg. Chem. *5*, 511 (1966)
209. Schlemper, E. O., Britton, D.: Inorg. Chem. *5*, 507 (1966)
210. Mironov, V. F., Gar, T. K.: Organicheskie Soedinenija Germaniya. Moscow: Nauka 1967
211. Lesbre, M., Mazerolles, P., Satge, J.: The Organic Compounds of Germanium. London: Wiley 1971
212. Sullivan, S. A., DePuy, C. H., Damrauer, R.: J. Am. Chem. Soc. *103*, 480 (1981)
213. Carlstrom, D., in: Ref. 9, p. 523
214. Shklover, V. E., Struchkov, Yu. T.: Uspekhi Khim. *49*, 518 (1980)
215. Baur, W. H.: Acta Crystallogr. *B36*, 2198 (1980)
216. Hoeve, W. T., Wynberg, H.: J. Org. Chem. *48*, 2925, 2930 (1980)
217. Dubchak, I. L., Dashevskii, V. G., Struchkov, Yu. T.: Zh. Strukt. Khim. *24*, 114 (1983)
218. Krogh-Jespersen, M.-B., Chandrasekhar, J., Wurthwein, E. U., Collins, J. B., Schleyer, P.: J. Am. Chem. Soc. *102*, 2263 (1980)
219. Wurthwein, E.-U., Schleyer, P. R.: Angew. Chem., Int. Ed. Engl. *18*, 553 (1979)
220. Minkin, V. I., Minyaev, R. M., in: Fizicheskaja Khimija, p. 180 (ed. Ya. M. Kolotyrkin). Moskow: Khimija 1983
221. Collins, J. B., Dill, J. D., Jemmis, E. D., Apeloig, Y., Schleyer, P. R., Seeger, R., Pople, J. A.: J. Am. Chem. Soc. *98*, 5419 (1976)
222. Shen, Q., Hilderbrandt, R. L., Burns, G. T., Barton, T. J.: J. Organometal. Chem. *195*, 39 (1980)
223. Schei, H., Shen, Q., Cunico, R. F., Hilderbrandt, R. L.: J. Mol. Struct. *63*, 59 (1980)
224. Corriu, R., Guerin, C.: J. Organometal. Chem. *195*, 261 (1980)
225. Cartledge, F. K.: J. Organometal. Chem. *255*, 131 (1982)
226. Meyer, H., Nogersen, G.: Angew. Chem., Int. Ed. Engl. *18*, 551 (1979)
227. Dunitz, J. D.: Angew. Chem., Int. Ed. Engl. *19*, 1034 (1980)
228. Nagorsen, G., Meyer, H.: Angew. Chem., Int. Ed. Engl. *19*, 1034 (1980)
229. Wilkinson, G.: Chem. Brit. *17*, 156 (1981)
230. Bibber, J. W., Barnes, C. L., Helm, D., Zuckerman, J. J.: Angew. Chem. Suppl. 668 (1983)
231. Schomburg, D.: Angew. Chem., Int. Ed. Engl. *22*, 65 (1983)
232. Muller, R., Heinrich, L.: Chem. Ber. *94*, 1943 (1961)
233. Frye, C. L.: J. Am. Chem. Soc. *86*, 3170 (1964); *92*, 1205 (1970)
234. Hawkins, C. J.: Absolute Configuration of Metal Complexes. New York: Wiley-Interscience 1971
235. Sheldrick, W. S.: Top. Curr. Chem. *73*, 1 (1978)

236. Holmes, R. R.: Pentacoordinated Phosphorus, Vol. 1 (ACS Monograph 175), Washington 1980
237. Martin, J. C., Perozzi, E. F.: Science *191*, 154 (1976)
238. Perozzi, E. F., Martin, J. C.: J. Am. Chem. Soc. *101*, 1591 (1979)
239. Perozzi, E. F., Michalak, R. S., Figuly, G. D., Sevenson, W. H., Dess, D. B., Ross, M. R., Martin, J. C.: J. Org. Chem. *46*, 1049 (1981)
240. Nguyen, T. T., Amey, R. L., Martin, J. C.: J. Org. Chem. *47*, 1024 (1982)
241. Kumok, V. N.: Zakonomernosti v. Ustoichiovosti Koordinats. Soedinenii v. Rastvorakh (Principles of the Stability of Coordination Compounds in Solutions). Tomsk: Tomskii Universitet 1977
242. Jong, F., Reinhoudt, D. N.: Adv. Phys. Org. Chem. *17*, 279 (1970)
243. Dashevskii, V. G., Baranov, A. P., Kabachnik, M. I.: Uspekhi Khim] *52*, 268 (1983)
244. Lamb, J. D., Izatt, R. M., Christensen, J. J., Eatough, D. J., in: Coordination Chemistry of Macrocyclic Compounds, p. 145 (ed. G. A. Melson). New York: Plenum 1979
245. Popov, A. I., Lehn, J.-M., in: Ref. 244, p. 537
246. Reibnegger, G. J., Rode, B. M.: Inorg. Chim. Acta *72*, 47 (1983)
247. Simmons, E. L.: J. Chem. Educ. *56*, 578 (1979)
248. Fransto da Silva, J. J. R.: J. Chem. Educ. *60*, 390 (1983)
249. Barbucci, R., Fabbrizzi, L., Paoletti, P.: Coord. Chem. Revs. *8*, 31 (1972)
250. Voronkov, M. G., Gubanova, L. I., Frolov, Yu. L., Chernov, N. F., Gavrilova, G. A., Chipanina, N. N.: J. Organometal. Chem. *271*, 169 (1984)
251. Albanov, A. I.: Dissertation Inst. Organic Chem. Irkutsk 1984
252. Lehn, J. M., Sauvage, J. P.: J. Am. Chem. Soc. *97*, 6700 (1975)
253. Sacconi, L., Mani, F.: Transition Metal Chem. *8*, 179 (1982)
254. Mani, F., Sacconi, L.: Comments Inorg. Chem. *2*, 157 (1983)
255. Lukevics, E., Popelis, Yu. Yu., Simchenko, L. I.: Zh. Obshch. Khim. *44*, 1750 (1974)
256. Pekhk, T., Lippmaa, E., Lukevics, E., Simchenko, L. I.: Zh. Obshch. Khim. *46*, 602 (1976)
257. Lukevics, E. J., Kovalev, I. F., Ignatova, V. A., Yankovskaya, I. S., Mazheika, I. P., Popelis, Yu. Yu., Simchenko, L. I.: Izv. Akad. Nauk Latv. SSR, Ser. Khim. 339 (1974)
258. Lukevics, E. J.: Izv. Akad. Nauk Latv. SSR, Ser. Khim. 351 (1974)
259. Potapov, V. M.: Stereokhimija, Ch. 5.6. Moskow: Khimija 1976
260. Naumov, A. D., Bashkirova, S. A., Komalenkova, N. G., Chernyshev, E. A., Popov, A. G., Antipova, V. V.: Zh. Obshch. Khim. *46*, 1808 (1976)
261. Tandura, S. N.: Dissertation Inst. Organic Chem. Irkutsk 1977
262. Voronkov, M. G., Pestunovich, V. A., Liepinsh, E. E., Tandura, S. N., Zelchan, G. I., Lukevics, E. J.: Izv. Akad. Nauk. Latv. SSR, Ser. Khim. 44 (1978)
263. Liepins, E., Popelis, J., Birgele, I., Urtane, I., Zelchan, G., Lukevics, E.: J. Organometal. Chem. *201*, 113 (1980)
264. Pestunovich, V. A., Shterenberg, B. Z., Tandura, S. N., Baryshok, V. P., Brodskaja, E. I., Komalenkova, N. G., Voronkov, M. G.: Dokl. Akad. Nauk SSSR *264*, 632 (1982)
265. Liepinsh, E. E., Birgele, I. S., Zelchan, G. I., Urtane, I. P., Lukevics, E.: Zh. Obshch. Khim. *53*, 1076 (1983)
266. Kupce, E., Liepinsh, E., Lukevics, E.: J. Organometal. Chem. *248*, 131 (1983)
267. Hegyes, P., Foldeak, S., Hencsei, P., Zsombok, G., Nagy, J.: J. Organometal. Chem. *251*, 289 (1983)
268. Urtane, I. P., Zelchan, G. I., Liepinsh, E. E., Yankovska, I. S., Lukevics, E.: Khimija i prakticheskoye primenenie kremnij i fosfororganicheskikh soedinenij, p. 24, Leningrad 1979
269. Birgele, I. S.: Dissertation Inst. Organic Synth. Riga 1981
270. Karcsev, G. N., Ignat'eva, S. I., Komalenkova, N. G., Bashkirova, S. A., Chernyshev, E. A.: Zh. Obshch. Khim. *53*, 2262 (1983)
271. Bochkarev, V. N., Sljusarenko, T. F., Silkina, N. N., Polivanov, A. N., Gar, T. K., Khromova, N. Yu.: Zh. Obshch. Khim. *50*, 1080 (1980)
272. Bochkarev, V. N., Bernadskij, A. A., Polivanov, A. N., Komalenkova, N. G., Bachkirova, S. A., Popov, A. G., Antipova, V. V., Chernyshev, E. A.: Zh. Obshch. Khim. *48*, 2700 (1978)
273. Penn, R. E., Birkenmeier, J. A.: J. Mol. Spectrosc. *62*, 416 (1976)
274. Penn, R. E., Olsen, R. J.: J. Mol. Spectrosc. *62*, 423 (1976)

275. Burgess, A. W., Shipman, L. L., Nemenoff, R. A., Scheraga, H. A.: J. Am. Chem. Soc. *98*, 23 (1976)
276. Rasanen, M., Aspiala, A., Homanen, L., Murto, J.: J. Mol. Struct. *96*, 81 (1982)
277. Alsenov, C., Scarsdale, J. N., Williams, J. O., Schafer, L.: J. Mol. Struct. *86*, 291 (1982)
278. Houriet, R., Rüfenacht, H., Carrupt, P.-A., Vogel, P., Tichy, M.: J. Am. Chem. Soc. *105*, 3417 (1983)
279. Wolfe, S.: Acc. Chem. Res. *5*, 102 (1972)
280. Kingsbury, C. A.: J. Chem. Educ. *56*, 431 (1979)
281. Spassov, S. L., Simeonov, M. F., Randall, E. W.: J. Mol. Struct. *77*, 289 (1981)
282. Randino, A., Millefiori, S., Zuccarello, F., Millefiori, A.: J. Mol. Struct. *51*, 295 (1979)
283. Hancock, R. D.: Inorg. Chim. Acta *49*, 145 (1981)
284. Brodalla, D., Mootz, D.: Angew. Chem., Int. Ed. Engl. *20*, 791 (1981)
285. Starova, G. L., Frank-Kamenetskaja, O. V., Fundamenskii, V. S., Semenova, N. V., Voronkov, M. G.: Dokl. Akad. Nauk SSSR *260*, 888 (1981)
286. Shklover, V. E., Gridunova, G. B., Struchkov, Yu. T., Voronkov, M. G., Kryuchkova, Yu. I., Mirskova, A. N.: Dokl. Akad. Nauk SSSR *269*, 387 (1983)
287. Tandura, S. N., Gurkova, S. N., Gusev, A. I., Alekseev, N. V.: Information Rev. Ser. Organoelement compounds and their application. Organogermanium compounds with extended coordinate sphere, p. 32, NIITEKhim, Moscow 1983
288. Cox, B. G., Schneider, H.: J. Chem. Soc., Perkin Trans. II 1293 (1979)
289. Lehn, J. M.: Pure Appl. Chem. *49*, 857 (1977); Acc. Chem. Res. *11*, 49 (1978)
290. Alder, R. W.: Acc. Chem. Res. *16*, 321 (1983)
291. Cheney, J., Kintzinger, J. P., Lehn, J. M.: Nouv. J. Chim. *2*, 411 (1978)
292. Smith, P. B., Dye, J. L., Cheney, J., Lehn, J.-M.: J. Am. Chem. Soc. *103*, 6044 (1981)
293. Metz, B., Moras, D., Weiss, R.: J. Chem. Soc., Perkin Trans. II 423 (1976)
294. Eaborn, C., Odell, K. J., Pidcock, A., Scollary, G. R.: J. Chem. Soc., Chem. Commun. 317 (1976)
295. Scollary, G. R.: Aust. J. Chem. *30*, 1007 (1977)
296. Clardy, J. C., Milbrath, D. S., Verkade, J. G.: J. Am. Chem. Soc. *99*, 631 (1977)
297. Shen, Q., Hilderbrandt, R. L.: J. Mol. Struct. *64*, 257 (1980)
298. D'yachenko, O. A., Atovmyan, L. O., Soboleva, S. V.: Fifth European Crystallographic Meeting, Abstracts, p. 289, Copenhagen 1979
299. Aldoshin, S. M., D'yachenko, O. A., Atovmyan, O. L., Chekhlov, A. N., Al'yanov, M. I.: Koord. Khim. *6*, 936 (1980)
300. Smith, K. M. (ed.): Porphyrins and Metalloporphyrins. Amsterdam: Elsevier 1975
301. Dolphin, D.: The Porphyrins. New York: Academic Press 1978
302. Berezin, B. D., Koifman, O. I.: Uspekhi Khim. *49*, 2389 (1980)
303. Berezin, B. D.: Coordination Compounds of Phorphyrins and Phthalocyanines. London: Wiley 1981
304. Sayer, P., Gouterman, M., Connell, C. R.: Acc. Chem. Res. *15*, 73 (1982)
305. Ugi, I., Ramirez, F.: Chem. Brit. *8*, 198 (1972)
306. Holmes, R. R.: Acc. Chem. Res. *12*, 257 (1979)
307. Denney, D. B., Denney, D. Z., Hammond, P. J., Hsu, Y. F.: J. Am. Chem. Soc. *103*, 2340 (1981)
308. Forbus, T. R., Martin, J. C.: J. Am. Chem. Soc. *101*, 5057 (1979)
309. Zubieta, J. A., Zuckerman, J. J.: Progr. Inorg. Chem. *24*, 251 (1978)
310. Pelizzi, C., Pelizzi, G., Predieri, G.: J. Organometal. Chem. *263*, 9 (1984)
311. Hargittai, M., Hargittai, I.: The Molecular Geometries of Coordination Compounds in Vapour Phase. Budapest: Akademiai Kiado 1975
312. Tijima, K., Shivata, S.: Bull. Chem. Soc. Jpn. *53*, 1908 (1980)
313. Boal, D., Ozin, G. A.: Can. J. Chem. *51*, 609 (1973)
314. Rami, T., Henzen, K.: J. Inorg. Nucl. Chem. *33*, 937 (1971)
315. Griffiths, J. E.: Spectrochim. Acta *A* 30, 169 (1974)
316. Long, C. A., Ewing, G. E.: J. Chem. Phys. *58*, 4824 (1973)
317. Voronkov, M. G., Deich, A. J.: Dokl. Akad. Nauk SSSR *168*, 337 (1966)
318. Voronkov, M. G., Pozdnyakova, M. V., Shagata, L. A.: Zh. Obshch. Khim. *40*, 1425 (1970)
319. Gundersen, G., Mayo, R. A., Rankin, D. W. H., Robertson, H. E.: to be published, cited in Ref. 156

320. Barrow, M. J., Ebsworth, E. A. V., Harding, M. M.: Acta Crystallogr. *B35*, 2093 (1979)
321. Barrow, M. J., Ebsworth, E. A. V., Harding, M. M.: J. Chem. Soc., Dalton Trans. 1838 (1980)
322. Frazer, M. J., Gerrard, W., Twaits, R.: J. Inorg. Nucl. Chem. *25*, 637 (1963)
323. Mehta, P., Zeldin, M.: Inorg. Chim. Acta *22*, L33 (1977)
324. Zeldin, M., Mehta, P., Vernon, W. D.: Inorg. Chem. *18*, 463 (1979)
325. Voronkov, M. G., Frolov, Yu. L., D'yakov, V. M., Chipanina, N. N., Gubanova, L. I., Gavrilova, G. A., Klyba, L. V., Aksamentova, T. N.: J. Organometal. Chem. *201*, 165 (1980)
326. Frolov, Yu. L., Aksamentove, T. N., Gavrilova, G. A., Chipanina, N. N., Modonov, V. B., Gubanova, L. I., D'yakov, V. M., Voronkov, M. G.: Dokl. Akad. Nauk SSSR *267*, 646 (1982)
327. Frolov, Yu. L., Voronkov, M. G., Gavrilova, G. A., Chipanina, N. N., Gubanova, L. I., D'yakov, V. M.: J. Organometal. Chem. *244*, 107 (1983)
328. Voronkov, M. G., Aksamentova, T. N., Modonov, V. B., Gubanova, L. I., Frolov, Yu. L., D'yakov, V. M.: Izv. Akad. Nauk SSSR, Ser. Khim. 685 (1984)
329. Marciniec, B., Gulinska, H.: J. Organometal. Chem. *146*, 1 (1978); *236*, 1 (1982)
330. Khorshev, S. Ya., Tsvekova, V. L., Egorochkin, A. N.: J. Organometal. Chem. *264*, 169 (1984)
331. Voronkov, M. G., Broskaya, E. I., Reich, P., Shevchenko, S. G., Baryshok, V. P., Frolov, Yu. L.: Dokl. Akad. Nauk SSSR *241*, 1117 (1978); J. Organometal. Chem. *164*, 35 (1979)
332. Helmer, B. J., West, R., Corriu, R. J. P., Poirier, M., Royo, G., DeSaxce, A.: J. Organometal. Chem. *251*, 295 (1983)
333. Hazen, R. M., Finger, L. W.: Comparative Crystal Chemistry. New York: Wiley 1982
334. Satchell, D. P. N., Satchell, R. S.: Chem. Rev. *69*, 251 (1969)
335. Pearson, R. G.: Hard and Soft Acids and Bases. Stroudsburg: Dowden, Hutchinson Ross 1973
336. Jensen, W. B.: Chem. Revs. *78*, 1 (1978)
337. Gutmann, V.: The Donor-Acceptor Approach to Molecular Interactions. New York: Plenum 1978
338. Jensen, W. B.: The Lewis Acid-Base Concept — an Overview. New York: Wiley 1980
339. Finston, H. L., Rychtman, A. C.: A New View of Current Acid-Base Theories. New York: Wiley 1982
340. Lewis, G. N., in: Valence and the Structure of Atoms and Molecules. New York: Dover 1966
341. Halgren, T. A., Brown, L. D., Kleier, D. A., Lipscomb, W. N.: J. Am. Chem. Soc. *99*, 6793 (1977)
342. Zandler, M. E., Talaty, E. R.: J. Chem. Educ. *61* 124 (1984)
343. Gianturco, F. A.: J. Chem. Soc. A *1969*, 1293
344. Smith, D. W.: J. Chem. Educ. *57*, 106 (1980)
345. Egorochkin, A. N., Vyazankin, N. S., Khorshev, S. Ya.: Uspekhi Khim. *41*, 828 (1972)
346. Egorochkin, A. N., Khorshev, S. Ya.: Uspekhi Khim. *49*, 1697 (1980)
347. Jørgensen, C. K.: Structure and Bonding *3*, 106 (1967)
348. Dyatkina, M. E., Klimenko, N. M.: Zh. Strukt. Khim. *14*, 173 (1973)
349. Charkin, O. P.: Zh. Strukt. Khim. *14*, 389 (1973)
350. Dyatkina, M. E., Klimenko, N. M., Rosenberg, E. L.: Pure Appl. Chem. *38*, 391 (1974)
351. Sadimenko, A. P., Kolodyazhny, Yu. V., Osipov, O. A.: Fiz. Molekul (Kiev) *6*, 77 (1978)
352. Bochvar, D. A., Gambaryan, N. P., Epshtein, L. M.: Uspekhi Khim. *45*, 1316 (1976)
353. Ratner, M. A., Sabin, J. R.: J. Am. Chem. Soc. *99*, 3954 (1977)
354. D'yachenko, O. A., Atovmyan, L. O.: Zh. Strukt. Khim. *24*, 144 (1983)
355. Rundle, R. E.: Survey Progr. Chem. *1*, 81 (1963)
356. Muscher, J. I.: Angew. Chem., Int. Ed. Engl. *8*, 54 (1969)
357. Halgren, T. A., Brown, L. D., Kleier, D. A., Lipscomb, W. N.: J. Am. Chem. Soc. *99*, 6793 (1977)
358. Kuznetsov, V. I. (ed.): Theory of Valency in Progress, Ch. 9. Moscow: MIR 1980
359. Harcourt, R. D.: Qualitative Valence-Bond Discriptions of Electron-Rich Molecules: Pauling "3-Electron Bonds" and "Increased-Valence" Theory. Berlin: Springer 1982
360. Sidorkin, V. F., Pestunovich, V. A., Voronkov, M. G.: Dokl. Akad. Nauk SSSR *235*, 1363 (1977)

361. Balakhchi, G. K., Keiko, V. V., Sidorkin, V. F., Pestunovich, V. A., Voronkov, M. G.: Dokl. Akad. Nauk SSSR *275*, 393 (1984)
362. Hall, M. B.: Inorg. Chem. *17*, 2261 (1978)
363. Schleyer, P. R., Wurthwein, E.-U.: J. Am. Chem. Soc. *104*, 5839 (1982)
364. Klimenko, N. M., Zakzhewskii, V. G., Charkin, O. P.: Koord. Khim. *8*, 903 (1982)
365. Minyaev, R. M., Orlova, G. V.: Zh. Obshch. Khim. *24*, 38 (1983)
366. Albano, V. G., Sansoni, M., Chini, P., Martinengo, S.: J. Chem. Soc., Dalton Trans. 651 (1973)
367. Albano, V. G., Chini, P., Martinengo, S., Sansoni, M., Strumolo, D.: J. Chem. Soc., Chem. Commun. 299 (1974)
368. Albano, V. G., Chini, P., Ciani, G., Sansoni, M., Strumolo, D., Heaton, B. T., Martinengo, S.: J. Am. Chem. Soc. *98*, 5027 (1976)
369. McLean, W., Schultz, J. A., Pedersen, L. G., Jarnagin, R. C.: J. Organometal. Chem. *175*, 1 (1979)
370. Jemmis, E. D., Chandrasekhar, J., Schleyer, P. R.: J. Am. Chem. Soc. *101*, 527 (1979)
371. Schwarz, H.: Angew. Chem., Int. Ed. Engl. *20*, 991 (1981)
372. Jemmis, E. D., Chandrasekhar, J., Wurthwein, E.-U., Schleyer, P. R., Chinn, J. W., Landro, F. J., Lagow, R. J., Luke, B., Pople, J. A.: J. Am. Chem. Soc. *104*, 4275 (1982)
373. Siebert, W., Edwin, J., Pritzkow, H.: Angew. Chem., Int. Ed. Engl. *21*, 148 (1982)
374. Gurak, J. A., Chinn, J. W., Lagow, R. J.: J. Am. Chem. Soc. *104*, 2637 (1982)
375. Braye, E. H., Dahl, L. F., Hubel, W., Wampler, D. L.: J. Am. Chem. Soc. *84*, 4633 (1962)
376. Sirigu, A., Bianchi, M., Benedetti, E.: J. Chem. Soc., Chem. Commun. 596 (1969)
377. Churchill, M. M., Wormald, M., Benedetti, E.: J. Chem. Soc., Chem. Commun. 596 (1969)
378. Dietrich, H., Rewicki, D.: J. Organometal. Chem. *205*, 281 (1981)
379. Jastrzebski, J. T. B. H., Koten, G., Konijn, M., Stam, C. H.: J. Am. Chem. Soc. *104*, 5490 (1982)
380. Koten, G., Noltes, J. G.: J. Organometal. Chem *84*, 129 (1975)
381. Schleyer, P. R., Wurthwein, E.-U., Kaufmann, E., Clark, T.: J. Am. Chem. Soc. *105*, 5930 (1983)
382. Schleyer, P. R.: Pure Appl. Chem. *56*, 151 (1984)
383. Lowdin, P.-O., Pullman, A. (eds.): New Horizons of Quantum Chemistry, p. 95, Dordrecht: Reidel 1983
384. Gillespie, R. J.: J. Chem. Soc. *1963*, 4672
385. Kepert, D. L.: Inorg. Chem. *12*, 1938 (1973)
386. Kepert, D. L.: Inorganic Stereochemistry. Berlin: Springer 1982
387. Holmes, R. R.: Progr. Inorg. Chem. *32*, (1984)
388. Sau, A. C., Day, R. O., Holmes, R. R.: J. Am. Chem. Soc. *102*, 7972 (1980)
389. Sau, A. C., Carpino, L. A., Holmes, R. R.: J. Organometal. Chem. *197*, 181 (1980)
390. Sau, A. C., Day, R. O., Holmes, R. R.: Inorg. Chem. *20*, 3076 (1981)
391. Durkin, T. R., Schram, E. P.: Inorg. Chem. *11*, 1048 (1972)
392. Dashevskii, V. J., Asatryan, R. S., Baranov, A. P.: Zh. Strukt. Khim. *19*, 465 (1978)
393. Marsden, C. J.: Inorg. Chem. *22*, 3177 (1983)
394. Adley, A. D., Bird, P. H., Fraser, A. R., Onyszchuk, M.: Inorg. Chem. *11*, 1402 (1972)
395. Luckenbach, R.: Dynamic Stereochemistra of Pentaco-ordinated Phosphorus and Related Elements. Stuttgart: Georg Thieme Verlag 1973
396. Favas, M. C., Kepert, D. L.: Progr. Inorg. Chem. *27*, 325 (1980)
397. Demuynck, J., Strich, A., Veillard, A.: Nouv. J. Chim. *1*, 217 (1977)
398. Klanberg, F., Muetterties, E. L.: Inorg. Chem. *7*, 155 (1968)
399. Ugi, I., Marguarding, D., Klusacek, H., Gillespie, P.: Acc. Chem. Res. *4*, 288 (1971)
400. Musher, J. I.: J. Am. Chem. Soc. *94*, 5662 (1972)
401. Jesson, J. P., Meakin, P.: J. Am. Chem. Soc. *96*, 5760 (1974)
402. Berry, R. S.: J. Chem. Phys. *32*, 933 (1960)
403. Spiridonov, V. P., Ischenko, A. A., Ivashkevich, L. S.: J. Mol. Struct. *72*, 153 (1981)
404. Wilhite, D. L., Spialter, L.: J. Am. Chem. Soc. *95*, 2100 (1973)
405. Holloway, C. E., Luongo, R. R., Phe, R. M.: J. Am. Chem. Soc. *88*, 2060 (1966)
406. Thompson, D. W.: Inorg. Chem. *8*, 2015 (1969)

407. Serpone, N., Hersh, K. A.: J. Organometal. Chem. *84*, 177 (1975)
408. Haworth, D. T., Lin, Gong-Yu, Wilkie, C. A.: Inorg. Chim. Acta *40*, 119 (1980)
409. Inoue, T.: Inorg. Chem. *22*, 2435 (1983)
410. Morokuma, K.: Acc. Chem. Res. *10*, 294 (1977)
411. Kollman, P., in: Modern Theoretical Chemistry, Vol. 4 (ed. H. F. Schaefer). New York: Plenum 1978
412. Kollman, P. A.: Acc. Chem. Res. *10*, 365 (1977)
413. Nalewajski, R. F.: J. Am. Chem. Soc. *106*, 944 (1984)
414. Tossell, J. A., Gibbs, G. V.: Acta Crystallogr. *A 34*, 463 (1978)
415. Oberhammer, H., Boggs, J. E.: J. Am. Chem. Soc. *102*, 7241 (1980)
416. Glidewell, C.: Inorg. Chim. Acta *12*, 219 (1975)
417. Bartell, L. S.: Inorg. Chem. *5*, 1635 (1966)
418. Burdett, J. K.: Structure and Bonding *31*, 67 (1976)
419. Gimarc, B. M.: J. Am. Chem. Soc. *100*, 2346 (1978)
420. Burdett, J. K.: Molecular Shapes. New York: Wiley 1980
421. Bills, J. L., Steed, S. P.: Inorg. Chem. *22*, 2401 (1983)
422. Day, M. C., Jr., Selbin, J.: Theoretical Inorganic Chemistry, 2nd Edition. New York: Reinhold 1969
423. Baybutt, P.: Mol. Phys. *29*, 389 (1975)
424. Keil, F., Ahirichs, R.: Chem. Phys. *8*, 384 (1975)
425. Kleboth, K., Rode, B. M.: Monatsh. Chem. *105*, 815 (1974)
426. Vitkovskaya, N. M., Mantsivoda, V. B., Moskovskaya, T. E., Voronkov, M. G.: In.. J. Quantum Chem. *17*, 299 (1970)
427. Vitkovskaya, N. M., Mantsivoda, V. B., Dolgunicheva, O. Yu., Voronkov, M.. G.: Dokl. Akad. Nauk SSSR *238* 1098 (1978)
428. Chehayber, J. M., Nagy, S. T., Lin, C. S.: Can. J. Chem. *62*, 27 (1984)
429. Glascow, L. C., Olbrich, G., Potzinger, P.: Chem. Phys. Lett. *14*, 466 (1972)
430. Chuiko, A. A., Tkachenko, K. I., Tertykh, V. A., Gorlov, Yu. I.: Koord. Khim. *2*, 1609 (1976)
431. Sidorkin, V. F., Pestunovich, V. A., Voronkov, M. G.: Uspekhi Khim. *49*, 789 (1980)
432. Mironov, S. L., Gorlov, Yu. I., Chuiko, A. A.: Teor. Eksp. Khim. *14*, 838 (1978)
433. Blandamer, M. J., Burgess, J., Hamshere, S. J., Peacock, R. D., Rogers, J. H., Jenkins, H. D. B.: J. Chem. Soc., Dalton Trans. 726 (1981)
434. Hencsei, P., Csonka, G.: Acta Chim. Acad. Sci. Hungr. *106*, 285 (1981)
435. El-Issa, B. D.: J. Chem. Soc., Faraday Trans. II *78*, 561 (1982)
436. Gucsev, G. L., Boldyrev, A. I.: Zh. Neorg. Khim. *27*, 868 (1982)
437. Payzant, J. D., Tanaka, K., Betowski, L. D., Bohme, D. K.: J. Am. Chem. Soc. *98*, 894 (1976)
438. Boyd, D. R. J.: J. Chem. Phys. *23*, 922 (1955)
439. Atoji, M., Lipscomb, W. W.: Acta Crystallogr. *7*, 597 (1954)
440. Kolbjorn, H., Kenneth, H.: J. Chem. Phys. *59*, 1549 (1973)
441. Muetterties, E. L.: Quart. Rev. *20*, 245 (1966); *21*, 109 (1967); Acc. Chem. Res. *3*, 266 (1970); Res. Chem. Progr. *31*, 51 (1970)
442. Berthier, G., David, D. J., Veillard, A.: Theor. Chim. Acta *14*, 329 (1969)
443. Mulder, J. J. C., Wright, J. S.: Chem. Phys. Lett. *5*, 445 (1970)
444. Ritchie, C. D., Chappell, G. A.: J. Am. Chem. Soc. *92*, 1819 (1970)
445. Cremaschi, P., Gamba, A., Simonetta, M.: Theoret. Chim. Acta *25*, 237 (1972)
446. Gutmann, V.: Rev. Chim. Roum. *22*, 679 (1977); Pure Appl. Chem. *51*, 2197 (1979)
447. Barrow, M. J., Cradock, S., Ebsworth, E. A. V., Rankin, D. W. H.: J. Chem. Soc., Dalton Trans. 1968 (1981)
448. Zefirov. Yu. V.: Zh. Struct. Khim. *22*, 194 (1981)
449. Sheldrick, W. S., Wolfsberger, W.: Z. Naturforsch. *32B*, 22 (1977)
450. Onan, K. D., McPhail, A. T., Yoder, C. H., Hillyard, R. W.: J. Chem. Soc., Chem. Commun. 209 (1978)
451. Hillyard, R. W., Jr., Ryan, C. M., Yoder, C. H.: J. Organometal. Chem. *153*, 369 (1978)
452. Blayden, H. E., Webster, M.: Inorg. Nucl. Chem. Lett. *6*, 703 (1970)
453. Bain, V. A., Killean, R. C. G., Webster, M.: Acta Crystallogr. *B25*, 156 (1969)

454. Sawitzki, G., Schnering, H. G.: Chem. Ber. *109*, 3728 (1976)
455. Klebe, G., Qui, D. T.: Acta Crystallogr. *C40*, 476 (1984)
456. Mooney, J. R., Choy, C. K., Knox, K., Kenney, M. E.: J. Am. Chem. Soc. *97*, 3033 (1975)
457. Swift, D. R.: Ph. D. Thesis Case Western Reserve Univ., Cleveland, 1970 quoted in Ref. 459
458. Kroenke, W. J., Sutton, L. E., Joyner, R. D., Kenney, M. E.: Inorg. Chem. *2*, 1064 (1963)
459. Dirk, C. W., Inabe, T., Schoch, K. F., Marks, T. J.: J. Am. Chem. Soc. *105*, 1539 (1983)
460. Bird, P., Harrod, J. F., Than, Khin Aye: J. Am. Chem. Soc. *96*, 1222 (1974)
461. Schomburg, D., Krebs, R.: Inorg. Chem. *23*, 1378 (1984)
462. Schomburg, D.: J. Organometal. Chem. *221*, 137 (1981)
463. Farnham, W. B., Harlow, R. L.: J. Am. Chem. Soc. *103*, 4608 (1981)
464. Boer, F. P., Flynn, J. J., Turley, J. W.: J. Am. Chem. Soc. *90*, 6973 (1968)
465. Schomburg, D.: Z. Naturforsch. *37B*, 195 (1982)
466. Schomburg, D.: Z. Naturforsch. *38B*, 938 (1983)
467. Flynn, J. J., Boer, F. P.: J. Am. Chem. Soc. *91*, 5756 (1969)
468. Durif, A., Averbuch-Pouchot, M. T., Guitel, J. C.: Acta Crystallogr. *B32*, 2957 (1976)
469. Hesse, K.-F.: Acta Crystallogr. *B35*, 724 (1979)
470. Loehlin, J. H.: Acta Crystallogr. *C40*, 570 (1984)
471. Deadmore, D. L., Bradley, W. F.: Acta Crystallogr. *15*, 186 (1962)
472. Cipriani, C., Curzio, A.: Rend. Soc. Min. Ital. *11*, 22 (1955)
473. Zalkin, A., Forrester, J. D., Templeton, D. H.: Acta Crystallogr. *17*, 1408 (1964)
474. Hoard, J. L., Vincent, W. B.: J. Am. Chem. Soc. *62*, 3126 (1940)
475. Hoard, J. L., Williams, M. B.: J. Am. Chem. Soc. *64*, 633 (1942)
476. Frlec, B., Gantar, D., Golic, L., Leban, I.: Acta Crystallogr. *B36*, 1917 (1980)
477. Cameron, T. S., Knop, O., MacDonald, L. A.: Can. J. Chem. *61*, 184 (1983)
478. Driessen, R. A. J., Hulsbergen, F. B., Vermin, W. J., Reedijk, J.: Inorg. Chem. *21*, 3594 (1982)
479. Stanko, J. A., Paul, I. C.: Inorg. Chem. *6*, 486 (1967)
480. Lynton, H., Siew, P.-Y.: Can. J. Chem. *51*, 227 (1973)
481. Ray, S., Zalkin, A., Templeton, D. H.: Acta Crystallogr. *B29*, 2741 (1973)
482. Clark, M. J. R., Fleming, J. E., Lynton, H.: Can. J. Chem. *47*, 3859 (1969)
483. Hunt, G. W., Terry, N. W., Amma, E. L.: Cryst. Struct. Commun. *3*, 523 (1974)
484. Hunt, G. W., Terry, N. W., Amma, E. L.: Acta Crystallogr. *B35*, 1235 (1979)
485. Schnering, H. G., Vu, D.: Angew. Chem., Int. Ed. Engl. *22*, 408 (1983)
486. Sawitzki, G., Schnering, H. G., Kummer, D., Seshadri, T.: Chem. Ber. *111*, 3705 (1978)
487. Corey, E. R.: unpublished results cited in Ref. 104
488. Holmes, R. R., Deiters, J. A.: J. Am. Chem. Soc. *99*, 3318 (1977)
489. Holmes, R. R.: J. Am. Chem. Soc. *106*, 3745 (1984)
490. Holmes, R. R.: Pentacoordinated Phosphorus, Vol. I, Structure and Spectroscopy (ACS Monogr. 175) Washington 1980
491. Bersuker, I. B.: Electronic Structure and Properties of Coordination Compounds. Leningrad: Leningr. Otd. 1976
492. Gabuda, S. P., Zemskov, S. V.: NMR in Complex Compounds. Novosibirsk: Nauka 1976
493. Gabuda, S. P., Gagarinskii, Yu. V., Polischuk, S. A.: NMR in Inorganic Fluorides: Structure and Chemical Bonding. Moscow: Atomizdat 1978
494. Borshch, S. A., Ogurtsov, I. Ya., Bersuker, I. B.: Zh. Strukt. Khim. *23*, 7 (1982)
495. Bissert, G., Liebau, F.: Naturwissenschaften *56*, 212 (1969)
496. Glasser, L. S. D., in: Molecular Structure by Diffraction Methods Vol. 6, p. 132 (1978)
497. Yagi, T., Akimoto, S.: Tectonophysics *35*, 259 (1976)
498. Marignac, Ch.: Annales de mines *15*, 223 (1859)
499. Voronkov, M. G., Kashaev, A. A., Zelbst, E. A., Frolov, Yu. L., D'yakov, V. M., Gubanova, L. I.: Dokl. Akad. Nauk SSSR *247*, 1147 (1979)
500. Zelbst, E. A., Shklover, V. E., Struchkov, Yu. T., Kashaev, A. A., Gubanova, L. I., D'yakov, V. M., Frolov, Yu. L., Voronkov, M. G.: Dokl. Akad. Nauk SSSR *259*, 1369 (1981)
501. Zelbst, E. A., Shklover, V. E., Struchkov, Yu. T., Kashaev, A. A., Demidov, M. P., Gubanova, L. I., Voronkov, M. G.: Dokl. Akad. Nauk SSSR *260*, 107 (1981)

502. Helbst, E. A., Shklover, B. E., Struchkov, Yu. T., Frolov, Yu. L., Kashaev, A. A., Cubanova, L. I., D'yakov, V. M., Voronkov, M. G.: Zn. Strukt. Khim. *22*, 82 (1981)
503. Schubert, U., Wiener, M., Kohler, F. H.: Chem. Ber. *112*, 708 (1979)
504. Klebe, G., Nix, M., Hensen, K.: Chem. Ber. *117*, 797 (1984)
505. Hensen, K., Klebe, G.: Fresenius Z. Analyt. Chem. *312*, 24 (1982)
506. Klebe, G., Hensen, K., Fuess, H.: Chem. Ber. *116*, 3125 (1983)
507. Klebe, G., Bats, J. W., Hensen, K.: Z. Naturforsch. *38 B*, 825 (1983)
508. Halder, T., Schwarz, W., Weidlein, J., Fischer, P.: J. Organometal. Chem. *246*, 29 (1983)
509. Schomburg, D.: Z. Anorg. Allg. Chem. *493*, 53 (1982)
510. Kemme, A. A., Bleidelis, J. J.: IV Vses. Sov. Organ. Kristallokhim., Zvenigorod 1984; Tezisy dokl., p. 160, Chernogolovka 1984
511. Bleidelis, J. J.: Izv. Akad. Nauk Latv. SSR, Ser. Khim. 259 (1983)
512. Kemme, A. A., Bleidelis, J. J.: Urtane, I. P., Zelchan, G. I., Lukevics, E. J.: Zh. Strukt. Khim. *25*, 165 (1984)
513. Kemme, A. A., Bleidelis, J. J., Urtane, I. G., Zelchan, G. I., Lukevics, E. J.: Izv. Akad. Nauk Latv. SSR, Khim. 486 (1982)
514. D'yachenko, O. A., Atovmyan, L. O., Aldoshin, S. M., Komalenkova, N. G., Popov, A. G., Antipova, V. V., Chernyshev, E. A.: Izv. Akad. Nauk SSSR, Ser. Khim. 1081 (1975)
515. D'yachenko, O. A., Atovmyan, L. O., Aldoshin, S. M., Krasnova, T. L., Stepanov, V. V., Chernyshev, E. A., Popov, A. G., Antipova, V. V.: Izv. Akad. Nauk SSSR, Ser. Khim. 2648 (1974)
516. Daly, J. J., Sanz, F.: J. Chem. Soc., Dalton Trans. 2051 (1974)
517. Kemme, A., Bleidelis, J., Urtane, I., Zelchan, G., Lukevies, E.: J. Organometal. Chem. *202*, 115 (1980)
518. Paton, W. F., Corey, E. R., Corey, J. Y., Glick, M. D.: Acta Crystallogr. *B33*, 3322 (1977)
519. Shklover, V. E., Struchkov, Yu. T., Rodin, O. G., Graven, V. F., Stepanov, B. I.: J. Organometal. Chem. *226*, 117 (1984)
520. Parkanyi, L., Bihatsi, L., Hencsei, P.: Cryst. Struct. Commun. *7*, 435 (1978)
521. Pudova, O. A.: Dissertation Inst. Organic Synth. Riga 1980
522. Kemme, A. A.: Dissertation Inst. Organic Synth. Riga 1977
523. Kemme, A. A., Bleidelis, J. J., D'yakov, V. M., Voronkov, M. G.: Izv. Akad. Nauk SSSR, Ser. Khim. 2400 (1976)
524. Kemme, A. A., Bleidelis, J. J., D'yakov, V. M., Voronkov, M. G.: Zh. Strukt. Khim. *16*, 914 (1975)
525. Demidov, M. P., Shklover, V. E., Frolov, Yu. L., Struchkov, Yu. T., Lukina, Yu. A. D'yakov, V. M., Voronkov, M. G.: IV Vses. Sov. Organ. Kristallokhim., Zvenigorod 1984; Tezisy dokl., p. 161, Chernogolovka 1984
526. Shklover, V. E., Struchkov, Yu. T., Sorokin, M. S., Voronkov, M. G.: Dokl. Akad. Nauk SSSR *274*, 615 (1984)
527. Turley, J. W., Boer, F. P.: J. Am. Chem. Soc. *90*, 4026 (1968)
528. Parkanyi, L., Simon, K., Nagy, J.: Acta Crystallogr. *B30*, 2328 (1974)
529. Parkanyi, L., Nagy, J., Simon, K.: J. Organometal. Chem. *101*, 11 (1975)
530. Turley, J. W., Boer, F. P.: J. Am. Chem. Soc. *91*, 4129 (1969)
531. Bleidelis, J. J., in: Uspekhi Khimii Furana, p. 7 (ed. E. Lukevics). Riga: Zinatne 1978
532. Parkanyi, L., Hencsei, P., Bihatsi, L.: J. Organometal. Chem. *232*, 315 (1982)
533. Parkanyi, L., Hencsei, P., Bihatsi, L., Muller, T.: J. Organometal. Chem. *269*, 1 (1984)
534. Kemme, A. A., Bleidelis, J. J., Pestunovich, V. A., Baryshok, V. P., Voronkov, M. G.: Dokl. Akad. Nauk SSSR *243*, 688 (1978)
535. Kemme, A. A., Bleidelis, J. J., Zelchan, G. I., Urtane, I. P., Lukevics, E. J.: Zh. Strukt. Khim. *18*, 343 (1977)
536. Boer, F. P., Turley, J. W.: J. Am. Chem. Soc. *91*, 4134 (1969)
537. Parkanyi, L., Hencsei, P., Popowski, E.: J. Organometal. Chem. *197*, 275 (1980)
538. Kemme, A. A., Bleidelis, J. J., Lapsinya, A. F., Lukevics, E. J.: in press
539. Boer, F. P., Turley, J. W., Flynn, J. J.: J. Am. Chem. Soc. *90*, 5102 (1968)

540. Voronkov, M. G., Demidov, M. P., Shklover, V. E., Baryshok, V. P., D'yakov, V. M., Frolov, Yu. L.: Zh. Strukt. Khim. *21*, 100 (1980)
541. Demidov, M. P., Zelbst, E. A., Kashaev, A. A., Lunev, I. L., Aleksandrov, G. G., Shklover, V. E., Baryshok, B. P., D'yakov, V. M., Frolov, Yu. L., Voronkov, M. G.: III Vses. Soveshch. Org. Kristallokhim., Tezisy dokl., p. 16, Gorky 1981
542. Demidov, M. P., Aleksandrov, G. G., Struchkov, Yu. T., Baryshok, B. P., D'yakov, V. M., Frolov, Yu. L., Voronkov, M. G.: Koord. Khim. *7*, 1262 (1981)
543. Gusev, A. I., Alekseev, N. V., Baryshok, B. P., Voronkov, M. G.: in press
544. Kemme, A., Bleidelis, J., Solomennikova, I., Zelchan, G., Lukevics, E.: J. Chem. Soc., Chem. Commun. 1041 (1976)
545. Kemme, A. A., Bleidelis, J. J., Zelchan, G. I., Lukevics, E. J.: I Vses. Simp. Stroenie Reakts. Sposobnost Kremneorganicheskikh Soedinenij, Tezisy dokl., p. 224, Irkutsk 1977
546. Grobe, J., Henkel, G., Krebs, B., Voulgarakis, N.: Z. Naturforsch. *39B*, 341 (1984)
547. Parkanyi, L.: 6th International symposium on Organosilicon Chemistry, Abstracts of Papers, p. 49, Budapest 1981
548. Hencsel, P., Parkanyi, L.: Kem. Kozl. (in press), quoted in Ref. 533
549. Legon, A. C.: Chem. Rev. *80*, 231 (1980)
550. Alekseev, N. V., Gurkova, S. N., Tandura, S. N., Nosova, V. M., Gusev, A. I., Gar, T. K., Sel'man, I. R., Khromova, N. Yu.: Zh. Strukt. Khim. *22*, 135 (1981)
551. Muller, E., Burgi, H.-B.: Helv. Chim. Acta, *67*, 399 (1984)
552. Tandura, S. N., Pestunovich, V. A., Voronkov, M. G., Zelchan, G. I., Baryshok, V. P., Lukina, Yu. A.: Dokl. Akad. Nauk SSSR *235*, 406 (1977)
553. Tandura, S. N., Pestunovich, V. A., Glukhikh, V. I., Baryshok, V. P., Voronkov, M. G.: Spectroscopy Lett. *10*, 163 (1977)
Spectroscopy Lett. *10*, 163 (1977)
554. Shatz, V. D., Belikov, V. A., Zelchan, G. I., Solomennikova, I. I., Lukevics, E.: J. Chromatography *174*, 83 (1979)
555. Keiko, V. V., Kuz'menko, L. P., Baryshok, V. P., D'yakov, V. M., Vitkovskii, V. Yu., Tandura, S. N., Voronkov, M. G.: Zh. Obshch. Khim. *50*, 703 (1980)
556. Voegel, J. C., Thirry, J. C., Weiss, R.: Acta Crystallogr. *B30*, 56 (1974)
557. Voegele, J. C., Fischer, J., Weiss, R.: Acta Crystallogr. *B30*, 62 (1974)
558. Voegele, J. C., Fischer, J., Weiss, R.: Acta Crystallogr. *B30*, 66 (1974)
559. Voegel, J. C., Thierry, J. C., Weiss, R.: Acta Crystallogr. *B30*, 70 (1974)
560. Taira, Z., Osaki, K.: Inorg. Nucl. Chem. Lett. *7*, 509 (1971)
561. Mattes, R., Fenske, D., Tebbe, K.-F.: Chem. Ber. *105*, 2089 (1972)
562. Follner, H.: Monatsh. Chem. *104*, 477 (1973)
563. Bonzeek, M., Follner, H.: Monatsh. Chem. *107*, 283 (1976)
564. Shklover, V. E., Struchkov, Yu. T., Voronkov, M. G., Ovchinnikova, Z. A., Baryshok, V. P.: Dokl. Akad. Nauk SSSR (1984).
565. Gurkova, S. N., Gusev, A. I., Sharapov, V. A., Alekseev, N. V., Gar, T. K., Chromova, N. Yu.: J. Organometal. Chem. *268*, 119 (1984)
566. Kemme, A. A., Ignatovich, L. M., Lukevics, E. J., Bleidelis, J. J.: Izv. Akad. Nauk Latv. SSR, Ser. Khim. 96 (1984)
567. Swisher, R. G., Day, R. O., Holmes, R. R.: Inorg. Chem. *22*, 3692 (1983)
568. Tzschach, A., Weichmann, H., Jurkschat, K.: Organometal. Chem. Revs. *12*, 293 (1981)
569. Tzschach, A., Jurkschat, K.: Comments Inorg. Chem. *3*, 35 (1983)
570. Milbrath, D. S., Verkade, J. G.: J. Am. Chem. Soc. *99*, 6607 (1977)
571. Aken, D., Merkelbach, I. I., Koster, A. S., Buck, H. M.: J. Chem. Soc., Chem. Commun. 1045 (1980)
572. Follner, H.: Acta Crystallogr. *B28*, 157 (1972)
573. Atovmyan, L. O., Krasochka, O. N.: J. Chem. Soc., Chem. Commun. 1670 (1970)
574. Kemme, A. A., Bleidelis, J. J.: Izv. Akad. Nauk Latv. SSR, Ser. Khim. 371; (1974); Izv. Akad. Nauk SSSR, Ser. Khim. 322 (1976)
575. Boer, F. P., Remoortere, F. P.: J. Am. Chem. Soc. *91*, 2377; (1969); *92* 801 (1970)
576. Bett, W., Gradock, S., Rankin, D. W. H.: J. Molec. Struct. *66*, 159 (1980)
577. Barrow, M. J., Ebsworth, E. A. V., Huntley, C. M., Rankin, D. W. H.: J. Chem. Soc., Dalton Trans. 1131 (1982)

578. Kamenar, B., Bruvo, M.: Z. Kristallogr. *141*, 97 (1975)
579. Sheludyakov, V. D., Kirilin, A. D., Gusev, A. I., Sharapov, V. A., Mironov, V. F.: Zh. Obshch. Khim. *46*, 2712 (1976)
580. Sheludyakov, V. D., Gusev, A. I., Dmitrieva, A. B., Los', M. G., Kirilin, A. D.: Zh. Obshch. Khim. *53*, 2276 (1983)
581. Gusev, A. I., Sharapov, V. A., Alekseev, N. V., Kozykov, V. P., Orlov, G. I.: Zh. Strukt. Khim. *24*, 112 (1983)
582. Colvin, E. W., Beck, A. K., Bastani, B., Seebach, D., Kai, Y., Dunitz, J. D.: Helv. Chim. Acta. *63*, 697 (1980)
583. Gurkova, S. N., Gusev, A. I., Alekseev, N. V., Fedotov, N. S., Ryasin, G. V., Polyakova, M. V., Sokolov, V. V.: Zh. Strukt. Khim. *20*, 160 (1979)
584. Gurkova, S. N., Gusev, A. I., Alekseev, N. V., Ryasin, G. V., Fedotov, N. S.: Zh. Strukt. Khim. *24*, 160 (1983)
585. Gurkova, S. N., Gusev, A. I., Alekseev, N. V., Los', M. G., Zavodnik, V. E., Belskii, V. K., Ryasin, F. V., Fedotov, N. S.: Zh. Strukt. Khim. *20*, 1059 (1979)
586. Käss, D., Oberhammer, H., Brandes, D., Blaschette, A.: J. Mol. Struct. *40*, 65 (1977)
587. Lebedev, V. A., Drozdov, Yu. N., Kuzmin, E. A., Ganyushkin, A. V., Yablokov, V. A., Belov, N. V.: Dokl. Akad. Nauk SSSR *246*, 601 (1979)
588. Shklover, V. E., Ganyushkin, A. V., Yablokov, V. A., Struchkov, Yu. T.: Cryst. Struct. Commun. *8*, 869 (1979)
589. Razuvaev, G. A., Yablokov, V. A., Ganyushkin, A. V., Shklover, V. E., Isynkier, I., Struchkov, Yu. T.: Dokl. Akad. Nauk SSSR *242*, 132 (1978)
590. Shklover, V. E., Adyasuren, P., Tsinker, I., Yablokov, V. A., Ganyushkin, A. V., Struchkov, Yu. T.: Zh. Strukt. Khim. *21*, 112 (1980)
591. Antipin, M. Yu., Kravers, M. A., Struchkov, Yu. T., Sturkovich, R. J., Lukevics, E. J.: Izv. Akad. Nauk Latv. SSR, Ser. Khim. 89 (1980)
592. Glidewell, C., Robiette, A. G.: Chem. Phys. Lett. *28*, 290 (1974)
593. Cradock, S., Ebsworth, E. A. V., Meikle, G. D., Rankin, D. W. H.: J. Chem. Soc., Dalton Trans. 805 (1975)
594. Glidewell, C., Rankin, D. W. H., Robiette, A. G., Sheldrick, G. M.: J. Chem. Soc. A 318 (1970)
595. Vajda, E., Szekely, T., Hargittai, I., Maltsev, A. K., Baskir, E. G., Nefedov, O. M., Brunvoll, J.: J. Organometal. Chem. *188*, 321 (1980)
596. Vajda, E., Szekely, T., Hargittai, I., Maltsev, A. K. Baskir, E. G., Nefedov, O. M.: J. Molec. Struct. **73**, 243 (1981)
597. Morino, Y., Hirota, E.: J. Chem. Phys. *28*, 185 (1958)
598. Fink, M. J., Haller, K. J., West, R., Michl, J.: J. Am. Chem. Soc. *106*, 822 (1984)
599. Ilsley, W. H., Schaaf, T. F., Glick, M. D., Oliver, J. P.: J. Am. Chem. Soc. *102*, 3769 (1980)
600. Albright, M. J., Schaaf, T. F., Butler, W. M., Hovland, A. K., Glick, M. D., Oliver, J. P.: J. Am. Chem. Soc. *97*, 6261 (1975)
601. Calabrese, J. C., Dahl, L. F.: J. Am. Chem. Soc. *93*, 6042 (1971)
602. Alcock, N. W., Tracy, V. M., Waddington, T. C.: J. Chem. Soc., Dalton Trans. 2238, 2243 (1976)
603. Glidewell, C., Rankin, D. W. H., Robiette, A. G., Sheldrick, G. M., Beagley, B., Freeman, J. M.: J. Molec. Struct. *5*, 417 (1970)
604. Wennerstrom, H., Forsen, S., Roos, B.: J. Phys. Chem. *76*, 2430 (1972)
605. Alesov, C., Scarsdale, J. N., Schafer, L.: J. Moloc. Struct. *90*, 297 (1982)
606. Bartell, L. S.: J. Chem. Phys. *32*, 827 (1960)
607. Bellama, J. M., MacDiarmid, A. G.: J. Organometal. Chem. *18*, 275 (1969); *24*, 91 (1970)
608. Pola, J., Jakoubkova, M., Chvalovsky, V.: Collect. Czech. Chem. Commun. *42*, 2914 (1977)
609. Bellama, J. M., Harmow, L. A.: Inorg. Chem. *17*, 482 (1978)
610. Pola, J., Chvalovsky, V.: Collect. Czech. Chem. Commun. *43*, 3192 (1978)
611. Pola, J., Chvalovsky, V.: Collect. Czech. Chem. Commun. *42*, 3581 (1977)
612. Voronkov, M. G., Feshin, V. P., Romanenko, L. S., Pola, J., Chvalovsky, V.: Collect. Czech. Chem. Commun. *41*, 2718 (1976)
613. Feshin, V. P., Voronkov, M. G., Romanenko, L. S., Ignat'eva, L. P., Dolgushin, G. V.: Zh. Obshch. Khim. *54*, 1312 (1984)
614. Feshin, V. P., Romanenko, L. S., Voronkov, M. G.: Uspekhi Khim. *50*, 460 (1981)

615. Egorochkin, A. N., Skobeleva, S. E.: Uspekhi Khim. *48*, 2216 (1979)
616. Semenov, V. V., Brevnova, T. N., Khorshev, S. Ya.: Zh. Obshch. Khim. *53*, 2085 (1983)
617. Voronkov, M. G., Kashik, T. V., Deriglazova, E. S., Kositsyna, E. I., Pestunovich, A. E., Lukevics, E. J.: Zh. Obshch. Khim. *51*, 375 (1981)
618. Ponec, R., Kucera, J., Chvalovsky, V.: Collect. Czech. Chem. Commun. *48*, 1602 (1983)
619. Grekov, A. P., Veselov, V. J.: Uspekhi Khim. *47*, 1200 (1978)
620. Ponec, R., Chvalovsky, V., Voronkov, M. G.: J. Organometal. Chem. *264*, 163 (1984)
621. Khorshev, S. Ya., Egorochkin, A. N., Sevast'yanova, E. I., Kuz'min, O. V.: Zh. Obshch. Khim. *46*, 1801, (1976); *48*, 1348 (1978); *49*, 795 (1979)
622. Feshin, V. P., Sapozhnikov, Yu. E., Dolgushin, G. V., Yasman, Ya. B., Voronkov, M. G.: Dokl. Akad. Nauk SSSR *247*, 158 (1979)
623. Reich, P.: Z. Anorg. Allg. Chem. *450*, 131 (1979)
624. Ponec, R., Dejmek, L., Chvalovsky, V.: Collect. Czech. Chem. Commun. *44*, 1434 (1979)
625. McKean, D. C., Torto, I., Morrison, A. R.: J. Organometal. Chem. *226*, C 47 (1982)
626. Ponec, R., Dejmek, L., Chvalovsky, V.: J. Organometal. Chem. *197*, 31 (1980)
627. Dejmek, L., Ponec, R., Chvalovsky, V.: Collect. Czech. Chem. Commun. *45*, 3510, 3518 (1980)
628. Ponec, R., Dejmek, L., Chvalovsky, V. V.: Collect. Czech. Chem. Commun. *42*, 1859 (1977)
629. Zingler, G., Kelling, H., Popowski, E.: Z. Anorg. Allg. Chem. *476*, 41 (1981)
630. Yoder, C. H., Cader, B. M.: J. Organometal. Chem. *233*, 275 (1982)
631. Yablokov, V. A., Tomadze, A. V., Yablokova, N. V., Aleksandrov, Yu. A.: Zh. Obshch. Khim. *49*, 1787 (1979)
632. Glidewell, C., Rankin, D. W. H., Robiette, A. G., Sheldrick, G. M.: J. Molec. Struct. *6*, 231 (1970)
633. Glidewell, C., Robiette, A. G., Sheldrick, G. M.: Chem. Phys. Lett. *16*, 526 (1972)
634. Wheatley, P. J.: J. Chem. Soc. *1962*, 1721
635. Chioccola, G., Daly, J. J.: J. Chem. Soc. A *1968*, 1658
636. Clegg, W., Hesse, M., Klingebiel, U., Sheldrick, G. M., Skoda, L.: Z. Naturforsch. *B35*, 1359 (1980)
637. Parkanyi, L., Dunaj-Jurco, M., Bihatsi, L., Hencsei, P.: Cryst. Struct. Commun. *9*, 1049 (1980)
638. Clegg, W., Klingebiel, U., Krampe, C., Sheldrick, G. M.: Z. Naturforsch. *B35*, 275 (1980); *B37* 423 (1982)
639. Clegg, W., Haase, M., Sheldrick, G. M., Vater, N.: Acta Crystallogr. *C40*, 871 (1984)
640. Yokoi, M., Nomura, T., Yamasaki, K.: J. Am. Chem. Soc. *77*, 4484 (1955); Bull. Chem. Soc. Japan. *30*, 100 (1957)
641. Vilkov, L. V., Kusakov, M. M., Nametkin, N. S., Oppengeim, V. D.: Dokl. Akad. Nauk SSSR *183*, 830 (1968)
642. Hseu, T. H., Chi, Y., Liu, C.-S.: Inorg. Chem. *20*, 199 (1981)
643. Parkanui, L., Sasvari, K., Barta, I.: Acta Crystallogr. *B34*, 883 (1978)
644. Kirichenko, E. A., Emarkov, A. I., Samsonova, I. N.: Zh. Fiz. Khim. *51*, 2506 (1977)
645. Schaaf, T. F., Butler, W., Glick, M. D., Oliver, J. P.: J. Am. Chem. Soc. *96*, 7593 (1974)
646. Tecle, B., Ilsley, W. H., Oliver, J. P.: Organometallics *1*, 875 (1982)
647. Ilsley, W. H., Albright, M. J., Anderson, T. J., Glick, M. D., Oliver, J. P.: Inorg. Chem. *19*, 3577 (1980)
648. Auburn, M., Ciriano, M., Howard, J. A. K., Murray, M. Pugh, N. J., Spencer, J. L., Stone, F. G. A., Woodward, P.: J. Chem. Soc., Dalton Trans. 659 (1980)
649. Graham, W. A. G., Bennett, M. J.: Chem. Eng. News *48*, No 24, 75 (1970)
650. Schubert, U., Ackermann, K., Kraft, G., Worle, B.: Z. Naturforsch. *38B*, 1488 (1983)
651. Simpson, K. A.: Ph. D. Thesis Univ. Alberta, Edmonton 1973, quoted in Ref. 650
652. Schubert, U., Ackermann, K., Worle, B.: J. Am. Chem. Soc. *104*, 7378 (1982)
653. Ackermann, K.: Dissertation Tech. Univ. München 1982, quoted in Ref. 662.
654. Smith, R. A., Bennett, M. J.: Acta Crystallogr. *B33* 1113 (1977)
655. Cowie, M., Bennett, M. J.: Inorg. Chem. *16*, 2321 (1977)
656. Elder, M.: Inorg. Chem. *9*, 762 (1970)
657. Cowie, M., Bennett, M. J.: Inorg. Chem. *16*, 2325 (1977)

658. Smith, R. A., Bennett, M. J.: Acta Crystallogr. *B33* 1118 (1977)
659. Manojlovie-Muir, L., Muir, K. W., Ibers, J. A.: Inorg. Chem. *9*, 447 (1970)
660. Greene, J., Curtis, M. D.: J. Am. Chem. Soc. *99*, 5176 (1977)
661. Curtis, M. D., Greene, J., Butler, W. M.: J. Organometal. Chem. *164*, 371 (1979)
662. Schubert, U., Kraft, G., Kalbas, C.: Transition Metal. Chem. *9*, 161 (1984)
663. Schubert, U., Ackermann, K., Goddard, R., Worle, B., Stansfield, R. F. D.: Organometallics in press
664. Colomer, E., Corriu, R. J. P., Marzin, C., Vioux, A.: Inorg. Chem. 21, 368 (1982)
665. Andrews, M. A., Kirtley, S. W., Kaesz, H. D.: Adv. Chem. Ser. *167*, 214 (1978)
666. Colomer, E., Corriu, R. J. P.: Top. Curr. Chem. *96*, 79 (1981)
667. Barrow, M. J.: Acta Crystallogr. *B37*, 2239 (1981)
668. Barrow, M. J., Ebsworth, E. A. V.: J. Chem. Soc., Dalton Trans. 211 (1982)
669. Britton, D., in: Perspectives in Structural Chemistry, Vol. 1, p. 109 (eds. J. D. Dunitz, J. A. Ibers) New York: 1967
670. Konnert, J., Britton, D., Chow, Y. M.: Acta Crystallogr. *B28*, 180 (1972)
671. Alcock, N. W.: Adv. Inorg. Chem. Radiochem. *15*, 1 (1972)
672. Arnold, D. E. J., Cradock, S., Ebsworth, E. A. V., Murdoch, J. D., Rankin, D. W. H., Shea, D. C. J., Harris, R. K., Kimber, B. J.: J. Chem. Soc., Dalton Trans. 1349 (1981)
673. Borthwick, P. W.: Acta Crystallogr. *B36*, 628 (1980)
674. Brown, I. D.: J. Chem. Soc., Dalton Trans. 1118 (1980)
675. Hossain, M. A., Hursthouse, M. B., Malik, K. M. A.: Acta Crystallogr. *B35*, 2258 (1979)
676. Domenicano, A., Mazzeo, P., Vaciago, A.: Tetrahedron Lett. 1029 (1976)
677. Domenicano, A., Murray-Rust, P.: Tetrahedron Lett. 2283 (1979)
678. Gergo, E., Hargittai, I., Schultz, G.: J. Organometal. Chem. *112*, 29 (1976)
679. Gergo, E., Hargittai, I.: 6th Intern. Symp. Organosilicon Chem. Abstracts of Papers, p. 139, Budapest 1981
680. Shevchenko, S. G., Elin, V. P., Dolenko, G. N., Baryshok, V. P., Feshin, V. P., Frolov, Yu. L., Mazalov, L. N., Voronkov, M. G.: Zh. Obshch. Khim. *23*, 43 (1982); Dokl. Akad. Nauk SSSR *264*, 373 (1982)
681. Bent, H. A.: Chem. Rev. *68*, 587 (1968)
682. Burgi, H. B.: Inorg. Chem. *12*, 2321 (1973)
683. Burgi, H. B., Dunitz, J. D.: Acc. Chem. Res. *16*, 153 (1983)
684. Allen, F. H., Kennard, O., Taylor, R.: Acc. Chem. Res. *16*, 146 (1983)
685. Burgi, H. B.: Angew. Chem., Int. Ed. Engl. *14*, 460 (1975)
686. Dunitz, J. D.: X-Ray Analysis and the Structure of Organic molecules. New York: Cornell University 1979
687. Pestunovich, V. A., Sidorkin, V. F., Dogaev, O. B., Voronkov, M. G.: Dokl. Akad. Nauk SSSR *251* 1440 (1980)
688. Burgi, H. B.: private communication cited in Ref. 686.
689. Britton, D., Dunitz, J. D.: J. Am. Chem. Soc. *103*, 2971 (1981)
690. Tandura, S. N., Gurkova, S. N., Gusev, A. I., Alekseev, N. V.: Zh. Strukt. Khim. *26*, (1985)
691. Gasteiger, J., Marsili, M.: Org. Magn. Reson. *15*, 353 (1981)
692. Gur'yanova, E. N., Gol'dshtein, I. P., Romm, I. P.: Donor-Acceptor Bond. New York: Wiley 1975
693. Tandura, S. N., Pestunovich, V. A., Voronkov, M. G., Zelchan, G. I., Solomennikova, I. I., Lukevics, E.: Khim. Geterotsikl. Soed. 1063 (1977)
694. Pestunovich, V. A., Tandura, S. N., Voronkov, M. G., Baryshok, V. P., Zelchan, G. I., Glukhikh, V. I., Engelhardt, G., Witanowski, M.: Spectroscopy Lett. *11*, 339 (1978)
695. Harris, R. K., Jones, J., Ng, S.: J. Magn. Reson. *30*, 521 (1978)
696. Tandura, S. N., Pestunovich, V. A., Zelchan, G. I., Baryshok, V. P., Lukina, Yu. A., Sorokin, M. S., Voronkov, M. G.: Izv. Akad. Nauk SSSR, Ser. Khim. 295 (1981)
697. Egorochkin, A. N., Pestunovich, V. A., Voronkov, M. G., Zelchan, G. I.: Khim. Geterotsikl. Soed. 300 (1965)
698. Pestunovich, V. A., Voronkov, M. G., Zelchan, G. I., Lukevics, E. J., Libert, L. I., Egorochkin, A. N., Burov, A. I.: Khim. Geterotsikl. Soed., Sb. 2 339 (1970)

699. Pestunovich, V. A., Voronkov, M. G., Zelchan, G. I., Lapsina, A. F., Lukevics, E. J., Libert, L. I.: Khim. Geterotsikl. Soed., Sb. 2 348 (1970)
700. Pestunovich, V. A., Popelis, Yu. Yu., Lukevics, E. J., Voronkov, M. G.: Izv. Akad. Nauk Latv. SSR, Ser. Khim. 365 (1973)
701. Popowski, E., Michalik, M., Kelling, H.: J. Organometal. Chem. *88*, 157 (1975)
702. Lukevics, E. J., Solomennikova, I. I. Zelchan, G. I., Yudeika, I. A., Liepinsh, E. E., Yankovska, I. S., Mazheika, I. B.: Zh. Obshch. Khim. *47*, 105 (1977)
703. Lukevics, E. J., Zelchan, G. I., Solomennikova, I. I. Liepinsh, E. E., Yankovska, I. S., Mazheika, I. B.: Zh. Obshech. Khim. *47*, 109 (1977)
704. Hencsei, P., Bihatsi, L., Kovacs, I., Szalay, E., Karsai, E. B., Szollosy, A., Gal, M.: Acta Chim. Hungarica *112*, 261 (1983)
705. Voronkov, M. G., Lukina, Yu. A., D'yakov, V. M., Frolov, Yu. L., Tandura, S. N.: Zh. Obshch. Khim. *52*, 349 (1982)
706. Voronkov, M. G., Tandura, S. N., Sorokin, M. S., Shterenberg, B. Z., Pestunovich, V. A.: Izv. Akad. Nauk SSSR, Ser. Khim. 464 (1979)
707. Voronkov, M. G., Tandura, S. N., Shterenberg, B. Z., Kuznetsov, A. L., Mirskov, R. G., Zelchan, G. I., Chromova, N. Yu., Gar, T. K., Mironov, V. F., Pestunovich, V. A.: Dokl. Akad. Nayk SSSR *248*, 134 (1979)
708. Leonard, N. J.: Acc. Chem. Res. *12*, 423 (1979)
709. Voronkov, M. G., Baryshok, V. P., Tandura, S. N., Vitkovskii, V. Yu., D'yakov, V. M., Pestunovich, V. A.: Zh. Obshch. Khim. *48*, 2238 (1978)
710. Kupće, E., Liepinsh, E., Lapsina, A., Zelchan, G., Lukevics, E.: J. Organometal. Chem. 15 (1983)
711. Fedorov, L. A., Nametkin, N. S., Kuzovkina, M. E., Perchenko, V. N.: Dokl. Akad. Nayk SSSR *195*, 1347 (1970)
712. Cook, D. I., Fields, R., Green, M., Haszeldine, R. N., Ilea, B. R., Jones, A., Newlands, M. J.: J. Chem. Soc. A *1966*, 887
713. Kowalski, J., Lasocki, Z.: J. Organometal. Chem. *116*, 75 (1976)
714. Yoder, C. H., Ryan, C. M., Martin, G. F., Ho, P. S.: J. Organometal. Chem. *190*, 1 (1980)
715. Sau, A. C., Holmes, R. R.: J. Organometal. Chem. **217**, 157 (1981)
716. Shapiro, B. L., Mohrmann, L. E.: J. Phys. Chem. Ref. Data *6*, 919 (1977)
717. Thompson, D. W.: J. Magn. Reson. *1*, 606 (1969)
718. Hester, R. E.: Chem. Ind. 1397 (1963)
719. Smith, J. A. S., Wilkins, E. J.: J. Chem. Soc. A *1967*, 1749
720. Fay, R. C., Serpone, N.: J. Am. Chem. Soc. *90*, 5701 (1968)
721. Srivastava, G., Mehrotra, S. K.: Reviews Silicon, Germanium, Tin and Lead Compaunds *2*, 307 (1977)
722. Menrotra, R. C., Bohra, R., Gaur, D. P.: Metal-Diketonates and Allied Derivatives. New York: Academic 1978
723. Esposito, J. N., Lloyd, J. E., Kenney, M. E.: Inorg. Chem. *5*, 1979 (1966)
724. Kane, A. R., Yalman, R. G., Kenney, M. E.: Inorg. Chem. *7*, 2588 (1968)
725. Janson, T. R., Kane, A. R., Sullivan, J. F., Knox, K., Kenney, M. E.: J. Am. Chem. Soc. *91*, 5210 (1969)
726. Kane, A. R., Sullivan, J. F., Kenng, D. H., Kenney, M. E.: Inorg. Chem. *9*, 1445 (1970)
727. Hanack, M., Mitulla, K., Pawlowski, G., Subramanian, L. R.: Angew. Chem., Int. Ed. Engl. *18*, 322 (1979)
728. Hanack, M., Mitulla, K., Pawlowski, G., Subramanian, L. R.: J. Organometal. Chem. *204*, 315 (1981)
729. Abraham, R. J., Plant, J., Bedford, G. R.: Org. Magn. Reson. *19*, 204 (1982)
730. Abraham, R. J.: J. Magn. Reson. *43*, 491 (1981) Orlov, V. V., Mamaev, V. M., Gloriozov, I. P.: Zh. Strukt. Khim. *22*, 64 (1981)
731. Vysotskii, Yu. B., Kuzmitzkii, V. A., Solov'ev, K. I.: Zh. Strukt. Khim. *22*, 22 (1981)
732. Abraham, R. J., Bedford, G. R., McNeillie, D., Wright, B.: Org. Magn. Reson. *14*, 418 (1980)
733. Tandon, J. P., Gupta, S. R., Prasad, R. N.: Monatsh. Chem. *107*, 1379 (1976)
734. Voronkov, M. G., Skvortsova, G. G., Domnina, E. S., Ivlev, Yu. N., Chernov, N. F., Chipanina, N. N., Voronov, V. K., Toryashinova, D. D.: Zh. Obshch. Khim. *46*, 311 (1976)

735. Dashora, R., Singh, R. V., Tandon, J. P.: Synth. React. Inorg. Met.-Org. Chem. *13*, 745 (1983)
736. Kummer, D., Gaisser, K. E., Seifert, J., Wagner, R.: Z. Anorg. Allg. Chem. 459, 145 (1979)
737. Strelenko, Yu. A., Alekseev, N. V. Ryasin, G. V., Fedotov, N. S., Gurkova, S. N., Gusev, A. I.: Dokl. Akad. Nauk SSSR *251* 159 (1980)
738. Rupani, P., Singh, A., Rai, A. K., Mehrotra, R. C.: J. Organometal. Chem. *185*, 209 (1980)
739. Ebsworth, F. A. V., Rocktaschel, G., Thompson, J. C.: J. Chem. Soc. *A 1967*, 362
740. West, R., Gornowicz, G. A.: J. Organometal. Chem. *25*, 385 (1970)
741. Corriu, R. J. P., Royo, G., Saxce, A.: J. Chem. Soc. Commun. 892 (1980)
742. Breliere, C., Carre, F., Corriu, R. J. P., Saxce, A., Poirier, M., Royo, G.: J. Organometal. Chem. *205*, C 1 (1981)
743. Koten, G., Noltes, J. G.: Adv. Chem. *157* 275 (1976)
744. Koten, G., Noltes, J. G.: J. Am. Chem. Soc. *98*, 5393 (1976)
745. Koten, G., Jastrzebski, J. T. B. H., Noltes, J. G., Pontenagel, W. M. G. F., Kroon, J., Spek, A. L.: J. Am. Chem. Soc. *100*, 5021 (1978)
746. Oki, M., Ohira, M.: Chem. Lett. 1267 (1982)
747. McFarlane, W., Seaby, J. M.: J. Chem. Soc., Perkin Trans. II 1561 (1972)
748. Khankhodzhaeva, D. A., Reikhsfeld, V. O., Saratov, I. E.: Khimija i prakticheskoye primenenie kremnij i fosfororganicheskikh soedinenij, p. 92, Leningrad 1977
749. Kummer, D., Koster, H.: Z. Anorg. Allg. Chem. *398*, 279, 402 (1973); 297 297 (1973)
750. Tandon, J. P., Gupta, S. R., Prased, R. N.: Monatsh. Chem. *107*, 1379 (1976)
751. Singh, R. V., Tandon, J. P.: J. Indian Chem. Soc. *55* 764 (1978); Ann. Soc. Sci. Bruxelles, Ser. 1 *93*, 143 (1979)
752. Kummer, D., Seshadri, T.: Z. Anorg. Allg. Chem. *425*, 236, 432 (1976); 147,432 (1977); 153 (1977)
753. Kummer, D., Gaiber, K.-E., Seshadri, T.: Chem. Ber. *110*, 1950 (1977)
754. Kummer, D., Seshadri, T.: Chem. Ber. *110*, 2355 (1977)
755. Kummer, D., Balkin, A., Koster, H.: J. Organometal. Chem. *178*, 29 (1979)
756. Singh, R. V., Tandon, J. P.: Synth. React. Inorg. Met.-Org. Chem. *11*, 109 (1981)
757. Dashora, R., Singh, R. V., Tandon, J. P.: Synth. React. Inorg. Met.-Org. Chem. *13*, 209 (1983)
758. Biradar, N. S., Karajagi, G. V., Aminabhavi, T. M., Rudzinski, W. E.: Inorg. Chim. Acta *82*, 211 (1984)
759. Voronkov, M. G., Vitkovskii, V. Yu., Tandura, S. N., Alekseev, N. V., Basenko, S. V., Mirskov, R. G.: Izv. Akad. Nauk SSSR, Ser. Khim. 2696 (1982)
760. Voronkov, M. G., Lukina, Yu. A., Tandura, S. N., Voronov, V. K., D'yakov, V. M.: Zh. Obshch. Khim. *49*, 1278 (1979)
761. Koacher, J. K., Tandon, J. P., Mehrotra, R. C.: Indian J. Chem. *A 18* 75 (1979); J. Inorg. Nucl. Chem. *41*, 1409 (1979)
762. Birgele, I. S., Mazheika, I. B., Liepinsh, E. E., Lukevics, E : Zh. Obshch. Khim. *50*, 882 (1980)
763. Voronkov, M. G., Lukina, Yu. A., D'yakov, V. M.. Sigalov, M. V.: Zh. Obshch. Khim. *53*, 803 (1983)
764. Grobe, J., Voulgarakis, N.: Z. Naturforsch. *38 B*, 269 (1983)
765. Lycka, A., Snobl, D., Vencl, J.: J. Organometal. Chem. *174*, 41 (1979)
766. Mazerolles, P., Faucher, A., Mauret, P., Mermillod-Blardet, D., Fayet, J.-P.: J. Chim. Phys. Phys.-Chim. Biol. *77*, 329 (1980)
767. Piekos, R., Sujecki, R., Sankowski, M.: Z. Anorg. Allg. Chem. *454*, 187 (1979)
768. Voronov, V. K., Keiko, V. V., Baryshok, V. P., D'yakov, V. M., Voronkov, M. G.: Dokl. Akad. Nauk SSSR *236*, 147 (1977)
769. Voronov, V. K., Voronkov, M. G., D'yakov, V. M., Baryshok, V. P.: Izv. Akad. Nauk SSSR, Ser Khim. 1457 (1978)
770. Voronov, V. K., Voronkov, M. G.: J. Mol. Struct. *67*, 385 (1980)
771. Kordova, I., Fomichev, A. A., Zvolinskii, V. P., Gusarov, A. I., Pleshakov, V. G., Prostakov, N. S.: Zh. Strukt. Khim. *22*, 27 (1981)

772. Kuz'menko, L. P., Voronkov, M. G., Keiko, V. V., Pestunovich, V. A.: Zh. Obshch. Khim. *53*, 105 (1983)
773. Sidorkin, V. F., Keiko, V. V., Pestunovich, V. A., Kalinina, N. A., Voronkov, M. G.: Zh. Obshch. Khim. *53*, 581 (1983)
774. Glukhikh, V. I., Tandura, S. N., Kuznetsova, G. A., Keiko, V. V., D'yakov, V. M., Voronkov, M. G.: Dokl. Akad. Nauk SSSR *239*, 1129 (1978)
775. Glukhikh, V. I., Voronkov, M. G., Yarosh, O. G., Tandura, S. N., Alekseev, N. V., Khromova, N. Yu., Gar, T. K.: Dokl. Akad. Nauk SSSR *258*, 387 (1981)
776. Glukhikh, V. I., Voronkov, M. G.: Dokl. Akad. Nauk SSSR *248*, 142 (1978)
777. Liepinsh, E. E., Birgele, I. S., Solomennikova, I. I., Lapsinya, A. F., Zelchan, G. I., Lukevics, E.: Zn. Obshch. Khim. *50*, 2462 (1980)
778. Tandura, S. N., Strelenko, Yu. A., Voronkov, M. G., Alekseev, N. V., Yarosh, O. G.: Dokl. Akad. Nauk SSSR *267*, 397 (1982)
779. Rakita, P. E., Srebro, J. P., Worsham, L. S.: J. Organometal. Chem. *104*, 27 (1976); Org. Magn. Reson. *8*, 310 (1976)
780. Rakita, P. E., Worsham, L. S.: J. Organometal. Chem. *137*, 145 (1977); *139* 135 (1977)
781. Mitchell, T. N.: Org. Magn. Reson. *8*, 34 (1976)
782. Levy, G. C., White, D. M., Cargioli, J. D.: J. Magn. Reson. *8*, 280 (1972)
783. Hearn, M. T. W.: Aust. J. Chem. *29*, 2315 (1976)
784. Schraml, J., Chvalovsky, V., Magi, M., Lippmaa, E.: Collect. Czech. Chem. Commun. *42*, 306 (1977)
785. Lippmaa, E., Magi, M., Chvalovsky, V., Schramb, J.: Collect. Czech. Chem. Commun. *42*, 318 (1977)
786. Kamienska-Trela, K.: J. Organometal. Chem. *159*, 15 (1978)
787. Sebald, A., Wrackmeyer, B.: Spectrochim. Acta *37A* 365 (1981)
788. Kamienska-Trela, K.: J. Mol. Struct. *78*, 121 (1982)
789. Craik, D. J., Ternai, B.: Org. Magn. Reson. *15*, 268 (1981)
790. Hegre, W. J., Taft, R. W., Topsom, R. D.: Progr. Phys. Org. Chem. *12*, 159 (1976)
791. Mishima, M., Fujio, M., Takeda, R., Tsuno, Y.: Mem. Fac. Sci. Kyushu Univ. *C11*, 97 (1978)
792. Ewing, D. F.: Org. Magn. Reson. *12*, 499 (1979)
793. Lipowitz, J.: J. Am. Chem. Soc. *94*, 1582 (1972)
794. Daneshrad, A., Eaborn, C., Walton, D. R. M.: J. Organometal. Chem. *85*, 35 (1975)
795. Voronkov, M. G., Tandura, S. N., Sorokin, M. S., Shterenberg, B. Z., Pestunovich, V. A.: Izv. Akad. Nauk SSSR, Ser. Khim. 464 (1979)
796. Voronkov, M. G., Tandura, S. N., Pestunovich, V. A., Sorokin, M. S., D'yakov, V. M.: Izv. Akad. Nauk SSSR, Ser. Khim. 1948 (1978)
797. Gar, T. K., Khromova, N. Yu., Tandura, S. N., Nosova, V. M., Kisin, A. V., Mironov, V. F.: Zh. Obshch. Khim. *52*, 622 (1982)
798. Harris, R. K., Kimber, B. J.: J. Magn. Reson. *17*, 174 (1975)
799. Pestunovich. V. A., Tsetlina, E. O., Voronkov, M. G., Liepinsh, E. E., Bogoradovskii, E. T., Zavgorodnii, V. S., Petrov, A. A.: II Vses. Simp. "Stroenie i Reakts. Sposobn. Kremneorg. Soed.", Tezisy dokl., p. 17, Irkutsk 1981
800. Wrackmeyer, B.: J. Organometal. Chem. *116*, 353 (1979)
801. Tandura, S. N., Voronkov, M. G., Kisin, A. V., Alekseev, N. V., Shestakov, E. E. Ovchinnikova, Z. A., Baryshok, V. P.: Zh. Obshch. Khim. *54*, 2012 (1984)
802. Sarneski, J. E., Surprenaut, H. L., Molen, F. K., Reilley, C. N.: Anal. Chem. *47*, 2116 (1975)
803. Stothers, J. B.: Carbon-13 NMR Spectroscopy. New York; Akademic Press 1972
804. Beguin, C. G., Deschamps, M.-N., Boubel, V., Delpuech, J.-J.: Org. Magn. Reson. *10*, 418 (1978)
805. Cralk, D. J., Levy, G. C., Lombardo, A.: J. Phys. Chem. *86*, 3893 (1982)
806. Grant, D. M., Cheney, B. V.: J. Am. Chem. Soc. *89*, 5315 (1967)
807. Batchelor, J. G.: J. Magn. Reson. *18*, 212 (1975)
808. Musker, W. K., Larson, G. L.: J. Organometal. Chem. *6*, 627 (1966)
809. Albanov, A. I., Voronkov, M. G., Gubanova, L. I., Larin, M. F., Liepinsh, E. E., Pestunovich, V. A.: Izv. Akad. Nauk SSSR, Ser. Khim. 2402 (1983)
810. Bar, E., Fehlhammer, W. P., Breitinger, D. K., Mink, J.: Inorg. Chim. Acta *82*, L17 (1984)

811. Matsubayashi, G., Tanaka, T.: Spectrochim. Acta *30A*, 869 (1974)
812. Muetterties, E. L.: J. Am. Chem. Soc. *82*, 1082 (1960)
813. Brownstein, S.: Can J. Chem. *58*, 1407 (1980)
814. Muller, R.: Organometal. Chem. Revs. *1*, 359 (1966)
815. Coyle, T. D., Johannesen, R. B., Brinckman, F. E., Farrar, T. C.: J. Phys. Chem. *70*, 1682 (1966)
816. Clark, H. C., Dixon, K. R., Jacobs, W. J.: J. Am. Chem. Soc. *90*, 2259 (1968)
817. Gibson, J. A., Ibbott, D. G., Janzen, A. F.: Can. J. Chem. *51*, 3203 (1973)
818. Marat, R. K., Janzen, A. F.: Can. J. Chem. *55*, 3845 (1977)
819. Brownstein, S.: J. Chem. Soc., Chem. Commun. 149 (1980)
820. Brownstein, S.: Can. J. Chem. *56*, 343 (1978)
821. Pestunovich, V. A., Larin, M. F., Sorokin, M. S., Albanov, A. I., Voronkov, M. G.: 7th Intern. Symp. Organosilicon Chem., Abstracts of Papers, p. 244, Kyoto 1984
822. Albanov, A. I., Gubanova, L. I., Larin, M. F., Pestunovich, V. A., Voronkov, M. G.: J. Organometal. Chem. *244*, 5 (1983)
823. Johannesen, R. B., Brinckman, F. E., Coyle, T. D.: J. Phys. Chem. *72*, 660 (1968)
824. Marsmann, H. C., Lower, R.: Chem. Ztg. *97*, 660 (1973)
825. Dean, P. A. W., Evans, D. F.: J. Chem. Soc. *A* 1970, 2569
826. Kummer, D., Seshadri, T.: Z. Anorg. Allg. Chem. *428*, 129 (1977)
827. Nefedov, V. I., Yarzhemsky, V. G., Tarasov, V. P.: Koord. Khim. *2*, 1443 (1976); Chem. Phys. *18*, 417 (1976)
828. Marat, R. K., Janzen, A. F.: Can J. Chem. *55*, 1167 (1977)
829. Kucheryaev, A. G., Lebedev, V. A., Ovchinnikov, I. M.: Zh. Strukt. Khim. *11*, 925 (1970)
830. Pestunovich, V. A., Larin, M. F., Albanov, A. I., Gubanova, L. I., Kopylov, V. M., Voronkov, M. G.: Izv. Akad. Nauk SSSR, Ser. Khim. 1931 (1983)
831. Corriu, R. J. P., Poirier, M., Royo, G.: J. Organometal. Chem. *233*, 165 (1982)
832. Marat, R. K., Janzen, A. F.: J. Chem. Soc., Chem. Commun. 671 (1977)
833. Adley, A. D., Gilson, D. F. R., Onyszchuk, M.: J. Chem. Soc., Chem. Commun. 813 (1968)
834. Ilyin, E. G., Buslaev, Yu. A.: Dokl. Akad. Nauk SSSR *246*, 1165 (1979)
835. Buslaev, Yu. A., Ilyin, E. G.: J. Fluor. Chem. *25*, 57 (1984)
836. Gofman, M. M., Nefedov, V. I.: Inorg. Chim. Acta *31*, 155 (1978)
837. Albanov, A. I.: Dissertation Inst. Organic Chem. Irkutsk 1984
838. Thompson, J. C., Wright, A. P. G., Reynolds, W. F.: J. Fluor. Chem. *17*, 509 (1981)
839. Wray, V.: Fluorine-19 Nuclear Magnetic Resonance Spectroscopy. (Ann. Rep. NMR Spectroscopy, Vol. 14) London: Academic Press 1983
840. Stevenson, W. H., Martin, J. C.: J. Am. Chem. Soc. *104*, 309 (1982)
841. Grobe, J., Martin, R., Moller, U.: Angew. Chem., Int. Ed. Engl. *16*, 248 (1977)
842. Shcherbakov, V. A., Chernyshov, B. N., Davidovich, R. L.: Zh. Strukt. Khim. *14*, 1109 (1973)
843. Moskvich, Yu. N., Polyakov, A. M., Dotsenko, G. I., Afanas'ev, M. A.: Zh. Neorg. Khim. *27*, 1972 (1982)
844. Gabuda, S. P., Goncharuk, V. K., Panich, A. M., Ippolitov, E. G.: Dokl. Akad. Nauk SSSR *265*, 622 (1982)
845. Sergienko, V. I., Pershin, V. L., Ignat'eva, L. N.: Zh. Strukt. Khim. *24*, 46 (1983)
846. Mackowiak, M., Brown, R. J. C.: J. Magn. Reson. *52*, 71 (1983)
847. Schramb, J., Bellama, J. M.: Determ. Org. Struct. Phys. Meth. *6*, 203 (1976)
848. Williams, E. A., Cargioli, J. D.: Ann. Rep. NMR Spectroscopy 9, 221 (1970)
849. Marsmann, H.: N.M.R., Basic Princ. and Progr. *17*, 65 (1981)
850. Cella, J. A., Cargioli, J. D., Williams, E. A.: J. Organometal. Chem. *186*, 13 (1980)
851. Pestunovich, V. A., Tandura, S. N., Voronkov, M. G., Engelhardt, G., Lippmaa, E., Pekhk, T., Sidorkin, V. F., Zelchan, G., Baryshok, V. P.: Dokl. Akad. Nauk SSSR *240*, 914 (1978)
852. Hencsei, P., Marsmann, H. C.: Acta Chim. Acad. Sci. Hung. *105*, 79 (1980)
853. Hencsei, P., Parkangi, L.: Kemiai Kozlemenyek *54*, 252 (1980)
854. Pestunovich, V. A., Albanov, A. I., Larin, M. F., Voronkov, M. G., Kramarova, E. P., Baukov, Yu. I.: Izv. Akad. Nauk SSSR, Ser. Khim. 2178 (1980)
855. Klebe, G., Hensen, K., Jouanne, J.: Organometal. Chem. *258*, 137 (1983)

856. Engelhardt, G., Radeglia, R., Jancke, H., Lippmaa, E., Magi, M.: Org. Magn. Reson. *5*, 561 (1973)
857. Wolff, R., Radeglia, R.: Z. Phys. Chem. (Leipzig *258*, 145 (1977); *261* 726 (1980)
858. Mitchell, T. N.: J. Organometal. Chem. *255*, 279 (1983)
859. Mai, L. A.: Izv. Akad. Nauk Latv. SSR, Ser. Khim. 273 (1981)
860. Poleshchuk, O. Kh., Maksyutin, Yu. K.: Uspekhi Khim. *45*, 2097 (1976)
861. Gutman, V., Resch, G.: Comments Inorg. Chem. *1*, 265 (1982)
862. Ernst, C. R., Spialter, L., Buell, G. R., Wilhite, D. L.: J. Am. Chem. Soc. *96*, 5375 (1974)
863. Lukevich, E., Pudova, O. A., Popelis, Yu., Erchak, N. P.: Zh. Obshch. Khim. *51*, 369 (1981)
864. Kupche, E.: Dissertation Inst. Organic Synth. Riga 1984
865. Pestunovich, V. A., Shterenberg, B. Z., Tandura, S. N., Zelchan, G. I., Baryshok, V. P., Solomennikova, I. I., Urtane, I. P., Lukevits, E. Ya., Voronkov, M. G.: Izv. Akad. Nauk SSSR, Ser. Khim. 467 (1981)
866. Gorenstein, D. G.: J. Am. Chem. Soc. *97*, 898 (1975)
867. Zschunke, A., Tzschach, A., Jurkschat, K.: J. Organometal. Chem. *112*, 273 (1976)
868. Albanov, A. I., Larin, M. F., Pestunovich, V. A. Voronkov, M. G., Kramarova, E. P., Baukov, Yu. A.: Zh. Obshch. Khim. *51*, 488 (1981)
869. Pestunovich, V. A., Larin, M. F., Albanov, A. I., Ignatieva, L. P., Voronkov, M. G.: Dokl. Akad. Nauk SSSR *247*, 393 (1979)
870. Pestunovich, V. A., Albanov, A. I., Larin, M. F., Ignatieva, L. P., Voronkov, M. G.: Izv. Akad. Nauk SSSR, Ser. Khim. 2185 (1978)
871. Krapivin, A. M.: Izv. Akad. Nauk SSSR, Ser. Khim. 1950 (1980)
872. Duboudin, F., Bourgeois, G., Faucher, A., Mazerolles, P.: J. Organometal. Chem. *133*, 29 (1977)
873. Pestunovich, V. A., Tandura, S. N., Shterenberg, B. Z., Khromova, N. Yu., Gar, T. K., Mironov, V. F., Voronkov, M. G.: Izv. Akad. Nauk SSSR, Ser. Khim. 959 (1980)
874. Kennedy, J. D., McFarlane, W.: Reviews Silicon, Germanium, Tin and Lead Compounds *1*, 235 (1974)
875. Harris, R. K., Kennedy, J. D., McFarlane, W., in: NMR and the Periodic Table, p. 309 (eds. Harris, R. K., Mann, B. E.). London: Academic Press 1978
876. Otera, J.: J. Organometal. Chem. *221*, 57 (1981)
877. Hani, R., Geanangel, R. A.: Coord. Chem. Rev. *44*, 229 (1982)
878. Holecek, J., Nadvornik, M., Handlir, K., Lycka, A.: J. Organometal. Chem. *241*, 177 (1983)
879. Witanowski, M., Webb, G. A. (eds.): Nitrogen NMR. London: Plenum 1973
880. Martin, G. J., Martin, M. L., Gonesnard, J. P.: ^{15}N-NMR Spectroscopy. Berlin: Springer 1981
881. Witanowski, M., Stefaniak, L., Webb, G. A.: Nitrogen NMR Spectroscopy. London: Academic Press 1981
882. Witanowski, M., Stefaniak, L., Januszewski, H., Voronkov, M. G., Tandura, S. N.: Bull. Acad. Polon. Science., Ser. Sci. Chem. *24*, 281 (1976)
883. Giger, W.: Dissertation ETH Zurich 1971
884. Liepinsh, E. E., Lapsinya, A. F., Zelchan, G. I., Lukevics, E. J.: Izv. Akad. Nauk Latv. SSR, Ser. Khim. 371 (1980)
885. Liepinsh, E. E., Birgele, I. S., Zelchan, G. I., Urtane, I. P., Lukevics, E.: Zh. Obshch. Khim. *50*, 2733 (1980)
886. Liepinsh, E. E., Birgele, I. S., Zelchan, G. I., Lukevics, E.: Zh. Obshch. Khim. *44*, 1537 (1979)
887. Pestunovich, V. A., Tandura, S. N., Shterenberg, B. Z., Baryshok, V. P., Voronkov, M. G.: Izv. Akad. Nauk SSSR, Ser. Khim. 2159 (1979)
888. Pestunovich, V. A., Shterenberg, B. Z., Voronkov, M. G., Myagi, M. Ya., Samoson, A. V.: Izv. Akad. Nauk SSSR, Ser. Khim. 1435 (1982)
889. Pestunovich, V. A., Tandura, S. N., Shterenberg, B. Z., Baryshok, V. P., Voronkov, M. G.: Izv. Akad. Nauk SSSR, Ser. Khim. 2653 (1978)
890. Pestunovich, V. A., Tandura, S. N., Shterenberg, B. Z., Baryshok, V. P., Voronkov, M. G.: Dokl. Akad. Nauk SSSR *253*, 400 (1980)

891. Pestunovich, V. A., Shterenberg, B. Z., Lippmaa, E. T., Miagi, M. J., Alla, M. A., Tandura, S. N., Baryshok, V. P., Petukhov, L. P., Voronkov, M. G.: Dokl. Akad. Nauk SSSR *258*, 1410 (1981)
892. Myagi, M. J., Samoson, A. V., Lippmaa, E. T., Pestunovich, V. A., Tandura, S. N., Shterenberg, B. Z., Voronkov, M. G.: Dokl. Akad. Nauk SSSR *252*, 140 (1980)
893. Koppel, I. A., Palm, V. A., in: Advances in Linear Free Energy Relationships (eds. Chapman, N. B., Shorter, J.). New York: Plenum Press 1972
894. Palm, V. A.: Basic Quantitative Theory of Organic Reactions. Leningrad: Khimija 1977
895. Kamlet, M. J., Abboud, J. L. M., Taft, R. W.: Progr. Phys. Org. Chem. *13*, 485 (1981)
896. Taft, R. W., Kamlet, M. J.: Org. Magn. Reson. *14*, 485 (1980)
897. Shterenberg, B. Z.: Dissertation Inst. Organic Chem. Irkutsk 1984
898. Liepinsh, E. E., Zitsmane, I. A., Zelchan, G. I., Lukevics, E.: Zh. Obshch. Khim. *53*, 245 (1983)
899. Pekhk, T., Kiirend, E., Lippmaa, E.: in Ref. 109, p. 23
900. Liepinsh, E. E., Zitsmane, I. A., Ignatovich, L. M., Lukevics, E., Gubanova, L. I., Voronkov, M. G.: Zh. Obshch. Khim. *53*, 1789 (1983)

8 Note Added in Proof

Section 1

901. Guanli, W., Kaijuan, L., Yexin, W., in: "Fundamental Research in Organometallic Chemistry", Eds. M. Tsutsui et al., Van Nostrand, Reinhold Co., N.Y., p. 737, 1982
902. Karcsev, G. N., in: "Khimicheska a Svyaz i Stroenie Molekul" (Chemical Bonding and Molecular Structure), Ed. V. I. Nefedov, Nauka, Moscow, p. 233, 1984
903. Voronkov, M. G., Gubanova, L. I., Pestunovich V. A., Frolov, Yu. L.: J. Fluor. Chem. *29*, 75 (1985)
904. Ault, B. S.: J. Mol. Struct. *129*, 287 (1985)
905. Ault, B. S.: J. Mol. Struct. *127*, 357 (1985)
906. Ault, B. S.: J. Mol. Struct. *130*, 215 (1985)
907. Imbenotte, M., Palavit, G., Legrand, P.: J. Raman Spectrosc. *15*, 293 (1984)
908. Kolditz, L., Nitzsche, V.: Z. anorg. allg. Chem. *526*, 48 (1985)
909. Ueyama, K., Matsubayashi, G.-E., Tanaka, T.: Inorg. Chim. Acta *87*, 143 (1984)
910. Krug, B., Legat, W., Gruehn, R.: Z. anorg. allg. Chem. *515*, 159 (1984)

Section 2

911. Peterson, M. R., Csizmadia, I. G.: J. Mol. Struct. (Theochem) *123*, 399 (1985)
912. Nyburg, S. C., Faerman, C. H.: Acta Cryst. *B41*, 274 (1985)

Section 2.1

913. Wojnowski, W., Peters, K., Böhm, M. C., Schnering, H. G. von: Z. anorg. allg. Chem. *523*, 169 (1985)
914. Jurkschat, K., Mugge, C., Schmidt, J., Tzschach, A.: J. Organomet. Chem. *287*, C1 (1985)
915. Chung, C.-S.: J. Chem. Educ. *61*, 1062 (1984)
916. Wipff, G., Kollman, P.: Nouv. J. Chim. *9*, 457 (1985)
917. Tarasyancs, R. R., Ogaidzhan, E. P., Belakova, Z. V., Shevchenko, V M , Sheludyakov, V. D., Chernyshev, E. A.: Zh. Obshch. Khim. *54*, 1569 (1984)

Section 2.2

918. Hamann, S. D., Mazerolles, P., Faucher, A., Manuel, G.: Aust. J. Chem. *37*, 23 (1984)
919. Parker, S. C., Catlow, C. R. A., Cormack, A. N.: Acta Cryst. *B40*, 200 (1984)
920. Angell, C. A., Cheeseman, P., Tamaddon, S.: Bull. Mineral. *106*, 87 (1983)

Section 3.1

921. Cruickshank, D. W. J.: J. Mol. Struct. *130*, 177 (1985)

Section 4.2

922. Willem, R., Gielen, M., Pepermans, H., Brocas, J., Fastenakel, D., Finocchiaro, P.: J. Am. Chem. Soc. *107*, 1146 (1985)
923. Willem, R., Gielen, M., Pepermans, H., Hallenga, K., Becca, A., Finocchiaro, P.: J. Am. Chem. Soc. *107*, 1153 (1985)
924. Michalak, R. S., Wilson, S. R., Martin, J. C.: J. Am. Chem. Soc. *106*, 7529 (1984)
925. Stevenson, W. H., Wilson, S., Martin, J. C., Farnham, W. B.: J. Am. Chem. Soc. *107*, 6340 (1985)
926. Stevenson, W. H., Martin, J. C.: J. Am. Chem. Soc. *107*, 6352 (1985)

Section 4.3

927. Brandemark, U., Siegbahn, P. E. M.: Theor. Chim. Acta *66*, 233 (1984)
928. Frolov, Yu. L., Shevchenko, S. G., Voronkov, M. G.: J. Organomet. Chem. *292*, 159 (1985)
929. Sheldon, J. C., Hayes, R. N., Bowie, J. H.: J. Am. Chem. Soc. *106*, 7711 (1984)
930. Anderson, A. B., Gordon, T. L., Kenney, M. E.: J. Am. Chem. Soc. *107*, 192 (1985)
931. Davis, L. P., Burggraf, L. W., Gordon, M. S., Baldridge, K. K.: J. Am. Chem. Soc. *107*, 4415 (1985)

Section 4.4.1

932. Farnham, W. B., Whitney, J. F.: J. Am. Chem. Soc. *106*, 3992 (1984)
933. Ciliberto, E., Doris, K. A., Pietro, W. J., Reisner, G. M., Ellis, D. E., Fragala, I., Herbstein, F. M., Ratner, M. A., Marks, T. J.: J. Am. Chem. Soc. *106*, 7748 (1984)
934. Dirk, C. W., Marks, T. J.: Inorg. Chem. *23*, 4325 (1984)

Section 4.4.2

935. Lieban, F.: "Structural Chemistry of Silicates" Springer-Verlag, Berlin, 1985
936. Eriks, K., Beno, M. A., Bechgaard, K., Williams, J. M.: Acta Cryst. *C40*, 1715 (1984)
937. Sackerer, D., Nagorsen, G.: Z. anorg. allg. Chem. *437*, 188 (1977)
938. Holmes, R. R., Day, R. O., Chandrasekhar, V., Holmes, J. M.: Inorg. Chem. *24*, 2009 (1985)
939. Holmes, R. R., Day, R. O., Chandrasekhar, V., Hazland, J. J.: Inorg. Chem. *24*, 2016 (1985)
940. Lieban, F.: Inorg. Chim. Acta *89*, 1 (1984)
941. Hoskins, B. F., Linden, A., Mulvaney, P. C., O'Donnell, T. A.: Inorg. Chim. Acta *88*, 217 (1984)
942. Depisch, B., Gladrow, B., Kummer, D.: Z. anorg. allg. Chem. *519*, 42 (1984)
943. Kummer, D., Deppisch, B., Gladrow, B., Seifert, J.: 7th Intern. Symp. Organosilicon Chem., Kyoto 1984, Abstracts of Papers, p. 119

Section 4.4.3

944. Kleve, G., Bats, J. W., Fuess, H.: J. Am. Chem. Soc. *106*, 5202 (1984)
945. Klebe, G., Hensen, K.: J. Chem. Soc., Dalton Trans, 5 (1985)
946. Parkanyi, L., Szöllösy, A., Bihatsi, L., Hencsei, P., Nagy, J.: J. Organomet. Chem. *256*, 235 (1983)
947. Klebe, G.: J. Organomet. Chem. *293*, 147 (1985)
948. Klebe, G., Bats, J. W., Hensen, K.: J. Chem. Soc., Dalton Trans., 1 (1985)
949. Shklover, V. E., Struchkov, Yu. T., Rodin, O. G., Traven, V. F., Stepanov, B. I.: J. Organomet. Chem. *266*, 117 (1984)
950. Shklover, V. E., Struchkov, Yu. T., Traven, V. F., Rodin, O. G., Rokitskaya, V. I., Eismont, M. Yu., Stepanov, B. I.: See Ref. 943, p. 121
951. Ovchinnikov, Yu. E., Shklover, V. E., Struchkov, Yu. T., Rokitskaya, V. I., Eismont M. Yu., Rodin, O. G., Traven, V. F., Stepanov, B. I.: III Vses. Simp. Stroenie Reakts. Sposobnost Kremneorganicheskikh Soedinenij, Tezisy dokl., Irkutsk, p. 13, 1985
952. Ovchinnikov, Yu. E., Shklover, V. E., Struchkov, Yu. T., Palynlin, V. A., Rokitskaya, V. I., Eismont, M. Yu., Traven, V. F.: J. Organomet. Chem. *290*, 25 (1985)
953. Parkanyi, L., Hencsei, P., Bihätsi, L., Kovacs, I., Szöllosy, A.: Polyhedron, *4*, 243 (1985)
954. Demidov, M. P., Shklover, V. E., Frolov, Yu. L., Lukina, Yu. A., D'yakov, V. M., Struchkov, Yu. T., Voronkov, M. G.: Zh. Strukt. Khim. *26*, 103 (1985)
955. Kemme, A. A., Bleidelis, J. J., Lapsinya, A. F., Flaisher, M. B., Zelchan, G. I., Lukevics, E. J.: Izv. Akad. Nauk Latv. SSR, Ser. Khim., 242 (1985)
956. Hencsei, P., Kovacs, I., Pazkanyi, L.: J. Organomet. Chem. *293*, 185 (1985)
957. Hencsei, P., Parkanyi, L.: Kem. Kozl. *61*, 319 (1984)

Section 4.4.4

958. Sheludyakov, V. D., Dmitrieva, A. B., Gusev, A. I., Apal'kova, G. M., Kirilin, A. D.: Zh. Obshch. Khim. *54*, 2298 (1984)
959. Shklover, V. E., Struchkov, Yu. T., Ganyushin, A. V.: Zh. Strukt. Khim. *26*, 180 (1985)
960. Slovokhotov, Yu. L., Timofeeva, T. V., Antipin, M. Yu., Struchkov, Yu. T.: J. Mol. Struct. *112*, 127 (1984)
961. D'yachenko, O. A., Sokolova, Yu. A., Atormyan, L. O., Ushakov, N. V.: Izv. Akad. Nauk SSSR, Ser. Khim., 1030 (1985)
962. Wojnowski, W., Peters, K., Weber, D., Schnering, H. G. von: Z. anorg. allg. Chem. *519*, 134 (1984)
963. Gergo, E., Schultz, G., Hargittai, I.: J. Organomet. Chem. *292*, 343 (1985)
964. Hargittai, I.: "The Structure of Volatile Sulphur Compounds", Akademiai Kiado, Budapest, 1985
965. Tikkanen, W. R., Egan, J. W., Ir., Petersen, J. L.: Organometallics *3*, 825 (1984)
966. Tikkanen, W. R., Egan, J. W., Jr., Petersen, J. L.: Organometallics *3*, 1646 (1984)
967. Martin, J. C., Stevenson, W. H., Lee, D. Y.: See Ref. 943, p. 112
968. McKean, D. C., Morrisson, A. R., Torto, I., Kelly, M. I.: Spectrochim. Acta *41A*, 25 (1985)
969. Kovalev, I. F., Dernova, V. S., Anoshkin, V. I., Voronkov, M. G.: Izv. Akad. Nauk SSSR, Ser. Khim., 109 (1985)
970. Yakovlev, I. P., Finogenov, Yu. S., Gindin, V. A., Feshin, V. P., Nikitin, P. A., Shchegolev, A. E., Ivin, B. A.: Zh. Obshch. Khim. *55*, 1093 (1985)
971. Schubert, U., Kraft, G., Walther, E.: Z. anorg. allg. Chem. *519*, 96 (1984)
972. Mackay, K. M., Nicholson, B. K., Robinson, W. T., Sims, A. W.: J. Chem. Soc., Chem. Commun., 1276 (1984)
973. Böhm, M. C., Ramirez, R., Nesper, R., Schnering, H. G. von: Ber. Bunsenges. Phys. Chem. *89*, 465 (1985)
974. Matsunaga, T., Kodera, E., Komura, Y.: Acta Cryst. *C40*, 1668 (1984)
975. Ramirez, R., Nesper, R., Schnering, H. G. von, Böhm, M. C.: Chem. Phys. *95*, 17 (1985)
976. Schubert, U., Kraft, G.: See Ref. 943, p. 125
977. Carre, F., Colomer, E., Corrin, R. J. P., Vioux, A.: Organometallics *3*, 1272 (1984)

Section 4.4.5

978. Blake, A. J., Ebsworth, E. A. V., Henderson, S. G. D., Welch, A. J.: Acta Cryst. *C41*, 1141 (1985)

Section 4.5.1

979. Aminabhavi, T. M., Biradar, N. S., Patil, S. B.: Boddabasanagoudar, V. L., Inorg. Chem. Acta *107*, 231 (1985)
980. Ghose, B. N.: Acta Chim. Hung. *118*, 191 (1985)
981. Lukasiak, J.: Acta Chim. Hung. *117*, 271 (1984)
982. Voronkov, M. G., Kuznecsov, A. L., Mirskov, R. G., Shterenberg, B. Z., Rakhlin, V. I.: Zh. Obshch. Khim. *55*, 1038 (1985)
983. Pestunovich, V. A., Petukhov, L. P., Larin, M. F., Shterenberg, B. Z., Balakhchi, G. K., Sidorkin, V. F., Barushok, V. P., Voronkov, M. G.: See Ref. 951, p. 74
984. Grobe, J., Voulgarakis, N.: Z. anorg. allg. Chem. *517*, 125 (1984)
985. Wheeler, B. L., Nagasubramanian, G., Bard, A. J., Schechtman, L. A., Dininny, D. R., Kenny, M. E.: J. Am. Chem. Soc. *106*, 7404 (1984)

Section 4.5.2

986. Voronkov, M. G., Gubanova, L. I., D'yakov, V. M., Glukhikh, V. I., Glukhikh, N. G., Shirchin, B.: Zh. Obsch. Khim. *55*, 1041 (1985)
987. Kalikhman, I. D., Bannikova, O. B., Belousova, L. I., Gostevskij, B. A., Vyazankina, O. A., Vyazankin, N. S.: Zh. Obshch. Khim. *54*, 2609 (1984)
988. Kalikhman, I. D., Bannikova, O. B., Gostevskij, B. A., Medvedeva, E. N., Lopyrev, V. A., Vyazankina, O. A., Vyazankin, N. S.: Izv. Akad. Nauk SSSR, Ser. Khim., 1395 (1985)
989. Kalikhman, I. D., Bannikova, O. B., Gostevskij, B. A., Vyazankina, O. A., Vyazankin, N. S., Pestunovich, V. A.: Izv. Akad. Nauk SSSR, Ser. Khim., 1688 (1985)

Section 4.5.3

990. Kalbandkeri, R. G., Syed, Mohamed K., Padma, D. K., Vasudeva, Murthy A.R.: Polyhedron *41*, 787 (1985)
991. Corriu, R. J. P., Kpoton, A., Poirier, M., Royo, G., Corey, J. Y.: J. Organomet. Chem. *277*, C25 (1984)
992. Pestunovich, V. A., Larin, M. F., Sorokin, M. S., Albanov, A. I., Voronkov, M. G.: J. Organomet. Chem. *280*, C17 (1985)
993. Krebs, R., Schomburg, D., Schmutzler, R.: Z. Naturforsch. *40B*, 282 (1985)

Section 4.5.4

994. Cragg, R. H., Lane, R. D.: J. Organomet. Chem. *291*, 153 (1985)
995. Bassindale, A. R., Stout, T.: J. Chem. Soc., Chem. Commun., 1387 (1984)
996. Smith, J. V., Blackwell, C. S.: Nature *303*, 223 (1983)
997. Sidorkin, V. F., Pestunovich, V. A., Voronkov, M. G.: Magn. Reson. Chem. *23*, 491 (1985)
998. Schlraml, J., Krapivin, A. M., Luzin, A. B., Kilesso, V. M., Pestunovich, V. A.: Coll. Czech. Chem. Commun. *49*, 2897 (1984)
999. Sjöberg, S., Ingri, N., Nenner, A.-M., Öhman, L.-O.: J. Inorg. Biochem. *24*, 267 (1985)
1000. Liepins E., Birgele, I., Tomsons, P., Lukevics, E.: Magn. Reson. Chem. *23*, 485 (1985)
1001. Mudrakovskii, I. L., Mastikhin, V. M., Shmachkova, V. P., Kotsarenko, N. S.: Chem. Phys. Lett. *120*, 424 (1985)

Section 4.5.5

1002. Pestunovich, V. A., Shterenberg, B. Z., Petukhov, L. P., Rakhlin, V. I., Baryshok, V. P., Mirskov, R. G., Voronkov, M. G.: Izv. Akad. Nauk SSSR, Ser. Khim., 1935 (1985)
1003. Suresh, B. S., Chandrasekhar, V., Padma, D. K.: J. Chem. Soc., Dalton Trans., 1787 (1984)
1004. Spencer J. N., Barton, S. W., Cader B. M., Corsico, Ch. D., Harrison, L. E., Mankuta, M. E., Yoder, C. H.: Organometallics, *4*, 394 (1985)

Section 4.5.6

1005. Kupce, E., Liepins, E., Lapsina, A., Urtane, I., Zelcans, G., Lukevics, E.: J. Organomet. Chem. *279*, 343 (1985)
1006. Liepins, E., Kupce, E., Birgele, I., Lukevics, E.: See Ref. 943, p. 114
1007. Kupce, E. L., Liepins, E. E., Lapsinya, A. F., Zelchan, G. I., Lukevics, E. J.: See Ref. 951, p. 75
1008. Taba, K. M., Dahlhoff, W. V.: J. Organomet. Chem. *280*, 27 (1985)

Author Index Volumes 101–131

Contents of Vols. 50–100 see Vol. 100
Author and Subject Index Vols. 26–50 see Vol. 50

The volume numbers are printed in italics

FSC
www.fsc.org
MIX
Papier | Fördert
gute Waldnutzung
FSC® C083411

Kolibri 360 GmbH
Ferdinand-Jühlke-Straße 7
99095 Erfurt, Deutschland
produktsicherheit@kolibri360.de